Mechanical Engineering Series

Frederick F. Ling
Series Editor

Advisory Board

Mechanical Engineering Series

Introductory Attitude Dynamics
F.P. Rimrott

Balancing of High-Speed Machinery
M.S. Darlow

Theory of Wire Rope
G.A. Costello

Theory of Vibration
  Vol. I Introduction
  Vol. II Discrete and Continuous Systems
A.A. Shabana

Laser Machining: Theory and Practice
G. Chryssolouris

Underconstrained Structural Systems
E.N. Kuznetsov

Principles of Heat Transfer in Porous Media
M. Kaviany

E.N. Kuznetsov

Underconstrained Structural Systems

With 99 Illustrations

Springer-Verlag
New York Berlin Heidelberg London
Paris Tokyo Hong Kong Barcelona

E.N. Kuznetsov
Department of General Engineering
University of Illinois
Urbana, IL 61801
USA

Series Editor
Frederick F. Ling
President, Institute for Productivity Research
New York, NY 10018
and
Distinguished William Howard Hart Professor Emeritus
Department of Mechanical Engineering,
   Aeronautical Engineering and Mechanics
Rensselaer Polytechnic Institute
Troy, NY 12180-3590
USA

Library of Congress Cataloging-in-Publication Data
Kuznetsov, Edward N.
   Underconstrained structural systems / Edward N. Kuznetsov.
      p.   cm. — (Mechanical engineering series)
   Includes bibliographical references and index.
   ISBN 0-387-97594-2 (Springer-Verlag New York). — ISBN
   3-540-97594-2 (Springer-Verlag Berlin)
      1. Structural analysis (Engineering) I. Title.   II. Series:
   Mechanical engineering series (Berlin, Germany)
   TA646.K89   1991
   624.1′7 — dc20   91-15293

Printed on acid-free paper.

Camera-ready copy prepared by the author.
Printed and bound by Edwards Brothers, Inc., Ann Arbor, MI.
Printed in the United States of America.

9 8 7 6 5 4 3 2 1

ISBN 0-387-97594-2 Springer-Verlag New York Berlin Heidelberg
ISBN 3-540-97594-2 Springer-Verlag Berlin Heidelberg New York

Series Preface

Mechanical engineering, an engineering discipline born of the needs of the industrial revolution, is once again asked to do its substantial share in the call for industrial renewal. The general call is urgent as we face profound issues of productivity and competitiveness that require engineering solutions, among others. The Mechanical Engineering Series is a new series, featuring graduate texts and research monographs, intended to address the need for information in contemporary areas of mechanical engineering.

The series is conceived as a comprehensive one that will cover a broad range of concentrations important to mechanical engineering graduate education and research. We are fortunate to have a distinguished roster of consulting editors, each an expert in one of the areas of concentration. The names of the consulting editors are listed on the first page of the volume. The areas of concentration are applied mechanics, biomechanics, computational mechanics, dynamic systems and control, energetics, mechanics of materials, processing, thermal science, and tribology.

Professor Marshek, the consulting editor for dynamic systems and control, and I are pleased to present this volume of the series: *Underconstrained Structural Systems* by Professor Kuznetsov. The selection of this volume underscores again the interest of the Mechanical Engineering Series to provide our readers with topical monographs as well as graduate texts.

New York, New York   Frederick F. Ling

Preface

In statics and in structural analysis it is traditional practice to classify structural systems as either geometrically stable or otherwise—and to pronounce the latter class unacceptable or tacitly exclude it from consideration. Yet this practice corresponds neither to the requirements nor to the possibilities of the mechanics of structures. Indeed, many good structural systems would not survive such a simplistic classification. These, often misunderstood, structural systems, defined later as underconstrained, are the subject of this book. They are not geometrically stable in the conventional sense, but nevertheless provide excellent solutions to a wide variety of engineering problems.

The inferior role traditionally ascribed to underconstrained structural systems, in comparison to invariant structures, has long been a psychological obstacle in their research, development, and application. In reality, this class of systems possesses some remarkable and unique properties making it a rich and diverse resource for many engineering applications, both conventional and novel, and especially for space applications.

Competent use of underconstrained structural systems requires a thorough comprehension of the underlying principles: the magic follows the logic. This book contains a concise and comprehensive theory of these systems and its application to their analysis, synthesis, and design. The theory—a combination of advanced analytical statics and nonlinear structural mechanics—is developed in Part I (Chapters 1–4). Part II (Chapters 5–10) provides applications to particular types of underconstrained structural systems.

Most theoretical ideas and concepts are illustrated by the simplest possible models, exposing the feature unobscured by irrelevant details. When appropriate, the corresponding design implications are shown and discussed. A prerequisite for this book is familiarity with linear algebra; Strang (1988) is the most suitable text for this purpose. Understanding the material on cable systems, nets, and membranes requires elements of differential geometry; see, for example, Struik (1961).

Acknowledgments. Chapters 1 and 10 of this book were written together with W. B. Hall, and I am pleased to acknowledge this co-authorship. I am also grateful to him for numerous suggestions and discussions. My sincere thanks go to the Department of General Engineering, University of Illinois/Urbana, where this book was written, for support and encouragement. It was a pleasure to work with Peggy Hills and Maureen Nickols who helped in typing the manuscript, and Brian McBride who helped with the graphics. Last and most, my gratitude goes to my family and, especially, to my wife Helen, to whom this book is dedicated.

Contents

Part I
Fundamentals

Cause and effect are two sides of one fact.

Emerson

Find out the cause of this effect,
or rather say, the cause of this defect,
for this effect defective comes by cause.

Shakespeare

1
Statical-Kinematic Analysis

The objective of this chapter is to modernize and further develop the basic concepts of statical-kinematic analysis, and to refine the classification of generic structural systems. This is done with a view to both a rigorous treatment of the topic and the development of analytical criteria amenable to computerized analysis.

Statical-kinematic analysis assumes the material of the system components to be perfectly rigid, and studies both general and system-specific relations among external loads, internal forces, equilibrium configurations, and kinematically possible displacements, i.e., all information attainable without invoking constitutive relations. At the conceptual level, the analysis deals with identifying and studying generic statical-kinematic types of structural and mechanical assemblies, their basic properties and response to external actions. In applications, it is concerned with solving particular, mostly geometrically nonlinear, coupled statical-kinematic problems, like establishing the final, deformed configuration and the corresponding force distribution in a general-chain system under a given load.

Conceptual sources of statical-kinematic analysis are structural topology, geometry, and the principle of virtual work. Structural topology and geometry underlie the system kinematics identifying the type of mobility of the system (no mobility, infinitesimal, or finite mobility). By postulating preservation of energy in small perturbations, the principle of virtual work, in conjunction with system kinematics, is the foundation of statics in a broad sense. It describes conditions of equilibrium identifying the statically possible states of a system and, beyond that, determines the quality of equilibrium (stable, neutral, or unstable).

Statical-kinematic analysis is often attributed to structural mechanics, but its bare essentials can be traced to analytical mechanics. The latter precedes structural mechanics from a methodological viewpoint and uses fewer basic concepts and hypotheses, especially, phenomenological ones, such as experimentally established material properties. Accordingly, the analysis formulated here will be presented in terms of analytical mechanics, with its higher level of rigorousness. This unavoidably implies the re-examination, re-evaluation, and sometimes modification of many concepts currently used in structural mechanics. As a reward, it provides a more thorough study of the range of problems in question, including some new insights and logical connections.

1.1   Basic Concepts and Definitions

Rigid and Underconstrained Structural Systems

Positional Constraints and Mobility. For the purpose of statical-kinematic analysis, a *structural system* is taken to mean an assemblage of idealized components made of a perfectly rigid material and linked by ideal positional constraints. Components usually are conventional models, such as three-dimensional solids, shells, membranes, plates, bars, wires, and material points. Two simple examples of structural systems are a conventional roof truss and a kinematic chain, which are at the opposite poles of statical-kinematic behavior.

In the analysis of a system, some components are designated as free bodies or material points, and others as *constraints,* which control the relative positions of the first group of components. The control is accomplished either by fixing (bilateral constraints) or limiting (unilateral, or boundary constraints) distances, angles, and other geometrical parameters of the system. A rigid bar connecting two nodes (material points) in a truss, and preserving the distance between them, is a typical *bilateral* constraint. A flexible inextensible wire connecting the same two nodes and incapable of supporting compression represents a typical *unilateral* constraint in tension. Thus, the term "constraint" represents both the physical constraint that exists in the form of a designated structural element or component and the corresponding mathematical constraint (an equation or an inequality condition) used in an analysis.

Taken without the constraints, the system components possess a certain mobility, conventionally quantified as their combined number of degrees of freedom. Appropriately introduced constraints reduce this number, ultimately, all the way to zero. Mobility of a structural system depends on (a) the *system composition* (layout) and (b) the *geometric configuration.* The system composition is formally described in terms of structural topology (component and constraint arrangement, or system connectivity), whereas the geometric configuration is determined by the position of each structural member with reference to a chosen coordinate system.

Structural Topology, Geometry, and Rigidity. An idealized structural system that has no mobility, either finite or infinitesimal, is called *geometrically invariant.* For a system to be invariant, both its topology and geometry must be of the right type. The system topology determines the *potential maximum* number of degrees of freedom that can be taken away by the constraints. The *actual* number taken away depends on the particular geometric configuration of the system.

The requirements on the topology and geometry of an invariant system are illustrated by the chart in Fig. 1. If the topology of a structural system is one for which an invariant configuration is possible, the system is said to be

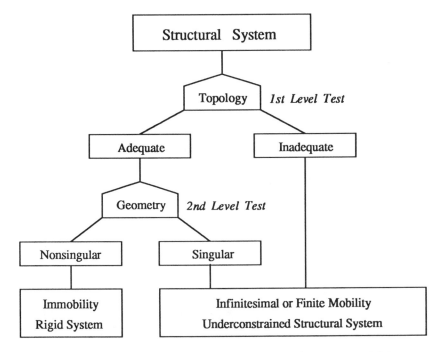

Figure 1.1. Topological and geometric requirements for structural rigidity.

topologically adequate for geometric invariance, or more simply, *topologically adequate*. A system having passed this first-level test is "almost always" invariant. However, to qualify as such, the system must pass a second-level, geometric test. Specifically, if the geometric configuration of a topologically adequate system is *nonsingular*, the system is geometrically invariant. The majority of conventional structures belong to this group. They are also called geometrically stable; statically rigid (reflecting their ability to support an arbitrary load); and "infinitesimally rigid" (this mathematical term, somewhat unusual for an engineer, indicates the impossibility of even infinitesimal displacements).

A topologically adequate system in a geometrically *singular, degenerate,* configuration is not geometrically invariant. These exceptional systems possess at least infinitesimal, and sometimes finite, mobility and are usually excluded from consideration for conventional structural applications. Detecting a singular configuration of a topologically adequate system may require a significant computational effort.

A *topologically inadequate* structural system is "almost always" kinematically mobile. Although such systems allow exceptional, and very interesting, singular geometric configurations with only infinitesimal mobility, they cannot be geometrically invariant; as a matter of principle, no variation in their geometry can remedy the deficient topology.

Underconstrained Structural Systems. This class of structural systems is the subject of this book.

Definition.   Structural systems that are topologically inadequate or geometrically singular are called *underconstrained structural systems.*

Thus, underconstrained systems include all structural systems that, for either topological or geometric reasons, are *not* geometrically invariant.

Geometrically invariant systems are sometimes called "just rigid" when the number of constraints is just sufficient for invariance, and overconstrained if the number is more than sufficient, as in redundant structures. For a system to be just rigid, the number of constraints, their arrangement, and the system geometry must all be correct. If the number of constraints, C, is smaller than the total number, N, of degrees of freedom for all of the system components, the system will be underconstrained regardless of its topology and geometry. Furthermore, an inadequate topology will cause a system to be underconstrained even when the number of constraints is excessive. And finally, even if both the number and arrangement of constraints are correct, a singular geometry will condemn the system to the underconstrained class.

Remark.   A relation between structural topology and geometry, whereby topology is defined first and is considered immutable, holds for structural and mechanical assemblies known as kinematic chains or general-chain systems. In fact, it is not unreasonable to take the existence of a fixed topology as a definition of a kinematic chain. However, in a broader context, the relation between structural topology and geometry is not that simple and straightforward.

The first complication is encountered when dealing with a system containing unilateral constraints. Although a topology reflecting the system composition and connectivity is well defined, it is not too useful. The reason is that the status of unilateral constraints (engaged or disengaged), is unknown until the system geometry is specified. The role of topology is blurred even more when considering granular media and problems of geometric packing. Here even an overall topology of the aggregate cannot be defined; instead, a set of feasible topologies is determined by distinct arrangements of the aggregate elements. The next step in this direction is an incompressibile liquid in hydrostatics where the only constraint, representing both topology and geometry, is the condition of volume preservation. This constraint is further relaxed in an ideal gas model.

Returning to structural systems, the statical-kinematic properties of a system are usually assessed either by inspection or by some analytical means like, for example, the well-known zero load test. In this test, geometric invariance of a structural system is ascertained by statical means, taking advantage of certain interrelations between the system statics and kinematics. The variety and sophistication of possible situations in the realm of underconstrained structural systems makes this state of the art unsatisfactory for the analysis, and especially synthesis, of structural systems. A more comprehensive statical-kinematic analysis of structural systems is presented below.

Intrinsic Properties of Structural Systems

Constraint Equations and Linearization. For systems with bilateral constraints, complete information on the kinematic properties of a system is contained in a set of simultaneous constraint equations,

$$F^i (X_1, \cdots, X_n, \cdots, X_N) = 0, \quad i = 1, 2, \cdots, C, \tag{1.1}$$

with at least one known solution, $X_n = X_n^0$, describing the reference geometric configuration. The C constraint functions F^i relate the N generalized coordinates X_n to the invariant geometric parameters of the system (member sizes, linear and angular distances, etc.). For example, constraint equations of a pin-bar assembly may be quadratic, of the form

$$\Delta X^2 + \Delta Y^2 + \Delta Z^2 - L^2 = 0, \tag{1.2}$$

with Δ denoting the coordinate difference of the bar ends and L the bar length.

Power expansion techniques are systematically used throughout the following analysis. It is therefore assumed that F^i can be expanded into power series at the solution point $X_n = X_n^0$:

$$F_n^i x_n + (1/2!)F_{mn}^i x_m x_n + \cdots = 0, \quad m, n = 1, 2, \cdots, N, \tag{1.3}$$

where x_n are infinitesimal increments of the respective coordinates. Here and below, a repeated subscript denotes summation over the indicated range.

The linear terms of the expansion appear in the linearized constraint equations obtained from (3),

$$F_n^i x_n = 0, \tag{1.4}$$

where the first derivatives,

$$F_n^i = \partial F^i / \partial X_n |_0, \tag{1.5}$$

are the elements of the constraint function *Jacobian matrix* at X_n^0.

Virtual and Kinematic Displacements. Infinitesimal coordinate increments x_n satisfying equations (4) are called *virtual displacements*. As is seen from their origin [cf. (5)], virtual displacements are tangent to the actual displacement path, which makes them local, that is, *configuration-specific*. Displacements that are found by solving the original constraint equations (1) are called *kinematically possible*, or simply *kinematic*. In contrast to virtual displacements, kinematic displacements are not necessarily infinitesimal. Another, more subtle yet crucial, distinction between virtual and kinematic displacements is best emphasized by Webster's definition of the word virtual as "being in effect but not in fact." Thus, virtual displacements are just formal, and only the kinematic ones are physically realizable. Yet, virtual displacements are more than a mathematical convenience: suffice it to say that virtual, rather than actual, displacements appear in the principle of virtual work.

Denoting the rank of the constraint Jacobian matrix as r, the solution of equations (4) can be found in terms of properly selected $(N - r)$ virtual displacements x_s designated as independent:

$$x_p = a_{ps} x_s; \quad p = 1, \cdots, r; \quad s = r+1, \cdots, N. \tag{1.6}$$

A useful expression for all displacements in terms of the independent ones can be written with the aid of Kronecker's δ:

$$x_n = a_{ns} x_s, \quad a_{ns} = \delta_{pn} a_{ps} + \delta_{ns}. \tag{1.7}$$

Generalized Constraint Reactions and Member Forces. According to the principle of virtual work, a system in a given configuration is in equilibrium under an external load P_n^* if and only if

$$P_n^* x_n = 0. \tag{1.8}$$

As is shown in analytical statics, the equilibrium equations can be obtained from the linearized constraint equations (4) and the equation of virtual work (8). In terms of matrix algebra, this development can be interpreted as follows. The equation of virtual work, being a necessary and sufficient condition for equilibrium, means that the subspaces of loads and virtual displacements are orthogonal. (This is why equilbrated loads do not perform mechanical work.) Since virtual displacements are in the nullspace of the constraint Jacobian matrix, the loads must be in its orthogonal complement, i.e., in the row space of the matrix. Hence, to satisfy (8), an external load, P_n^*, must be a linear combination of the Jacobian matrix rows:

$$F_n^i \Lambda_i = P_n^*. \tag{1.9}$$

The obtained equations (9) are the equilibrium equations of the system and their coefficient matrix is the transpose of the constraint Jacobian matrix, as is seen from the respective subscript patterns in (4) and (9). The multipliers Λ_i are called *generalized constraint reactions.* They do not always coincide with the actual member forces and depend on the particular form of the constraint functions (1). For example, if a constraint is described by equation (2), the constraint variation is

$$f_i = 2L_i l_i \quad \text{(no summation)}, \tag{1.10}$$

where l_i is a change in the member length. The member force N_i is evaluated by equating two expressions for virtual work produced if the constraint is released,

$$\Lambda_i f_i = N_i l_i \quad \text{(no summation)}, \tag{1.11}$$

wherefrom, upon eliminating f_i,

$$N_i = 2L_i \Lambda_i \quad \text{(no summation)}. \tag{1.12}$$

Thus, the generalized constraint reaction and bar force (or support reaction) are related but not identical. This difference is a direct result of choosing the quadratic expression (2) to represent the positional constraint imposed by a bar. The result would be quite different for alternative expressions. For example, if a constraint were made in terms of the distance between the nodes, rather than its square, the constraint equation would be equally correct but mathematically inconvenient, since it would involve a square root of terms. On the other hand, as is seen from (12), the generalized reactions corresponding to the chosen form (2), however convenient, have the unusual dimension (force/length). For this reason, they are sometimes called force densities. In what follows, the term constraint reactions refers to all generalized constraint reactions of the system, including those associated with support forces.

Constraint Dependence and Self-Stress. A rank r of the constraint Jacobian matrix that is less than the number of constraint equations C indicates *constraint dependence*. This means that some of the constraints deprive the system of certain degrees of freedom more than once (redundancy), while possibly leaving other degrees of freedom unconstrained. The constraint dependence is local if it occurs only in the reference configuration and global if it holds in any kinematically possible configuration. In the first case the configuration is singular and the rank restores upon exiting from it, whereas in the second case the rank does not restore.

Although the concept of constraint dependence is related closely to structural redundancy, it is far from being synonymous with it. As will be seen later (Chapter 3), dependent unilateral constraints are not always redundant and then removal or failure of one of them would result in the collapse of the system.

A statical consequence of constraint dependence is the potential counteraction of the corresponding constraint reactions. That is, reactions in dependent constraints can exist in the absence of external loads as a self-equilibrated force pattern called self-stress. Its anlytical manifestation is that, at $r < C$, the system of homogeneous equilibrium equations,

$$F_n^i \, \Lambda_i = 0, \tag{1.13}$$

can be solved in terms of properly selected $(C - r)$ constraint reactions Λ_k:

$$\Lambda_p = b_{pk} \, \Lambda_k; \quad p = 1, \cdots, r; \quad k = r + 1, \cdots, C. \tag{1.14}$$

It is convenient to express all reactions in terms of independent ones:

$$\Lambda_i = b_{ik} \Lambda_k; \quad b_{ik} = \delta_{ip} b_{kp} + \delta_{ik}. \tag{1.15}$$

Each solution Λ_i describes a pattern of statically possible initial (self-stress) reactions. Thus, constraint dependence makes self-stress statically possible and, conversely, the statical possibility of self-stress is a sign of constraint dependence. It is also true that a constraint is independent if and only if it cannot have an initial force.

1.2 External Loads and Constraint Variations. Statical-Kinematic Duality

Applied loads and small variations in constraint parameters (structural member sizes) are external actions that are represented by the right-hand sides of the equilibrium and linearized constraint equations. Both the admissibility and the analytical treatment of these inputs are predicated by the properties of the respective matrices, specifically, by the relations among the matrix dimensions, C and N, and the rank, r. A certain parallelism of the two analyses is not related to the fact that both of them are based on one and the same matrix and is simply a consequence of the underlying linear-algebraic relations.

Equilibrium and Perturbation Loads

Equilibrium Loads and Configurations. In any given configuration, a system can balance only those loads P_n^* that lie in the column space of the equilibrium matrix (the transposed Jacobian matrix). These statically possible loads are called *equilibrium loads* and are representable as linear combinations of the matrix columns with arbitrary coefficients Λ_i. The number of equilibrium loads equals the rank r and is, therefore, the number of independent constraints in the given configuration.

There is no one-to-one relation between equilibrium loads and configurations: there are r linearly independent equilibrium loads for a given configuration, whereas, generally, for a given load there is only *one* statically possible configuration called an *equilibrium configuration*. (Nonuniqueness of an equilibrium configuration raises the question of stability of equilibrium.) For a system with Jacobian matrix rank $r = N$, and only for such a system, the equilibrium equations are always compatible, so that every load is an equilibrium load.

Perturbation Loads. When the Jacobian matrix rank is smaller than N, there exist $N - r$ linearly independent loads that belong to the orthogonal complement of the column space of the equilibrium matrix. These loads are of the form (Kuznetsov, 1972b)

$$p_n = a_{ns}\, p_s, \tag{1.16}$$

with arbitrary p_s and are called *perturbation loads*; they cannot be balanced in the original configuration of the system and, if applied, would produce either displacements or infinite constraint reactions.

Comparison with (7) shows that perturbation loads are related to the virtual displacements of the system: both span the nullspace of the constraint Jacobian matrix and, therefore, are orthogonal to its row space, i.e., the column space of the equilibrium matrix. Accordingly, the number of arbitrary components p_s, which is the number of linearly independent perturbation loads, equals the number of independent virtual displacements. Systems with $r = N$, and only such systems, have no perturbation loads or virtual displacements; these systems have adequate topology and nonsingular geometry.

For a system subjected to a general load, P_n, the equation of virtual work (8) can be transformed with the aid of relations (6) and (16) as follows:

$$P_n^* x_n = (P_n - p_n) x_n = (P_n - a_{nt} p_t) a_{ns} x_s = 0, \quad s, t = r+1, \cdots, N. \quad (1.17)$$

Since this equation must be satisfied identically with respect to all independent displacements x_s, it yields $V = (N - r)$ equations in the unknown independent components p_t of the perturbation load p_n :

$$a_{ns} a_{nt} p_t = a_{ns} P_n. \quad (1.18)$$

This system of equations can also serve as a check for an equilibrium load: if

$$a_{ns} P_n = 0, \quad (1.19)$$

then a perturbation component in the load is absent and $P_n^* = P_n$, thereby producing a compatible system of equilibrium equations.

Orthogonal Load Resolution. The significance of the orthogonality condition,

$$P_n^* p_n = 0, \quad (1.20)$$

lies in the fact that an arbitrary load *can always be uniquely resolved into equilibrium and perturbation components.* This orthogonal resolution is the key to statical-kinematic analysis; its basic properties can be summed up as follows:

(1) For a system in a given geometric configuration, an arbitrary load resolves uniquely into mutually orthogonal equilibrium and perturbation components.

(2) An equilibrium load, including the special kind, self-stress, generates internal forces in the system but no displacements. A perturbation load produces displacements but, having no component in the column space of the equilibrium matrix, does not affect the member forces. (More precisely, a perturbation load cannot be accounted for in the equilibrium equations for the given configuration; it requires a nonlinear displacement analysis.)

(3) For equilibrium loads, superposition of the respective force distributions is valid. For perturbation loads, superposition of displacements holds.

(4) Both the resolution of a general load and the described properties of its components are local, i.e., related to a given configuration of the system and valid only for a linearized analysis.

The concepts of equilibrium and perturbation loads are illustrated in the following example.

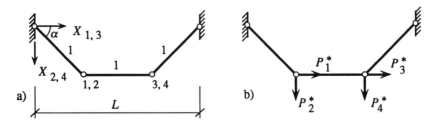

Figure 1.2. A three-bar structural system: a) a symmetric configuration; b) four load components.

Example 1.1. The pin-bar system in Fig. 2a (called a "four-bar" mechanism in mechanical engineering, but referred to here as a "three-bar" structural system) consists of three identical bars of unit length. In the Cartesian coordinate system shown, the constraint equations are

$$X_1^2 + X_2^2 = 1, \quad (X_3-X_1)^2 + (X_4-X_2)^2 = 1, \quad (L-X_3)^2 + X_4^2 = 1.$$

The shown symmetric configuration of the system is conveniently specified by the angle α: $X_1 = \cos \alpha$, $X_2 = X_4 = \sin \alpha$, $X_3 = 1 + \cos \alpha$. This description accounts for the fact that all three bar lengths are $L_i = 1$ and thus are present in the following calculations only implicitly. This is important to remember when verifying the dimensions of the related parameters and variables, both here and in many subsequent examples utilizing the same system. Denoting

$$\cos \alpha = c, \quad \sin \alpha = s, \tag{a}$$

the linearized constraint equations for this configuration can be written

$$F_n^i x_n = 2 \begin{bmatrix} c & s & 0 & 0 \\ -1 & 0 & 1 & 0 \\ 0 & 0 & -c & s \end{bmatrix} \begin{bmatrix} x_1 \\ x_2 \\ x_3 \\ x_4 \end{bmatrix} = 0. \tag{b}$$

Obviously, the given configuration is an equilibrium configuration for any symmetric load; it is intuitively less obvious (but true) that there are many asymmetric equilibrium loads as well. This can be easily verified by considering the system of equilibrium equations obtained by transposing the above constraint Jacobian matrix

$$2 \begin{bmatrix} c & -1 & 0 \\ s & 0 & 0 \\ 0 & 1 & -c \\ 0 & 0 & s \end{bmatrix} \begin{bmatrix} \Lambda_1 \\ \Lambda_2 \\ \Lambda_3 \end{bmatrix} = \begin{bmatrix} P_1^* \\ P_2^* \\ P_3^* \\ P_4^* \end{bmatrix}. \tag{c}$$

The matrix is of rank $r = 3$, so that only three out of the four equilibrium load components can be arbitrary. Suppose the first three components are of unit magnitude (Fig. 2b). Then the constraint reactions Λ_i are found by solving the first three equations:

$$\Lambda_1 = 1/2s, \quad \Lambda_2 = (c/s - 1)/2, \quad \Lambda_3 = (c/s - 2)/2c. \qquad (d)$$

According to (12), with $L_i = 1$, the corresponding bar forces are simply double the above reactions:

$$N_1 = 1/\sin \alpha, \quad N_2 = \operatorname{ctn} \alpha - 1, \quad N_3 = 1/\sin \alpha - 2/\cos \alpha.$$

The fourth component, P_4^*, of the equilibrium load is imposed by the equilibrium conditions of the given configuration and is evaluated from the last of equations (c):

$$P_4^* = (c - 2s)/2c = 1/2 - \tan \alpha. \qquad (e)$$

As is readily seen, this load depends on the chosen independent load components and on the system geometry represented in this case by the angle α. A special case occurs at a value of $\tan \alpha = 1/2$. At this angle the magnitude of P_4^* is zero, while at steeper angles its direction is downward, and at shallower angles, upward.

It is interesting (and may be contrary to intuition) that for this symmetric configuration of the system there exist many more independent asymmetric equilibrium loads than symmetric ones. Indeed, general, i.e., asymmetric, equilibrium loads are determined by three arbitrarily chosen components, whereas symmetric ones are determined by just two such components.

Since the rank of the constraint Jacobian matrix is $r = 3 < N = 4$, the system has one perturbation load. It coincides with the pattern of virtual displacements which is also determined by one independent component. In both cases, the remaining components can be evaluated with the aid of the matrix a_{ns} defined in (7). Designating x_2 as the only independent variable, the system (b) is solved to give

$$x_n = a_{n2}\, x_2, \quad a_{n2} = [-t, \; 1, -t, \; -1]^{\mathrm{T}} \quad (t = \tan \alpha), \qquad (f)$$

wherefrom the perturbation load is found as

$$p_n = a_{n2}\, p_2 \qquad (g)$$

with an arbitrary p_2.

For a rectilinear configuration of the system, $\sin \alpha = s = 0$ and the Jacobian matrix rank drops to $r = 2$. Accordingly, the number of independent components of the equilibrium load reduces to two, P_1^* and P_3^*; the dependent ones are $P_2^* = P_4^* = 0$. The system has two perturbation loads given by (16) with arbitrary p_s and

$$a_{ns} = \begin{bmatrix} 0 & 1 & 0 & 0 \\ 0 & 0 & 0 & 1 \end{bmatrix}^{\mathrm{T}}. \qquad (h)$$

Constraint Variations

Compatible and Anticompatible Constraint Variations. Formal observations and interpretations paralleling those made on the equations of equilibrium are valid for the inhomogeneous linearized constraint equations

$$F_n^i\, x_n = f_i. \tag{1.21}$$

Here the *constraint variations*, f_i, are infinitesimally small (of the same order as displacements x_n), and local (configuration-specific). They represent small changes in the geometric parameters of system components, due to deformations of the structural members of any origin (elastic or thermal deformations; lack of fit of components; or even computer experiments with member sizes). Mutually *compatible variations*, f_i^*, are in the column space of the constraint Jacobian matrix and produce infinitesimal displacements of the same order as the constraint variations themselves. When the rank is $r = C$, and only in this case, every variation of constraints is compatible. Otherwise, there exist $C - r$ linearly independent constraint variations that reside in the orthogonal complement of the column space (i.e., in the left nullspace) of the constraint Jacobian matrix and are of the form

$$f_i^{\perp} = b_{ik}\, f_k \tag{1.22}$$

with coefficients from (15) and arbitrary f_k. It is logical to call such a variation *anticompatible* since it is not just incompatible, but "pure": it does not contain any compatible component. A similar concept in the theory of elasticity is the tensor $Ink\,(\varepsilon)$ (from the German) which extracts the purely incompatible component of a general strain tensor. Anticompatible variations are impossible without either a constraint discontinuity (which amounts to the system disintegration) or large displacements. In an elastic system, these variations would produce self-stress. Since the left nullspace of the Jacobian matrix is the space of self-stress patterns, it is only natural that anticompatible variations correspond to the statical counteraction of constraints, i.e., self-stress.

Orthogonal Resolution of Constraint Variations. The orthogonality condition for the subspaces of compatible constraint variations and self-stress,

$$f_i^*\, \Lambda_i = 0, \tag{1.23}$$

can be transformed using relations (15) and (22):

$$f_i^*\, \Lambda_i = (f_i - f_i^{\perp})\, \Lambda_i = (f_i - b_{il}\, f_l^{\perp})\, b_{ik}\, \Lambda_k = 0, \quad k, l = r+1, \cdots, C. \tag{1.24}$$

Since this equation must be satisfied identically with respect to all independent constraint reactions Λ_k, it yields $S = (C - r)$ equations in the unknown independent components f_k of the anticompatible constraint variation $f_i^{\perp}$:

$$b_{ik}\, b_{il}\, f_l^{\perp} = b_{ik}\, f_i. \tag{1.25}$$

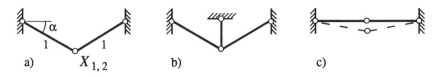

Figure 1.3. Constraint variations in simple trusses: a) all constraint variations are compatible; b) and c) examples of systems with anticompatible variations.

This system of equations can also serve as a check for a compatible variation: if

$$b_{ik} f_i = 0, \tag{1.26}$$

then an anticompatible component in the variation is absent and $f_i = f_i^*$, thereby producing a compatible system of linearized constraint equations. Thus, the matrix b_{ik} is a discrete analog of the above-mentioned operator $Ink\,(\,)$ from the theory of elasticity.

The orthogonality condition for compatible and anti-compatible variations,

$$f_i^* f_i^\perp = 0, \tag{1.27}$$

means that an arbitrary constraint variation resolves uniquely into the two components, f_i^*, and $f_i^\perp$. The basic properties of this resolution parallel those of the orthogonal load resolution:

(1)   For a system in a given geometric configuration, an arbitrary variation of constraints resolves uniquely into mutually orthogonal compatible and anticompatible components.

(2)   A compatible variation, including the special kind, a trivial variation ($f_i = 0$), generates displacements in the system but no internal forces. An anti-compatible variation produces forces but, having no component in the column space of the constraint Jacobian matrix, does not affect the system displacements. (More precisely, an anticompatible variation cannot be accounted for in the linearized constraint equations for the configuration; it requires a nonlinear displacement analysis.)

(3)   For compatible constraint variations, superposition of the respective displacements is valid. For anticompatible variations, superposition of self-stress patterns holds.

(4)   Both the resolution of a general constraint variation and the described properties of its orthogonal components are local, i.e., related to a given configuration of the system and valid only for a linearized analysis.

Example 1.2. Using notation (a) of Example 1, the constraint equations for a two-bar truss with unit length bars, shown in Fig. 3a, are

$$X_1^2 + X_2^2 = 1, \quad (2c - X_1)^2 + X_2^2 = 1. \tag{a}$$

The inhomogeneous linearized constraint equations (18) can be written

$$2 \begin{bmatrix} c & s \\ -c & s \end{bmatrix} \begin{bmatrix} x_1 \\ x_2 \end{bmatrix} = \begin{bmatrix} f_1^* \\ f_2^* \end{bmatrix}. \tag{b}$$

The Jacobian matrix rank is $r = 2$ and equals the number of constraints, $C = 2$. Accordingly, any infinitesimal constraint variation in this system is compatible and produces the corresponding infinitesimal displacements found from (b). The displacements are of the same order as the constraint variations.

Introducing a third bar (Fig. 3b) gives rise to an additional constraint equation,

$$(X_1 - c)^2 + X_2^2 = s^2. \tag{c}$$

As a result, the linearized constraint equations become

$$2 \begin{bmatrix} c & s \\ -c & s \\ 0 & s \end{bmatrix} \begin{bmatrix} x_1 \\ x_2 \end{bmatrix} = \begin{bmatrix} f_1^* \\ f_2^* \\ f_3^* \end{bmatrix}. \tag{d}$$

Here, $C = 3$, but the rank is still $r = 2$, so that only two components in any compatible constraint variation are independent. The remaining one must conform to the first two, as determined by equations (d); otherwise, it is incompatible. If, for instance, the first two components are of one and the same magnitude, f, the compatible third one must be $f_3^* = f$ so that

$$f_i^* = f[1, \ 1, \ 1]^{\mathrm{T}}. \tag{e}$$

The anticompatible constraint variation $f_i^{\perp}$ mimics the self-stress pattern of the system and, therefore, can be found by solving the homogeneous equilibrium equations

$$2 \begin{bmatrix} c & -c & 0 \\ s & s & s \end{bmatrix} \begin{bmatrix} \Lambda_1 \\ \Lambda_2 \\ \Lambda_3 \end{bmatrix} = 0. \tag{f}$$

With Λ_3 as independent,

$$\Lambda_1 = \Lambda_2 = -\Lambda_3/2 \tag{g}$$

and, according to (17), anticompatible variations are determined by one arbitrary component, f:

$$f_i^{\perp} = b_{i3} f, \qquad b_{i3} = [-1/2, \ -1/2, \ 1]^{\mathrm{T}}. \tag{h}$$

In a degenerate system in Fig. 3c, $s = 0$, so that $r = 1$. There is only one independent component in the compatible constraint variation, say, $f_1^* = f$, and then the second component must be $f_2^* = -f$. A discontinuity would result if $f_2^* < -f$, whereas $f_2^* > -f$ would produce a very large displacement (in comparison to f).

Statical-Kinematic Duality

Statical-kinematic duality is a set of formal interrelations between corresponding statical and kinematic concepts, their respective analytical attributes, and their numerical characterizations. The duality stems from the role of kinematics in the very origin of statics as a description of equilibrium in structural and mechanical systems. The duality relations are not only illuminating but also very useful in establishing the basic properties of a structural system associated with the type of mobility (finite, infinitesimal, or zero). Using these relations, the kinematic properties of a system can be revealed by means of intuitively more appealing and easily verifiable statical analysis. An example of such utility is the zero load test, which is simply a check of equilibrium matrix rank by statical means. Other examples will be considered in Chapter 2, after exploring nonlinear manifestations of statical-kinematic duality.

Conceptual Sources of Statical-Kinematic Analysis. The definition of equilibrium as expressed by the principle of virtual work is universal: it applies to both discrete and continuous systems and to any particular form of equilibrium equations (algebraic, differential, integral, etc.). Of course, it also applies to systems with deformable, flexible components, if the elastic displacements are treated as virtual.

Although the principle of virtual work is analytically equivalent to the equations of equilibrium, it conceptually precedes them and, in a sense, is more powerful. The principle of virtual work defines equilibrium as the condition of orthogonality between any linear combination of constraint reactions (such a combination is an equilibrium load) and any virtual displacement of the system in a given configuration. This orthogonality is, in fact, the reason why the system is static: an equilibrium load is what it is precisely because it cannot perform work over the virtual displacements, lowering in the process the potential energy of the system. In other words, in the context of systems with ideal positional constraints, the principle of virtual work is a statement of conservation of mechanical energy under infinitesimal reversible perturbations.

When superimposed on kinematics, the principle of virtual work engenders the entire field of statics, including the stability aspect of equilibrium, as a *derivative* conception. Moreover, it does it in two ways: logically (virtual displacements are needed as a prerequisite for implementing the principle), and literally (virtual displacements are a manifold tangent to kinematic displacements). Thus, the very way the principle of virtual work links statics to kinematics clearly demonstrates the local and derivative nature of statics and the precedence of kinematics.

The described relation is schematized in Fig. 4. The starting point is the global kinematics of a system with ideal positional constraints, as formally expressed by constraint equations. Their linearization engenders virtual displacements as infinitesimal increments of generalized coordinates and describes

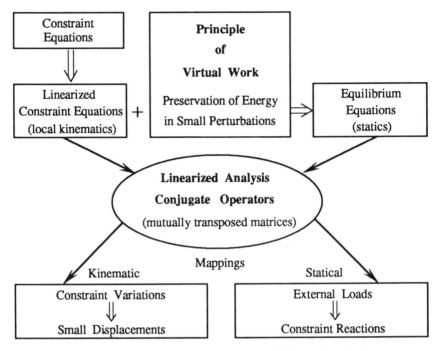

Figure 1.4. Conceptual sources of statical-kinematic analysis.

the local kinematics of the system by means of the linearized constraint equations. From here it is obvious that the kinematic aspect of the problem exists, and is analytically formulated, absolutely independently of any possible subsequent considerations related to the system energy, statics, dynamics, stability, and so on.

Only after virtual displacements are formally introduced can the principle of virtual work be implemented. Within the framework of a first-order (linear) analysis, this principle brings forward the force, the energy, and the ensuing equilibrium aspects that accompany kinematically possible transformations of the system. Thereby, the system kinematics is linked with such new and extraneous factors as mechanical work, external loads, and constraint reactions. The resulting equilibrium equations provide the means for identifying, among all kinematically possible configurations of the system, those configurations that are statically possible as well. In a second- or higher-order analysis employing the respective higher-order coordinate increments, the principle describes the stability aspect of equilibrium.

Thus, there is no conceptual symmetry or parity between statics and kinematics: kinematics can (and does) exist without any relation to statics, but not vice versa. In other words, kinematics is precedent to statics, in spite of the ostensibly opposite connotation implied by such traditional terms as statical-kinematic analysis and duality.

Statical and Kinematic Mappings. In contrast to the analysis of constraint equations (1), which describe the global kinematics of the system, a statical-kinematic analysis based on the linearized constraint equations and equilibrium equations imparts a local character to all related considerations and analytical results. Therefore, the validity of equations (4) and (9) is confined to a certain fixed (albeit not necessarily known in advance) state of the system. This involves both the geometric configuration at which the linearization was carried out, and the status of each constraint, such as yielding, buckling, or (in the case of unilateral constraints) disengagement. Obviously, different geometric configurations and states of one and the same system that arise from arbitrary constraint variations may result in different statical-kinematic properties. Accordingly, these properties characterize the particular states and configurations and cannot be attributed to the system per se unless the configuration is unique and the constraints are perfectly rigid.

For a system in a given configuration, the statical and local kinematic analyses can be interpreted as mappings (Kuznetsov, 1972a). The solution of the equilibrium equations (9) maps the load space into the constraint reaction space. Investigation of the properties of this mapping (its feasibility, domains in both spaces, conditions for single- or multiple-valuedness, the relevant arbitrary elements, etc.) is the object of statical analysis. For instance, the mapping is infeasible if the external load is not an equilibrium load for the given configuration. If it is, the mapping is nonunique when homogeneous equations allow nontrivial solutions. Then the number of such solutions equals the equilibrium matrix nullity,

$$S = C - r, \qquad (1.28)$$

which is the number of independent self-stress patterns and of anticompatible constraint variations.

Analogously, the object of local kinematic analysis based on equations (21) is the investigation of a mapping of the space of constraint variations into the displacement space, or more precisely, into the tangential hyperplane of the coordinate space corresponding to a particular reference configuration of the system. This mapping also may be nonunique, impossible, or multivalued. The latter takes place when the nullspace of the constraint Jacobian matrix is of a nonzero dimension, in which case the matrix nullity,

$$V = N - r, \qquad (1.29)$$

is the number of independent virtual displacements and of perturbation loads for the configuration considered.

The fact that the matrices of equations (9) and (21) are mutually transposed is the reason for the properties of the two mappings being closely related. Obviously, similar relations exist for continuous problems as well. In this case the statical and kinematic aspects are represented by boundary value problems with conjugate operators and natural boundary conditions. This connection between the statical and local kinematic analyses (and between the two associated

Table 1.1.   Statical-Kinematic Duality: Summary of Dual Concepts.

Local parameter	Kinematically is the number of independent	Statically is the number of independent
r	compatible constraint variations	equilibrium loads
$V = N - r$	virtual displacements	perturbation loads
$S = C - r$	anticompatible constraint variations	self-stress patterns

mappings) gives rise to a system of interrelations called the *statical-kinematic duality*. Within this framework, every pertinent concept, parameter, or equation in kinematics has a counterpart in statics, and vice versa. Furthermore, even relations of a mixed, statical-kinematic, nature are mirrored by their kinematic-statical counterparts. For example, the equation of virtual work (8), clearly of a mixed nature, has a counterpart (23),

$$f_i^* \, \Lambda_i = 0. \tag{1.30}$$

Of course, this condition of orthogonality between the spaces of compatible constraint variations and self-stress does not represent a fundamental independent law like the principle of virtual work and is, in fact, a consequence of the latter. Nevertheless, it is a valid and useful equation. Moreover, if analytically combined with independently (axiomatically) established equations of statics, relation (30) would produce the linear constraint equations and, thereby, the entire linear local kinematics.

Summary of Statical-Kinematic Interrelations. For a discrete system, the conjugate kinematic and statical operators are, respectively, the constraint Jacobian matrix and its transpose. As a result, the pivotal role in statical-kinematic duality belongs to the matrix rank r. Its relationship to the number of coordinates, N, and the number of constraints, C, has both kinematic and statical implications summarized in Table 1.

The apparent symmetry among the concepts in the table should not be construed as an indication of equal standing between statical and kinematic properties of a system. Even the symmetry itself is somewhat superficial. For example, among the objects figuring in the duality relations, only two statical objects—equilibrium loads and constraint reactions—are of finite magnitudes, whereas their kinematical counterparts, along with all other duality objects, are infinitesimal. The adopted notation emphasizes this disparity by consistently denoting the respective objects by capital or lowercase letters. Of course, the infinitesimal quantities, such as displacements or perturbation loads, are extrapolated in practical calculations to small but finite magnitudes.

All of the concepts and parameters in Table 1 are local, i.e., configuration-specific, due to the presence of the Jacobian matrix rank. However, not all statical-kinematic properties of a system are of a local nature. Some are determined by the system composition, that is, by its topology rather than geometry. In particular, a few global invariants are obtained below (section 4) by eliminating the rank r from several statical-kinematic relations.

At the same time, the fact that the most important local properties of a system are determined by the constraint Jacobian matrix does not mean that a local statical-kinematic analysis is confined to the study of equilibrium equations and linearized constraint equations. The energy aspect introduced by the principle of virtual work makes possible a more versatile and comprehensive local analysis. In it, the second- or higher-order infinitesimal vicinity of the reference configuration of the system is investigated using power-series expansions. This is necessary, for instance, in assessing the stability of equilibrium which requires an evaluation of the lowest-order nonvanishing virtual work. This analysis, presented later in Chapter 2, reveals several new local statical-kinematic interrelations and extends the concept of statical-kinenatic duality beyond the conventional first-order analysis.

1.3 Statical-Kinematic Types of Structural Systems

Relations Among Statical and Kinematic Properties

Statical Criterion of Invariance. For the purpose of classification, structural systems can be divided into two basic classes: those that are capable of balancing an arbitrary load in the given configuration and those that are not. The two classes are respectively denoted by S^+ and S^-. This statical division has a kinematic counterpart that separates systems into those that possess a unique geometric configuration (i.e., lack kinematic mobility), denoted by K^-, and those that allow displacements, denoted by K^+. Despite the ostensible symmetry between these two classifying aspects, there is no parity between them. The kinematic division is primal and global, whereas the statical one is derivative and local.

For a structural system, the ability to balance an arbitrary load in its original configuration means that every load is an equilibrium load. This is equivalent to S^+, so that, in terms of rank r of equilibrium equations (9),

$$r = N \Leftrightarrow S^+. \tag{1.31}$$

It follows immediately from equations (4) that (31) entails the absence of nontrivial virtual displacements and, hence, of kinematic displacements as well. Thus, the outcome S^+ is a necessary and sufficient condition for geometric invariance of the system (statical criterion of invariance) and yields the result K^-. This may be written in the following form

$$S^+ \Leftrightarrow \text{invariance} \Rightarrow K^-. \tag{1.32}$$

Obviously, K^+ is a criterion of a geometrically variant system and yields S^-:

$$K^+ \Leftrightarrow \text{invariance} \Rightarrow S^-. \tag{1.33}$$

Thus the combinations $(S^+ K^-)$ and $(S^- K^+)$ characterize the two main statical-kinematic types of systems—invariant and variant, respectively. The combination $(S^+ K^+)$ is ruled out by virtue of (32), but the case of $(S^- K^-)$, covering other possible types, remains to be examined.

It is interesting to observe how the perceived symmetry of the statical and kinematic analyses and their corresponding classifications breaks down. On the statical side, the equilibrium equations (9) are the only basis for the statical division and their analysis yields a definite result, either S^+ or S^-. On the kinematic side, however, the corresponding result is not always so easy to obtain, because analyzing the linearized homogeneous constraint equations (4) may at best answer only the question on *virtual*, not kinematic, mobility.

The absence of virtual mobility, denoted by V^-, produces a conclusive result. It entails both the uniqueness of the system configuration and its ability to support an arbitrary external load. That is,

$$K^- \Leftarrow V^- \Leftrightarrow S^+ \Leftrightarrow \text{invariance}. \tag{1.34}$$

On the other hand, the presence of virtual mobility, V^+, is conclusive only for a nonsingular configuration of the system (the Jacobian matrix is of full rank), in which case

$$K^+ \Leftarrow V^+ \Leftrightarrow S^- \Leftrightarrow \text{variance}. \tag{1.35}$$

Singular Configurations.   A major complication arises when virtual mobility is accompanied by system degeneration (a drop in the Jacobian matrix rank relative to adjacent configurations), producing a singular configuration. In this case an analysis based on the linear approximation (4) is inadequate, and higher-order terms of expanded constraint equations (3) must be considered. It can happen that a degenerate system, while allowing virtual displacements, nevertheless lacks kinematic mobility. In this case the situation $(S^- K^-)$ does arise, indicating that even though the system is not geometrically invariant (it cannot balance an arbitrary load), it possesses a unique configuration. More precisely, such a system allows infinitesimally small displacements requiring only second- or, perhaps, higher-order variations of constraints; this property qualifies the system as one with infinitesimal mobility.

Systems with infinitesimal mobility belong to two distinct statical-kinematic types. The difference between them is that of principle and is rooted in structural topology. A system of the first type has an adequate number and arrangement of constraints (e.g., bars) to be geometrically invariant, but is an infinitesimal mechanism because of its singular, degenerate geometric configuration. In other words it is *topologically adequate*, but *geometrically singular*. There always exists a constraint variation (a change in the structural component sizes) that would take the system out of degeneration and make it geometrically invariant. For this reason, systems of this type are called *quasi-invariant*.

In contrast, an infinitesimally mobile system of the second type is *topologically inadequate*; its structural topology is such that no constraint variation would render it invariant. Such systems are almost always variant but allow exceptional, *geometrically singular*, immobile configurations. Here, constraint variations can only restore kinematic mobility, thereby reducing the system to its generic type—geometrically variant. Therefore, singular systems of this type are called *quasi-variant*. Thus, in both quasi-invariant and quasi-variant systems, infinitesimal mobility has its source in geometric singularity.

It should be borne in mind that not every singular configuration of a variant system is characterized by infinitesimal mobility; a typical example is a "dead-center" configuration of a four-bar mechanism. Here the singularity has some profound consequences, both statical and kinematic, but kinematic immobility is not one of them. On the other hand, degeneration of an invariant system may lead not only to infinitesimal, but to finite mobility as well.

A Statical-Kinematic Classification

Four Statical-Kinematic Types of Structural Systems. The above considerations underlie a comprehensive classification of structural systems. Adopting the point of view of a structural engineer, this classification arranges the four (and only four) existing statical-kinematic types of systems in order of a step-by-step degradation of stiffness. This is reflected in the successive relaxation of the respective criteria for the four types. The classification reflects the already apparent bias favoring the kinematic aspect of system description. The above-mentioned precedence of the kinematic properties of a structural system over the statical ones is seen in the fact that the basic *kinematic properties* (virtual and kinematic mobility or immobility, constraint dependence) *control the statical properties* (such as the system ability to support all or only selected external loads, uniqueness or nonuniqueness of statically possible states under these loads).

On the other hand, the alternative, statical, characterization of a system is in many cases intuitively appealing and easily obtained (recall the unit load test). In such cases, the statical-kinematic interrelations become helpful in revealing the less obvious kinematic properties of the system (examples of this kind are discussed on many occasions below). Furthermore, because of practical limitations on geometric precision, desired singular configurations can be realized only by statical means.

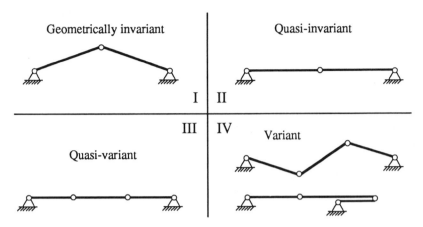

Figure 1.5. The simplest prototypes of the four statical-kinematic types of structural systems.

Definitions. The following definitions of the four statical-kinematic types of structural systems provide equivalent dual formulations, while putting first the kinematic aspect.

Type I: A *geometrically invariant* system is one devoid of even virtual mobility; perturbation loads do not exist (any load is an equilibrium load).

Type II: A *quasi-invariant* system is one possessing only virtual mobility and admitting a constraint variation that renders it invariant; perturbation loads exist and would produce infinite constraint reactions.

Type III: A *quasi-variant* system is one possessing only virtual mobility and *not* admitting a constraint variation that would render it invariant; perturbation loads exist and would produce infinite constraint reactions.

Type IV: A *geometrically variant* system is one possessing kinematic mobility; perturbation loads exist and produce a change in configuration.

On the basis of these definitions, the statical-kinematic classification of structural systems can be presented in the form of a four-quadrant Classification Table shown in Fig. 5. The main diagonal of the table (quadrants I and IV) contains the main types—invariant and variant—and the secondary diagonal (quadrants II and III) contains the singular, degenerate types—quasi-invariant and quasi-variant. The table also shows the simplest prototypes of the four types.

For brevity, the Roman numerals will sometimes be used to designate systems, in place of the above names for the four statical-kinematic types. Note, for example, that types I, II, and III possess a unique geometric configuration (they lack kinematic mobility); types II, III, and IV cannot balance an arbitrary load in their original configurations (perturbation loads exist along with virtual displacements); types II and III are infinitesimal mechanisms (they

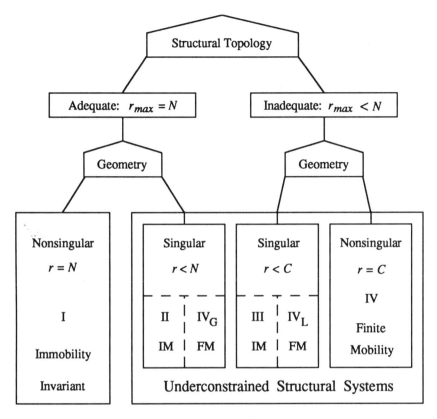

Figure 1.6. Combinations of structural topology and geometry producing the four types of structural systems. G and L mean global and local singularity, IM and FM—infinitesimal and finite mobility.

possess virtual, but not kinematic, mobility). Systems of type IV are the only ones that allow kinematic displacements. It is convenient to attribute the latter to trivial constraint variations ($f_i = 0$). Obviously, these variations cannot change the system type; only *nontrivial* constraint variations can do so. A familiar example is a constraint variation bringing a singular system II or III in and out of degeneration.

Another view of the four types of systems is given in Fig. 6, showing their topological and geometric origins in detail. Summarizing, a topologically adequate structural system is almost always invariant, but under nontrivial constraint variations may degenerate and acquire virtual or even kinematic mobility. In other words, a system of type I may degenerate into type II or even a singular type IV with globally dependent constraints (this is illustrated in Example 3 later in this section). On the other hand, a topologically inadequate structural system is almost always variant (type IV), although some nontrivial variations of constraints may bring it into a singular, kinematically immobile configuration (type III), where the system possesses only virtual mobility.

It is clear from the logic underlying the definitions of the four statical-kinematic types of systems that no structural system falls outside the developed classification. However, the introduced classification is more than just a matter of convenient and comprehensive labeling of the structural types: it provides valuable information on what to expect and count on, and what to beware and avoid in structural design.

Available Analytical Criteria. Topological adequacy or inadequacy is an immutable characteristic of a structural system: no constraint variation can turn an inadequate system into an adequate one, and vice versa. The reason is simple: structural topology is inherently blind to constraint variations. An analytical criterion for a *topologically adequate* structural system is

$$r_{max} = N, \tag{1.36}$$

where r_{max} is the maximum rank of the constraint Jacobian matrix attained in nontrivial constraint variations of the system. A *topologically inadequate* system is one with

$$r_{max} < N. \tag{1.37}$$

As to geometry, a *geometrically nonsingular* system has

$$r = r_{max}, \tag{1.38}$$

and a *geometrically singular* system is one with

$$r < r_{max}. \tag{1.39}$$

These observations pave the way to formulating explicit analytical criteria for each of the four statical-kinematic types. At this stage, such criteria can be given only for the two main types of systems, I and IV, and even then only the criterion for geometrically invariant systems can be given completely. To fully specify the criteria for the remaining system types, it is necessary to deal first with the subject of singular geometry. This requires an investigation of the second (or, perhaps, a higher) infinitesimal vicinity of the reference configuration of a system in the coordinate space, which is done in the next chapter.

A necessary and sufficient analytical criterion for a *geometrically invariant* system (type I) is straightforward and comprehensive. Conditions (36) on topology and (38) on geometry must be satisfied simultaneously, yielding

$$r = N. \tag{1.40}$$

This means that a system is both topologically adequate and geometrically nonsingular, a formal description of systems of type I.

With the other main type—*variant* systems—the situation is simple only for the nonsingular case. Applying conditions (37) and (38) leads to the sufficient analytical criterion:

$$C = r < N. \tag{1.41}$$

This criterion is not all-inclusive, as it omits systems of type IV in singular configurations (like the previously mentioned dead-center position), when $r < C$. Here a higher-order analysis is required to discriminate between singular configurations of kinematically mobile (singular type IV) systems and kinematically immobile (type III) systems.

1.4   Structural Indeterminacies

Statical Determinacy and Indeterminacy

Insufficiency of Conventional Concepts.   The conventional concepts of statical determinacy and indeterminacy (redundancy) in structural mechanics have been derived, respectively, from the sufficiency or insufficiency of static equilibrium conditions to determine constraint reactions, while the degree of statical indeterminacy is evaluated by counting redundant constraints. These concepts originated in the context of geometrically invariant systems, analyzed as undeformable (or linearly deformable) models, and were fully adequate so long as consideration was confined to such systems. Cautious authors have avoided answering the question on statical determinacy or indeterminacy for systems other than invariant, and imply or insist that the very question is inappropriate unless invariance has been ascertained.

Attempts to apply these concepts, together with their traditional definitions, to noninvariant systems, has led to a loss of both clarity and substance. For example, in accordance with current definitions, a three-hinged arch is statically determinate when analyzed as an undeformable model, but must be recognized as statically indeterminate under finite displacements. (Equilibrium conditions are insufficient for determining its internal forces.) The purely qualitative basic concept thus becomes conditional on a quantitative—and highly subjective— factor (the displacement magnitude). This situation is further complicated by the fact that in the latter case there would be no satisfactory answer to the question of degree of indeterminacy. Further examples are a cable (regarded by some authors as statically indeterminate), a cable net (which at first glance is multiply indeterminate), the large class of the so-called *tensegrity* systems, membranes, and many other structural systems.

The situation with systems involving unilateral constraints is especially confusing. As is known (and will be shown again below), an invariant system involving unilateral constraints, even just one, is statically indeterminate. But if so, how can one answer the question as to the very existence of a statically determinate system with unilateral constraints? Since such a system cannot be invariant, the very formulation of the question in this manner is ambiguous.

Thus, the conventional concepts and definitions are consistent and meaningful only within the confines of a linear analysis of geometrically invariant systems. The questions arise, first, Why are these concepts not directly applicable to other types of systems? And, second, Are they substantive enough to be worthy of extending to all types of systems? Whereas answering the first question requires some elaboration, the answer to the second question is beyond doubt. The feature of statical determinacy or indeterminacy bears upon the system reliability and its behavior under yielding (or complete failure) of constraints, temperature changes, support settlement, or lack of fit of components. It also determines the possibility of prestressing, redistribution or adjustment of internal forces, and other important structural attributes. Ideally an extended concept should:

(i)   cover all types of systems;
(ii)   yield a clear-cut, unambiguous, and universal measure for the degree of statical indeterminacy; and
(iii)   coincide with the existing concepts when applied to a geometrically invariant system.

Extended Concept. Two factors have proved to be helpful in understanding and overcoming the difficulties with the conventional definitions when extending them to underconstrained structural systems (Kuznetsov, 1971). The two factors are: the purely statical character of the concepts in question, and the local nature of equilibrium equations. If both observations are taken into consideration, *statical determinacy* can be interpreted as the capability of the system in a given configuration to balance each of its equilibrium loads in a unique way in terms of the internal forces. It follows then that *statical indeterminacy* would mean the capability of a system to balance its equilibrium loads in a nonunique way. However, in spite of its simplicity and clarity, such a definition is weak in that the equilibrium loads figure in it while they are only a means for detecting this property of the configuration. The following definitions are more precise and yet provide a direct reference to uniqueness or nonuniqueness of the statically possible states:

Statical determinacy is the *absence* of nontrivial solutions to the homogeneous system of equilibrium equations. *Statical indeterminacy* is the *exsistence* of such solutions. The *number* of these solutions, S, is the *degree of statical indeterminacy*. It is also the number of independent self-stress patterns, the number of dependent constraints, and the equilibrium matrix nullity given by (28).

These definitions are based solely on the properties of the analytical model of the system, specifically, on the operator of the homogeneous statical problem which can be a matrix, a differential, integral, or some other type of operator. This reflects the fact that determinacy or indeterminacy is a fundamental intrinsic property of a given configuration of the system. Although this property reveals itself in the behavior of a system under many internal or external actions (for example, constraint variations, thermal or force loads, misfit of components, rheological phenomena, and so on), it depends neither on these actions nor even on their presence or absence. Only an implicit connection with such actions might exist: they determine both the configuration and the status of the constraints in the system, and thereby may affect all of its statical-kinematic properties, including the degree of statical indeterminacy.

It is readily seen that all three requirements (i)–(iii) posed above are now satisfied. First, by divorcing the concept of statical indeterminacy from the notion of redundant constraints and reverting to the purely statical nature of the concept, the new definition establishes its local, configuration-specific character and extends the concept to all types of systems. Second, the measure of degree of indeterminacy is rigorously, yet naturally, generalized to the same extent. Finally, since an invariant system with an undeformable model has a unique configuration, the extended concepts not only fully retain their original meanings but also coincide with their conventional counterparts.

It is interesting to reconsider, in this new light, the examples of a three-hinged arch, a flexible cable, and a prestressed net. Clearly, a three-hinged arch is geometrically invariant and statically determinate in all configurations, except for a singular one where the three hinges are collinear. In the latter case, the system degenerates into type II and admits a nontrivial solution of the homogeneous statical problem. As a result, the solution of the corresponding inhomogeneous problem for equilibrium loads is not unique (the only equilibrium load for this configuration is a force acting at the middle pin and collinear with the bars).

Both a cable and a cable net are topologically inadequate systems. All curvilinear configurations of a cable are statically determinate, whereas the rectilinear ones are indeterminate to at least the first degree. (The actual degree depends on the number of folds.) As regards a contour-supported cable net, it is statically determinate in any nonsingular configuration, but admits exceptional configurations (most interesting for applications) in which it is indeterminate to the first degree. Multiple statical indeterminacy is also possible; as demonstrated below, it occurs in a net arranged along asymptotic lines of a surface of negative total curvature.

Kinematic and Virtual Indeterminacies

Kinematic Determinacy and Indeterminacy. Whereas the concept of statical determinacy has lent itself to a natural and rather painless extrapolation to all statical-kinematic types of structural systems, the existing concept of kinematic determinacy needs a thorough overhaul. The number of independent virtual displacements, V [see (29)], obtained in a linearized analysis, is often perceived as the number of degrees of freedom and is sometimes called the degree of kinematic indeterminacy. One of the unsettling features of this approach is the fact that some or even all of the V degrees of freedom might be realizable only as infinitesimal displacements. In this case, the indiscriminate lumping together of virtual and kinematic displacements is inconsistent and may be misleading.

A convenient starting point is the following conceptualization of kinematic determinacy and indeterminacy as applied to underconstrained systems:

Kinematic determinacy is the *absence* of nontrivial solutions to the system of constraint equations. *Kinematic indeterminacy* is the *existence* of such solutions. The *number* of these solutions, K, is the *degree of kinematic indeterminacy*. It is also the number of independent kinematic displacements and the number of kinematic degrees of freedom. For any *nonsingular* configuration of the system this number is the parameter V given by (29).

Thus, kinematic determinacy means kinematic immobility and uniqueness of the system configuration whereas indeterminacy implies the opposite. It should be noted that both the concepts and the definitions are rigorous and unambiguous only within the context of statical-kinematic analysis. However, the same terminology is conventionally used with respect to displacements and degrees of freedom of a deformable geometrically invariant system. This is done, for example, with respect to the number of elastic displacements and the dimension of the stiffness matrix of the system. Such a usage is consistent only as long as genuinely kinematic displacements are absent and would be ambiguous, say, for an underconstrained structural system with flexible members. Here mobility comes from two sources: the system geometry and the component flexibility. Only the first source is reflected in the above conceptualization of kinematic determinacy and, in application to underconstrained systems, it will be used only in this proper, natural, sense.

Virtual Determinacy and Indeterminacy. A parallel set of definitions dealing with virtual displacements and the corresponding degrees of freedom evolves naturally.

Virtual determinacy is the *absence* of nontrivial solutions to the system of *linearized* constraint equations. *Virtual indeterminacy* is the *existence* of such solutions. The *number* of these solutions, V, is the *degree of virtual indeterminacy*. It is also the number of independent virtual displacements, the number of virtual degrees of freedom, and the constraint Jacobian matrix nullity, given by (29).

It is easy to see from the above definitions that virtual determinacy (i.e., virtual immobility) entails kinematic determinacy (kinematic immobility), but *not* vice versa. The definitions reaffirm the local and infinitesimal character of virtual displacements (in contrast, on both counts, to kinematic displacements). Virtual displacements are local with reference to the configuration at which linearization of the constraint equations was carried out; and are infinitesimal regarding the extent of validity of the solution to the linearized equations. Once again, the difference between virtual and kinematic displacements is not a matter of scale or extrapolation, but a matter of principle.

Table 1.2. Indeterminacies Present in Each Type of Structural System.

Type of System	I	II	III	IV	$IV_{G, L}$
Indeterminacies Present	$S*$	S, V	S, V	K, V	S, K, V

*Type I may be statically determinate as well; type IV excludes compound systems with statically indeterminate components; G—global, L—local.

The four statical-kinematic types of systems and the three types of structural indeterminacies are intertwined. In each system type, the indeterminacies may, and in most case must, be present in certain combinations. Table 2 summarizes the combinations of indeterminacies (denoted $S, K,$ and V) characterizing each of the four system types.

The significance of the presented definitions of structural indeterminacies is in their general nature, rigorous analytical basis and easily quantifiable degrees. According to these definitions, virtual determinacy is equivalent to geometric invariance and virtual indeterminacy—to being underconstrained. A system that is virtually indeterminate but kinematically determinate is a system with infinitesimal mobility; it is the subject of the following chapter.

Degree of Singularity. Changes in the configuration of a system caused by trivial ($f_i = 0$) or nontrivial constraint variations may lead to degeneration. The resulting singular configuration is characterized by a drop in the Jacobian matrix rank r compared to its maximum value, r_{max}, which is attained at adjacent nonsingular configurations of the system. The drop in the rank, which is the *degree of degeneracy* (or degree of singularity), is

$$D = r_{max} - r. \tag{1.42}$$

It is an important local parameter of the system. In particular, it establishes the differences between the local and global degrees of indeterminacy:

$$D = S - S_G = V - K. \tag{1.43}$$

Here S_G is the degree of global (topological) statical indeterminacy, present, for example, in a finite mechanism with statically indeterminate components. Note that although both V and S are local, their differrence

$$V - S = K - S_G = N - C \tag{1.44}$$

is a global invariant determined by the system topology. The reason is that the system degeneration, accompanied by a drop in the rank, affects the virtual and statical degrees of indeterminacy equally. From here, it follows, for instance, that a system that has $r = C = N$ is statically, virtually, and kinematically determinate $(S = V = K = 0)$. As is seen from (31) and (32), this is a geometrically invariant system. For this system, and only for it, every load is an equilibrium load while, at the same time, every constraint variation is

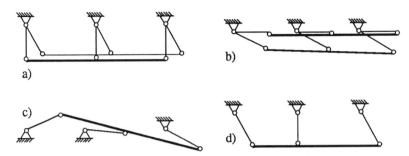

Figure 1.7. Transformations of a topologically adequate system: a) a finite mechanism with globally dependent constraints; b) a doubly degenerate (folded) configuration; c) and d) geometrically invariant configurations.

compatible—the familiar properties of a statically determinate system of type I. Thus, a topologically adequate system with $N = C$, possessing *one* of the three properties, statical determinacy, virtual determinacy, and geometric invariance, possesses *all* three of them and, in addition, is kinematically determinate as well.

Example 1.3. The assembly in Fig. 7 is a rigid beam with three parallel support bars. It is a simple, yet surprisingly multifaceted example of possible situations. This is a topologically adequate, but geometrically degenerate, finite mechanism with globally dependent constraints (type IV_G in Fig. 6). In any ordinary configuration it has one statically possible self-stress pattern and one kinematically possible displacement:

$$S = V = K = 1: \Rightarrow \text{type } 1. \tag{a}$$

The number of equilibrium loads is 2—a force parallel to the support bars and a moment—both applied to the rigid beam at an arbitrary location. The folded configuration in Fig. 7b, resulting from trivial variations of the constraints, is singular: the Jacobian matrix rank drops, so that

$$S = V = 2, \quad K = 1. \tag{b}$$

Accordingly, in this configuration the system further degenerates, increasing the degrees of virtual and statical indeterminacies. As a result, it allows two linearly independent self-stress patterns (initial forces in any two bars can be assigned arbitrarily), and two independent displacements (the horizontal bar can undergo a vertical translation and an infinitesimal tilt). The number of equilibrium loads reduces to 1 (a horizontal force); the moment can no longer be balanced because of the acquired virtual rotation (tilt).

Another interesting feature of this system, following from its characterizations (*a*) and (*b*), in conjunction with (44), is

$$N = C, \tag{c}$$

meaning that the system is topologically adequate. This entails the possibility

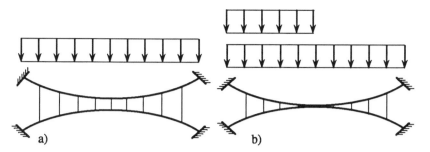

Figure 1.8. Prestressed cable systems: a) with one vertical equilibrium load and b) with two such loads.

of its conversion into type I, resulting from nontrivial constraint variations. Two of the geometrically invariant, statically determinate systems that can be obtained by changing the bar lengths are shown in Fig. 7c and d. More dangerous is the possibility of the opposite conversion: from immobility to kinematic—worse than virtual—mobility.

Design Implications. The above observations on equilibrium loads and statical, kinematic, and virtual indeterminacies are helpful in the conceptual design of underconstrained systems. Generally, the structural composition and geometric configuration of a system should be such that all, or at least the major, external loads are equilibrium loads. Then the undesirable kinematic displacements are either eliminated or minimized. For example, for the cable system in Fig. 8a, the uniform load shown is the only vertical equilibrium load. If it is also necessary to accommodate a half-span load, the system can be modified as shown in Fig. 8b.

One of the ways of optimizing the distribution of member forces in an underconstrained system is to employ a statically indeterminate configuration and to take advantage of nonuniqueness of the statically possible state. As an example, consider a suspended system of an elliptic planform, depicted in Fig. 9. An elliptic contour ring is known to be momentless (i.e., to be in pure compression) when the loads parallel to the axes of the ellipse are in the ratio

$$H_1 : H_2 = a^2 : b^2.$$

This force ratio can be realized in a cable net having the form of an elliptic parabolid. It happens that a cable net is a variant system (type IV) and an elliptic paraboloid is a singular, hence, statically indeterminate, configuration of this system with $S = D = 1$. The desired cable force ratio can be achieved in one of the statically possible force patterns under a uniform vertical load, W:

$$W = 2s(H_1/a^2 + H_2/b^2),$$

where s is the sag of the net. The above two equations uniquely determine the cable forces in a paraboloidal net with given sizes a, b, and s.

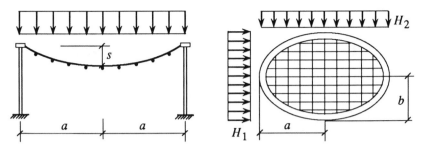

Figure 1.9. Advantageous use of statical indeterminacy in a variant system.

After the forces are determined, the required cross sections of the cables can be evaluated so as to have them fully stressed. In this case, only two types of cables are necessary, one for each of the two crossing arrays. Note that the cables being uniformly stressed is an independent design consideration, and not a requirement for achieving the desired cable force ratio. The latter can be realized (perhaps, somewhat contrary to intuition) with any cable cross sections, say, with all of the cables being identical.

With the cable forces and cross section areas known, the elastic elongations and the corresponding initial lengths of the cables can be evaluated, thereby providing complete information on the cable system geometry. Although a direct implementation of this geometry would entail the described advantageous stress state, it is impractical to use the individual cable lengths as a means for attaining it. A more practical solution is to "tune-up" the system by adjusting the tension level in the cables.

2
Systems with Infinitesimal Mobility

For a structural engineer, infinitesimal mobility has traditionally been an obscure side issue. In a typical text on structural analysis, systems with infinitesimal mobility are discussed very briefly, labeled geometrically unstable, hence unsuitable for practical applications, and abandoned from further consideration. For a mechanical engineer, infinitesimal mechanisms have been even less interesting, as they lack the quintessential feature of mobility. One of very few applications of such systems has been their use as a "force amplifier".

Recently, however, structural and mechanical assemblies with infinitesimal mobility have attracted renewed interest. These systems are unique in that they open a fundamentally different way of obtaining structural stiffness, acquiring it by means of prestress, at the expense of the material strength of a structural member rather than an increased size. Advances in materials science, providing an ever-increasing abundance of strength, underlie both the success and the potential of the concept. Moreover, since an increase in strength is usually not accompanied by a gain in stiffness, prestress may be the only way of capitalizing on the material strength. This is best exemplified by the *tensegrity* structures of Buckminster Fuller, which are multifreedom systems with infinitesimal mobility, stiffened by prestress. More recently, the concept has attracted the attention of structural, mechanical, and control theory specialists in the area of large, lightweight, deployable structures.

An analytical criterion for a geometrically invariant (stiff) three-dimensional system was formulated by Möbius (1837) for an assembly of n solids constrained in their motion by six external supports, and interacting at p points where mutual normal contact forces develop. Such an assembly is stiff when $p \geq 6(n-1)$. Stating immediately that this condition is necessary but not sufficient, Möbius proceeded with a thorough investigation of exceptional systems that satisfy his criterion but are not stiff. He revealed three interrelated properties of such systems: (a) they possess only infinitesimal mobility; (b) when the equations of equilibrium allow a solution, it is not unique; and (c) each structural member has a maximum or minimum size compatible with other members of the assembly.

When formulating a stiffness criterion for pin-bar systems, Maxwell (1890) reversed the roles of joints and bars by considering the former as material points and the latter as constraints. More interestingly, he recognized the existence of exceptional systems of the opposite nature: having fewer constraints than necessary for invariance, these systems are topologically inadequate and yet kinematically immobile. Like Möbius, Maxwell attributed the singular nature of his exceptional systems to the presence of maximum or minimum length bars. Mohr (1885) found this extremal feature to be a general condition for the statical possibility of self-stress in structural systems. Kinematic properties of singular systems were studied in more detail by Kötter (1912) and Levi-Civita and Amaldi (1930). The latter developed the most general approach, which is a basis and starting point for this work (Kuznetsov, 1991).

Definition. A system with *infinitesimal mobility* is a system allowing virtual displacements but no kinematic displacements.

Such a system, also called an infinitesimal mechanism, is described by

$$V > K = 0. \tag{2.1}$$

Thus, an infinitesimal mechanism possesses virtual degrees of freedom but no actual, kinematic ones and, therefore, is kinematically immobile. Statically, an infinitesimal mechanism is characterized by that perturbation loads exist and, in theory, produce infinite constraint reactions—the "force amplification" effect.

When the rank r of the constraint Jacobian matrix is full, that is, equal to the smaller of the matrix sizes, it follows from equations (1.4) that the system is nonsingular and belongs to one of the two main types, I or IV. Indeed, at $r = N \leq C$, all virtual displacements are zero and the system is of type I, while at $r = C < N$, a nonzero solution for x_n exists, so that displacements are possible and the system is of type IV. Thus, a unique configuration with infinitesimal mobility can occur only when r is smaller than both C and N, i.e., only in singular systems.

2.1 Simple First-Order Mechanisms

Analytical Criterion for a Simple First-Order Mechanism

For a singular system, $D > 0$ and the homogeneous equilibrium equations admit at least one nontrivial solution, $\Lambda_i \neq 0$. This can be used to form a linear combination of expanded constraint equations:

$$\Lambda_i f^i = \Lambda_i [F_n^i x_n + (1/2!) F_{mn}^i x_m x_n + ...] = 0. \tag{2.2}$$

Physically, this construct represents the incremental work of constraint reactions of self-stress over infinitesimal constraint variations produced by virtual displacements. Note that for a pin-bar system, the power expansions of the constraint functions contain only linear and quadratic terms. On multiplying out, the first product vanishes by virtue of (1.15) and the kinematic properties of the system are determined by the remaining quadratic or higher-order forms.

It is convenient to begin the analysis with a class of systems, hereafter called *simple infinitesimal mechanisms*, satisfying the following assumptions:

(1)   the degree of singularity (of degeneracy) is $D = 1$ and the self-stress pattern is comprehensive, so that all Λ_i are different from zero; and

(2)   the leading (lowest-order) nonvanishing form in (2) is quadratic, making the second-order analysis meaningful, if not always conclusive.

The assumption on a comprehensive self-stress pattern means that all of the system constraints are mutually dependent: an independent constraint in a prestressed system would be force-free.

Complete and Reduced Quadratic Form. The quadratic form in (2) is a form in all variables x_n; it is the *complete* quadratic form. It can be reduced by eliminating the dependent displacements with the aid of solution (1.7). The result is the *reduced* quadratic form $Q = Q(x_s, x_t, \dots, x_N)$ in V independent virtual displacements. If this form is definite, all displacements x_s must be zero. This means that only the linear forms in expansions (1.3) are compatible and vanish with arbitrary x_s, whereas the quadratic forms can simultaneously turn into zero only in a trivial way, i.e., together with all the variables. This prevents any nonzero virtual displacement from satisfying the constraint equations (1.1) and, consequently, prohibits any kinematic displacements. Thus, definiteness of the quadratic form Q is a sufficient condition for the system to be an infinitesimal mechanism.

To demonstrate the necessity of this condition, recall that the left-hand side in (2) is work performed by the self-equilibrated initial forces over constraint variations. The equilibrium state of the system does not change if one, say the j-th, constraint is cut and the force associated with Λ_j is applied at both sides of the cut. With the rest of the constraints intact, any set of displacements will produce a discontinuity f^j at the cut. (Were displacements possible without either a discontinuity or a constraint variation, the system would be kinematically mobile.) The displacement process requires the work $\Lambda_j f^j \neq 0$. In the absence of other constraint variations, this work equals Q; for it to be different from zero for any nonzero set of virtual displacements requires that the quadratic form Q be definite.

Thus, for a system satisfying the above two assumptions, definiteness of the reduced quadratic form Q constitutes both a necessary and sufficient condition for the system to be an infinitesimal mechanism.

In more formal terms, the existence of a nontrivial solution $\Lambda_j \neq 0$ is a necessary condition for any one of the constraint functions F^i to have an extremum subject to the fixed values of all the remaining functions. The type of the extremum depends on the character of the leading form in expansion (2). A strict constrained minimum or maximum of just one of the functions F^i entails a similar property for each of them and constitutes a necessary and sufficient analytical criterion for a simple infinitesimal mechanism.

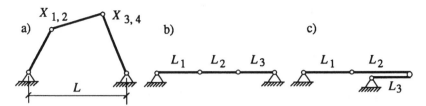

Figure 2.1. Configurations of a three-bar system: a) variant (finite mechanism); b) infinitesimal mechanism; and c) finite mechanism in a singular configuration.

For a pin-bar assembly, constructing the reduced quadratic form is especially easy. Since the constraint functions are homogeneous quadratic polynomials, the elementary quadratic forms in their power expansions are obtained instantly. Technically, this is done by replacing the coordinates, denoted by capital letters, with the coordinate increments, i.e., displacements, denoted by the same lowercase letters. The complete quadratic form is the sum of these elementary forms weighted by the respective constraint reactions Λ_i. The only remaining step in evaluating the reduced quadratic form is the elimination of r dependent displacements with the aid of the linearized constraint equations.

Quasi-Invariant and Quasi-Variant Systems. As defined in Chapter 1, a system with infinitesimal mobility is quasi-invariant (type II) if it is topologically adequate. According to (1.36), upon exiting from the degenerate configuration, the Jacobian matrix rank for such a system restores to $r = N$; the system looses its virtual mobility and becomes invariant. Thus, a quasi-invariant system is formally characterized by

$$V = D, \quad K = 0. \tag{2.3}$$

In contrast, a quasi-variant system (type III), as topologically inadequate, has

$$V > D, \quad K = 0, \tag{2.4}$$

and retains virtual mobility even after exiting from the degenerate configuration, since the rank r can never equal N.

Example 2.1. The familiar three-bar system (Fig. 1) is a ready probestone when dealing with infinitesimal mobility. The constraint equations are

$$X_1^2 + X_2^2 = L_1^2, \quad (X_3 - X_1)^2 + (X_4 - X_2)^2 = L_2^2, \quad (L - X_3)^2 + X_4^2 = L_3^2. \tag{2.5}$$

Upon omitting the factor 2, the constraint Jacobian matrices for an ordinary configuration and the two singular ones are, respectively,

$$
\begin{bmatrix}
X_1 & X_2 & 0 & 0 \\
X_1 - X_3 & X_2 - X_4 & X_3 - X_1 & X_4 - X_2 \\
0 & 0 & X_3 - L & X_4
\end{bmatrix},
\begin{bmatrix}
L_1 & 0 & 0 & 0 \\
-L_2 & 0 & L_2 & 0 \\
0 & 0 & \overline{+}L_3 & 0
\end{bmatrix}, \tag{2.6}
$$

where the double sign corresponds to the two respective singular configurations. As is readily seen, for a pin-bar assembly the Jacobian matrix elements are simply the horizontal and vertical projections of the bar lengths taken with a proper sign: positive for the displacements of the "far" end of the bar (in the global reference system), and negative for the "near" end. The corresponding equilibrium matrices are obtained by transposition:

$$
\begin{bmatrix}
X_1 & X_1-X_3 & 0 \\
X_2 & X_2-X_4 & 0 \\
0 & X_3-X_1 & X_3-L \\
0 & X_4-X_2 & X_4
\end{bmatrix}
,
\begin{bmatrix}
L_1 & -L_2 & 0 \\
0 & 0 & 0 \\
0 & L_2 & \overline{+}L_3 \\
0 & 0 & 0
\end{bmatrix}
. \tag{2.7}
$$

For an ordinary configuration of the system (Fig. 1a) the Jacobian matrix rank is full: $r = r_{max} = C = 3 < N = 4$. Hence,

$$V = K = 1,$$

and the linearized constraint equations given by the first matrix in (7) allow a nonzero solution; the system is kinematically mobile (variant, type IV).

For the two singular configurations shown, the rank drops to 2 so that

$$V = N - r = 2.$$

The equilibrium equations for these configurations given by the second matrix in (7), admit the respective nontrivial solutions

$$\Lambda_1 = \Lambda/L_1, \quad \Lambda_2 = \Lambda/L_2, \quad \Lambda_3 = \pm\,\Lambda/L_3, \tag{2.8}$$

differing only in the sign of the third reaction. The complete quadratic form is the sum of the constraint functions in the left-hand sides of equations (5) weighted by the corresponding reactions from (8), with displacements x_n replacing the respective coordinates X_n. The linearized constraint equations given by the second of the matrices (6) serve for the evaluation of dependent displacements. The choice of two independent displacements is in this case confined to x_2 and x_4, and the obtained values of the dependent displacements are $x_1 = x_3 = 0$. Their elimination from the complete quadratic form yields

$$Q = \Lambda\,[x_2^2/L_1 + (x_4-x_2)^2/L_2 \pm x_4^2/L_3]. \tag{2.9}$$

For the configuration in Fig. 1b the form is definite, hence $x_2 = x_4 = 0$ and

$$V = 2, \quad K = 0.$$

The system lacks kinematic mobility and is of type III. For the other singular configuration (Fig. 1c) the form is definite only at $L_3 > L_1 + L_2$; otherwise, it is indefinite, so that

$$V = 2, \quad K = 1,$$

and the system is a finite mechanism in a singular ("dead center") position.

Matrix Test for First-Order Infinitesimal Mobility

Definiteness of a Quadratic Form Subject to Linear Constraints.
However simple, the construction of the reduced quadratic form Q and the
required conversion to a matrix format in order to test for sign definiteness
involve some "manual" operations. An alternative method, more suited to a
computer-aided analysis, is based on a theorem due to Mann (1943). Adapted to
the problem under consideration, the theorem states that the complete quadratic
form in (2) is positive for all nonzero x_n satisfying the linearized constraint
equations (1.4), if and only if

$$(-1)^r \det \begin{bmatrix} K_{mn} & F^m{}_p \\ F^p{}_n & 0 \end{bmatrix} > 0 \tag{2.10}$$

for each of $n = r + 1, \ldots, N$, that is, for $V = N - r$ determinants. Here

$$K_{mn} = \Lambda_i F^i_{mn} \tag{2.11}$$

is the sum of constraint function Hessian matrices weighted by the respective
reactions, and $F^p{}_n$ $(p = 1, 2, \ldots, r)$ is the Jacobian matrix with dependent
rows removed.

At $r = C - 1$, condition (10) is sufficient for a strict minimum of one of the
constraint functions subject to fixed values of the remaining functions. In this
interpretation, the constraint function attaining a minimum is the one
represented by that row in the Jacobian matrix which was selected as dependent
and omitted in (10). Recall that for the system to be kinematically immobile,
this function (physically, a bar length squared) must be a minimum if the bar is
in tension, and a maximum if it is in compression. The condition for a strict
constrained maximum is obtained from (10) by replacing $(-1)^r$ with $(-1)^n$.

The matrix test is especially attractive for a computer-aided analysis of pin-bar
assemblies because of the utter simplicity and immediate availability of the
necessary matrices. Indeed, the constraint Jacobian matrices have been shown to
involve only the coordinate differences of the bar ends. Quite serendipitously,
the Hessian matrices corresponding to quadratic constraint functions are even
more readily available. They are determined solely by the system connectivity,
described by the system incidence matrix.

Example 2.1 (continued). To illustrate the matrix test (10), consider again
the previous example. The Hessian matrices of constraint functions (5) are

$$F^1_{mn} = \begin{bmatrix} 1 & 0 & 0 & 0 \\ 0 & 1 & 0 & 0 \\ 0 & 0 & 0 & 0 \\ 0 & 0 & 0 & 0 \end{bmatrix}, \quad F^2_{mn} = \begin{bmatrix} 1 & 0 & -1 & 0 \\ 0 & 1 & 0 & -1 \\ -1 & 0 & 1 & 0 \\ 0 & -1 & 0 & 1 \end{bmatrix}, \quad F^3_{mn} = \begin{bmatrix} 0 & 0 & 0 & 0 \\ 0 & 0 & 0 & 0 \\ 0 & 0 & 1 & 0 \\ 0 & 0 & 0 & 1 \end{bmatrix}. \tag{2.12}$$

With all the necessary data in place, the test matrix in (10) can be formed. It is symmetric, with maximum size 6×6 corresponding to $s = N = 4$:

$$
\begin{bmatrix}
1/L_1+1/L_2 & 0 & -1/L_2 & 0 & L_1 & 0 \\
 & 1/L_1+1/L_2 & 0 & -1/L_2 & 0 & 0 \\
 & & 1/L_2\pm1/L_3 & 0 & 0 & \mp L_3 \\
\vdots & & & 1/L_2\pm1/L_3 & 0 & 0 \\
 & & & & 0 & 0 \\
 & \cdot & & \cdots & & 0
\end{bmatrix}.
$$

The first four columns represent the sum of the above Hessian matrices weighted by the first set of Λ_i from (8), with the common scale factor Λ taken as 1. The last two columns are the transposed first and third rows of the Jacobian matrix. Any two of the three rows could be selected, according to the rank $r = 2$. As before, the double sign corresponds to the two respective singular configurations in Fig. 1b and c.

Two determinants are to be evaluated, D_3 and D_4. (When dealing with systems that are not as simple, the matrix test can be implemented with the evaluation of matrix pivots instead of the determinants.) Determinant D_3, corresponding to $n = r + 1 = 3$, is obtained after deleting from the above matrix the fourth row and fourth column:

$$D_3 = (1+L_1/L_2)(\pm L_3)^2, \quad D_4 = (L_3 \pm L_1 \pm L_2) L_3/L_2. \tag{2.13}$$

Since $D_3 > 0$ for both of the singular systems, the outcome of the test depends on the sign of D_4. For the system in Fig. 1b, $D_4 > 0$ and it passes the matrix test as type III. The second degenerate configuration (Fig. 1c) is of type III only at $L_3 > L_1 + L_2$; otherwise, it is a system of type IV.

Example 2.2. The top and bottom chord polygons of the system in Fig. 2 are affine with the respective ordinates differing by a factor of 2. It is easily found by inspection that the system is topologically inadequate (the number of constraints is insufficient for invariance). It seems to be statically indeterminate (the affinity of the chords entails the statical possibility of self-stress). Beyond that, the statical-kinematic properties of this system, including its type, are not as clear as, say, for a double-convex version of the system or for the system in the previous example. This makes the analysis somewhat more challenging, both conceptually and analytically.

The system will be analyzed first by applying the matrix test (10) and then by constructing and investigating the reduced quadratic form Q. Besides a mutual verification of the two results, this provides an opportunity for comparing the computational efforts of the two approaches. As has been noticed, the Jacobian matrix elements are simply the horizontal and vertical projections of the bar lengths taken with a proper sign. Both the necessary sizes and the matrix itself are shown in the figure.

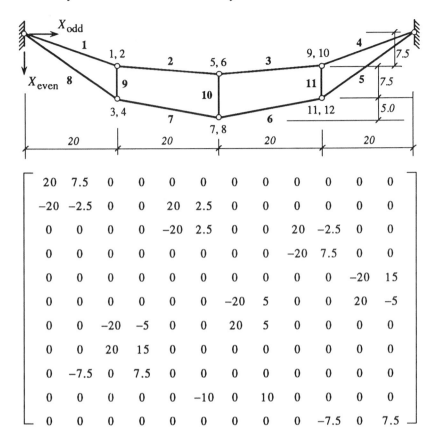

Figure 2.2. A system with affine polygon chords and its constraint Jacobian matrix.

Because of the apparent statical indeterminacy, this 11×12 matrix must be singular. Indeed, the matrix rank is $r = 10$, confirming the supposition. The constraint reactions of self-stress can be found by transposing the matrix, transferring any column to the right-hand side, and solving the obtained system of ten inhomogeneous equilibrium equations. The solution is

$$\Lambda_i = [\ 2,\ 2,\ 2,\ 2,\ -1,\ -1,\ -1,\ -1,\ 4/3,\ 1,\ 4/3\ ]^T. \qquad (a)$$

All the necessary data are now available for the construction of the test matrix (10). Its upper left block is a 12×12 matrix sum of the bar incidence matrices weighted by the respective Λ_i from (a). The 10×12 bottom left block is the Jacobian matrix with one row (say, the last, eleventh) removed. According to (a), bar 11 is in tension, so that for the system to be kinematically immobile, the bar length must be minimum. Since $V = N - r = 2$, only two determinants must be evaluated, of which the first is obtained after deleting from the test matrix the twelfth row and column. Both of the determinants turn out to be positive, so that the system passes the test as kinematically immobile. Thus, somewhat contrary to intuition, the system is quasi-variant (type III), like its double-convex variant. This conclusion holds regardless of the number of bays in the system, as long as the distance between the end pins is a maximum.

Comparison with the Alternative Analysis. In the analysis based on the construction of the reduced quadratic form, $r = 10$ dependent displacements are to be evaluated in terms of $V = 2$ independent ones. For a better visualization of results, displacements x_7 and x_8 of the center pin are chosen as independent. After the corresponding two columns of the Jacobian matrix are transferred to the right-hand side, the solution is obtained in the form (1.7) with

$$a_{ns} = \begin{bmatrix} 3/4 & -2 & 3/2 & -2 & 1/2 & 0 & 1 & 0 & 3/4 & 2 & 3/2 & 2 \\ 3/16 & -1/2 & 3/8 & -1/2 & 0 & 1 & 0 & 1 & -3/16 & -1/2 & -3/8 & -1/2 \end{bmatrix}^T.$$

The next step in this procedure, in contrast to the matrix test, requires writing the constraint functions in an explicit form, in order to obtain the quadratic terms in their power expansions. This operation, performed here manually, is technically simple and can be computerized. With the aid of the above solution, the dependent displacements were eliminated in the elementary quadratic forms before forming their weighted sum. The resulting reduced quadratic form is

$$Q = 15.25\,x_7^2 + 4.8125\,x_8^2.$$

The cross-product term in this form is absent due to the antisymmetry and symmetry of the respective displacement patterns corresponding to the chosen independent virtual displacements x_7 and x_8. The form is positive definite, thereby confirming the earlier conclusion on the system type.

Comparing the two approaches, the construction of the reduced quadratic form involves working with smaller matrices, $C \times N$ and $V \times V$, but requires an evaluation and elimination of dependent displacements. The matrix test, although dealing with a larger matrix size, $(N + r) \times (N + r)$, requires fewer "manual" operations and is better suited for fully computerized treatment.

The terse formalism of the matrix test (10) is quite remarkable. The test first extracts from the constraint functions the necessary information on the system topology and geometry, in the form of the Hessian and Jacobian matrices, respectively, and then arranges it in a matrix format entirely and uniquely suitable for computer treatment. Moreover, for the analysis of pin-bar systems there is not even any need for the constraint equations per se; the required input involves only the set of nodal coordinates and the system connectivity, which is the input of a routine truss analysis.

With all their efficiency, both of the presented methods are applicable only to simple first-order infinitesimal mechanisms. Their generalization to more complex systems, which involve multiple singularity or higher-order mobility, seems unlikely: finding conditions for a real solution, $X_n = X_n^0$, of constraint equations (1.1) to be isolated, is beyond conventional matrix means.

Second-Order Statical-Kinematic Duality

Virtual Self-Stress and Prestress. The existence of a nontrivial solution to the homogeneous equilibrium equations indicates the statical possibility of initial (self-stress) forces in the system. The leading form Q in expansion (2) is the lowest-order virtual work done by the initial forces Λ_i over the constraint variations f^i, resulting from displacements x_n. For the state of self-stress to be stable, it is necessary and sufficient that this work be strictly positive. Note that mere definiteness of Q amounts in this case to positive definiteness; if sign reversal is needed, it is achieved by simultaneously changing the sign of all initial forces. Thus, a definite form Q means in statical terms that the system allows stable self-stress (a stable state with initial forces), and in kinematic terms that all displacements are zero, i.e., the system is kinematically immobile.

The nature of the solution point stationarity, described by the reduced quadratic form, plays in the second-order analysis a dual role similar to that of the Jacobian matrix rank in the linear analysis. On the kinematic side, a definite or indefinite quadratic form, like the two cases of full rank, provide a conclusive outcome. When Q is definite, the system is kinematically immobile (type II or III), whereas an indefinite Q indicates a singular configuration of a variant system (type IV). A semidefinite Q indicates the possibility of a rigid motion.

Remark. In mathematical literature, a kinematically immobile system is called rigid and, if it also lacks virtual mobility, infinitesimally rigid. The conventional engineering terminology is more convenient in that it allows infinitesimal mobility, rather than rigidity, of a system, to be quantified in terms of its order. For example, simple mechanisms defined above, possess first-order infinitesimal mobility if the quadratic form Q is definite.

On the statical side, a definite Q, as has been mentioned, is a sign of stability of self-stress. An indefinite form means unstable, hence, physically unrealizable self-stress, like that of a dead-center configuration. A semidefinite Q indicates the possibility of neutral equilibrium with initial forces, as in a tensioned chain with one end pinned and the other sliding along a circular guide.

Remark. The stability of a self-stressed equilibrium state, as considered here, is of kinematic, or geometric (as opposed to elastic) nature. With the system constraints, i.e., structural members, assumed ideal, the entire development has been devoid of any notion of elasticity.

A certain symmetry is now apparent between the concepts of virtual and kinematic displacements on the one hand, and the concepts of statically possible and physically realizable self-stress on the other. The following definitions are helpful in pursuing this parallel.

Virtual self-stress is a statically possible pattern of initial forces determined by a nontrivial solution of the homogeneous equilibrium equations.

Prestress is a stable state of the system with initial (self-equilibrated) forces.

The term *virtual* reflects the formal analytical origin of the concept and emphasizes its effectual, rather than actual, nature. The effect of virtual self-stress is in the nonuniqueness of statically possible states of the system (statical indeterminacy).

The concepts of virtual displacement and virtual self-stress have much in common: both are only formal local effects reflecting the existence of nontrivial solutions of simultaneous homogeneous equations; both are engendered by one and the same constraint Jacobian matrix and belong, respectively, to its nullspace and left nullspace; both may be realizable and become the respective prerequisites for an actual, kinematic displacement or a state of prestress; both originate in the first-order analysis and, therefore, represent inherently linear concepts; and, finally, the realizability of each depends on certain higher-order conditions.

Remarkably, these realizability conditions reside with one and the same higher-order construct—the leading (i.e., the lowest-order, nonvanishing) form in expansion (2); moreover, the two conditions are *mutually exclusive*.

Mutual Exclusion of Kinematic Mobility and Prestressability. The statical possibility of prestress, hereafter called *prestressability*, requires a definite Q, thereby precluding the virtual displacements from realization, that is, from becoming kinematic. Conversely, kinematic mobility requires an indefinite Q, thereby ruling out stable self-stress. A semidefinite Q, as has been mentioned, corresponds to the statical possibility of neutral self-stress and rigid mobility. Since rigid motion (with or without self-stress) is of little interest for structural applications, the term "kinematic mobility" in this book always implies a change in the geometric configuration of a system, thereby excluding rigid motion. With this in mind, the following concise statement holds.

Exclusion Condition. In any structural system, kinematic mobility and prestressability are mutually exclusive.

In other words, the presence of one of these two features rules out the other by rendering it unrealizable: either mobility or self-stress must remain only virtual. However, the opposite statement does *not* hold: the absence of one of the two features does not cause the other to be present, so that both of them can be absent at the same time. This is obvious in the case of a geometrically invariant and statically determinate system, which has neither of these features. But it is not so obvious for systems with infinitesimal mobility.

The exclusion condition is a second-order manifestation of the statical-kinematic interrelationship. As before, the apparent symmetry of the condition with respect to mobility and prestressability does not imply parity between the statical and kinematic aspects; rather, the kinematic aspect, being the source of the relationship, has primacy. Note that in the same way as kinematic mobility is merely the possibility of motion, prestressability is only a prerequisite for the realization of stable self-stress.

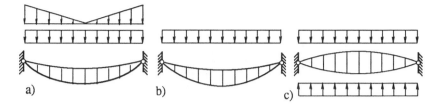

Figure 2.3. Funicular trusses: a) nonsingular; b) and c) singular; and their vertical equilibrium loads.

statical and kinematic aspects; rather, the kinematic aspect, being the source of the relationship, has primacy. Note that in the same way as kinematic mobility is merely the possibility of motion, prestressability is only a prerequisite for the realization of stable self-stress.

Two symmetric corollaries of the exclusion condition are:

Statical Criterion of Immobility: A prestressable system possesses a unique geometric configuration, i.e., is kinematically immobile; and

Kinematic Criterion of Impossibility of Prestress: A kinematically mobile (variant) system is unprestressable.

It is important to emphasize that it is just the *statical possibility* (as opposed to the physical presence) of prestress, that constitutes the statical criterion of immobility. Indeed, just the formal existence of a solution $\Lambda_i \neq 0$, as a set of numbers engendering a definite Q, leads to $x_s = 0$ and indicates immobility; the actual presence or absence of initial forces is irrelevant. At the same time, definiteness of Q, ensuring stability of this statically possible state, makes the system prestressable.

The significance of the statical criterion is two-fold: first, it relates the issue of infinitesimal mobility to the vast knowledge base in stability; and second, prestressability is in many cases evident and does not require any analysis. Such is the case, for example, of a comprehensive tensile self-stress, when even the complete quadratic form is positive: it is a combination of the squares of coordinate differences with positive Λ_i. Also, prestressability quite often is assured by actually prestressing the system.

The kinematic criterion seems intuitively obvious. In any nonsingular configuration, a variant system is statically determinate and does not allow even virtual self-stress. In a singular configuration, like dead-center, self-stress exists only as virtual and is physically unrealizable. (Recall that the trivial case of neutral equilibrium with rigid mobility has been excluded.)

The above criteria are useful for a quick assessment of system properties.

Example 2.3. All three variants of a funicular truss presented in Fig. 3 are underconstrained (they miss diagonal bars). The first truss (Fig. 3a) has its top and bottom chords in the form of two different funicular curves, one each

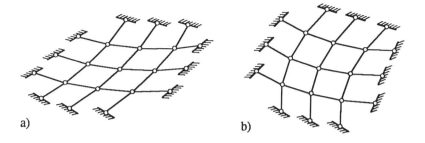

Figure 2.4. Singular hinge-bar systems: a) unprestressable; b) prestressable.

Being underconstrained and statically determinate, this must be a nonsingular configuration of a variant system (type IV). It has two vertical equilibrium loads, and the distribution of internal forces under any combination of these loads is unique: each cord will support only the load component that shaped it, and "filter" through the other component. There are also many other (not vertical) equilibrium loads, and under each of them the statically possible state is unique.

The profiles of the upper and bottom chords in the second truss (Fig. 3b) *are* affine, suggesting the existence of virtual self-stress, with one of the chords in tension and the other in compression. This configuration is statically indeterminate, has only one vertical equilibrium load, and the corresponding statically possible state is not unique. In the light of Example 2, the system is quasi-variant (type III) and admits prestress with tension in the upper chord. Although virtual self-stress with the upper chord in compression, by itself, is unstable, a suitable equilibrium load (like the one shown in the figure) may induce sufficient tension in the upper chord to overcome the compression, thereby making the combined force distribution stable.

Finally, the third truss (Fig. 3c) is doubly convex, with the chord profiles affine. The system is statically indeterminate and its virtual self-stress (with both chords in tension) is stable. Thus, this is a system of type III with just one equilibrium vertical load; its statically possible state under this (or any other) equilibrium load is not unique.

Comparing the three systems, note that a decrease by one in the number of equilibrium loads for each of the two singular configurations is accompanied by a corresponding increase in the number of virtual self-stress patterns, i.e., in the degree of statical indeterminacy.

Example 2.4. Each of the space hinge-bar systems presented in Fig. 4 consists of two arrays of identical funicular polygons located in mutually perpendicular vertical planes. In the first system, the directions of convexity of the two arrays are coincident, while in the second they are opposite. Both systems possess virtual self-stress. In the system in Fig. 4a, the initial forces are of opposite signs; hence, one of the arrays must be in compression, the self-stress is unstable, and the system is variant. As in the preceding case of funicular trusses, depending on the direction (sign) of an applied equilibrium load, the final force distribution may be stable and realizable. Its nonuniqueness, due to the statical indeterminacy, can be used advantageously, for example, by selecting and implementing a statically possible state most favorable for the supporting structure, as was the case for the system in Fig. 1.9.

The system in Fig. 4b has entirely tensile, therefore stable, virtual self-stress. The system is quasi-variant, hence, statically indeterminate, and prestressable. Upon being prestressed, it will support an equilibrium load of either sign, at least until the prestressing force is reduced to zero in some member, making it susceptible to buckling or disengagement.

2.2 Multiple Singularity. Compound Systems

Complex and Compound Systems

Dependent Versus Independent Constraints.   Further advance of the foregoing analysis is achieved by relaxing, one at a time, the simplifying assumptions used. The first assumption to be revised is the one restricting the degree of singularity to $D = 1$. A structural system may be comprised of several singular subsystems, each involving a set of dependent constraints. Then each subsystem has virtual self-stress, thereby producing $D > 1$ simultaneous reduced quadratic forms Q_d. Among these forms, definite and indefinite ones have quite different implications for kinematic mobility. The role of a definite form is straightforward: it immobilizes the subsystem by preventing all of its virtual displacements from becoming kinematic. The effect of an indefinite Q_d is more subtle. It also constrains kinematic mobility of the subsystem, but only to the extent of mutually relating the virtual displacements figuring in the form. The result is a reduction by one in the number of independent virtual displacements, and a corresponding increase in the number of dependent ones.

When considering multiple singularity, two basic classes of systems are encountered: systems with all of the constraints mutually dependent, and those containing independent constraints as well. The two classes will be referred to as *complex* and *compound* systems; they are respectively characterized by overlapping and nonoverlapping subsystem self-stress patterns. Each of the two complex systems shown in Fig. 5 admits comprehensive self-stress representing a combination of the subsystem self-stress patterns. As in the case of a simple system $(D = 1)$ considered earlier, analyzing kinematic mobility of a complex system requires the construction and investigation of a reduced quadratic form. The only difference is that a definite quadratic form Q in all independent displacements is sought as a linear combination of the forms Q_d engendered by the D independent self-stress patterns. An iterative algorithm for identifying such a linear combination has been described by Calladine and Pellegrino (1991).

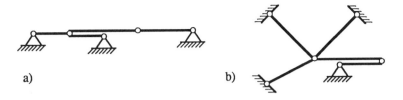

Figure 2.5. Complex systems: a) one-dimensional; b) two-dimensional.

In a compound system, the presence of independent constraints precludes overlapping of the self-stress patterns of the singular subsystems (recall that independent constraints cannot develop initial forces). As a result, the existence of a definite linear combination Q as a condition for infinitesimal mobility is only sufficient and excessively strong; in other words, the absence of a definite Q does not necessarily entail kinematic mobility. Accordingly, in assessing mobility of a compound system, the question is whether there exists a set of independent virtual displacements that turns all D forms Q_d into zero simultaneously. When there is no such set, the conclusion is unequivocal: the system is kinematically immobile and possesses only virtual (in this case, first-order infinitesimal) mobility. However, if such a set of virtual displacements exists, the outcome is ambiguous and requires further analysis; it only means that the system constraints are compatible at least to the second order but its mobility may be either infinitesimal or finite. It is convenient at this point to illustrate the situation with a simple example.

Example 2.5. The system in Fig. 6a involves two singular subsystems, each comprised of three horizontal bars. Their bar lengths, shown to scale, are either 1, or 2, except for the bottom bar, which has length L_6. The two subsystems are connected by a vertical bar 7 of length 1 and an inclined bar 8 as shown. The numbers at pins indicate coordinates; for example, the position of the first pin is given by (X_1, X_2). The constraint equations of the system are

$$(1) \; X_1^2 + X_2^2 = 2^2 \qquad\qquad (4) \; (X_5-1)^2 + (X_6-1)^2 = 1^2$$

$$(2) \; (X_3-1)^2 + X_4^2 = 2^2 \qquad (5) \; (X_7-X_5)^2 + (X_8-X_6)^2 = 2^2$$

$$(3) \; (X_3-X_1)^2 + (X_4-X_2)^2 = 1^2 \quad (6) \; (X_7-4+L_6)^2 + (X_8-1)^2 = L_6^2 \qquad (a)$$

$$(7) \; (X_5-X_1)^2 + (X_6-X_2)^2 = 1^2 \quad (8) \; (X_7-X_3)^2 + (X_8-X_4)^2 = 2.$$

In the given configuration,

$$X_1 = X_5 = 2, \quad X_2 = X_4 = 0, \quad X_3 = 3, \quad X_6 = X_8 = 1; \quad X_7 = 4. \qquad (b)$$

Deriving the linearized constraint equations for this configuration, and solving them with virtual displacements x_2 and x_4 as independent, yields

$$x_6 = x_2, \quad x_8 = x_4, \quad x_1 = x_3 = x_5 = x_7 = 0. \qquad (c)$$

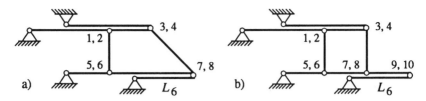

Figure 2.6. Compound systems: a) topologically adequate; b) inadequate.

The complete quadratic forms for the two subsystems are obtained from the first two triplets of constraint equations (a) by multiplying them by $\Lambda_i = \Lambda/L_i$ as described earlier. Eliminating the dependent displacements and omitting the Λ's leads to the reduced forms, which differ from (9) only in the bar lengths:

$$Q_1 = x_2^2/2 + (x_2 - x_4)^2 - x_4^2/2, \qquad (2.14)$$

$$Q_2 = x_2^2 + (x_2 - x_4)^2/2 - x_4^2/L_6. \qquad (2.15)$$

Each form is indefinite and constrains the corresponding subsystem by imposing a certain relation between the virtual displacements which were independent in the first-order analysis. Thus, taken separately, each subsystem possesses one kinematic and two virtual degrees of freedom.

The simplicity of the above forms enables the question of their compatibility to be resolved explicitly. Setting Q_1 to zero gives

$$x_4 = (2 \pm 1)\, x_2. \qquad (d)$$

This double-valued solution exemplifies a phenomenon that Tarnai (1988) called "bifurcation of compatibility." When the two solutions are introduced, one at a time, into the form Q_2, the resulting monomial form vanishes at $L_6 = 3$ and $L_6 = 1$. At any different bar length there must be $x_2 = 0$ leading to $x_4 = 0$ and, according to (c), all virtual displacements are zero. Hence,

$$V = D = 2, \qquad K = 0,$$

the system is kinematically immobile and, by virtue of (1.24), it is quasi-invariant (type II). The latter conclusion is verified by noticing that the number of constraints, $C = N = 8$, is sufficient for invariance; the system is topologically adequate and the constraints are dependent only because of the degenerate configuration. Upon exiting from it, the Jacobian matrix rank restores to $r = 8$ and the system becomes invariant.

Geometric Interpretation. Quasi-Variant Compound Systems. It appears from the above example that compound systems should almost always possess infinitesimal mobility, while constraint compatibility and kinematic mobility must be exceptional. A geometric interpretation illustrates and confirms this observation. The indefinite form Q_1, expressing the work of initial forces over constraint variations, represents in coordinates (x_2, x_4, Q) a hyperbolic paraboloid, with the plane $Q = 0$ intersecting it along two linear generators. The generators give the two directions [displacement ratios from (d)] of zero work, hence, zero constraint variations. The same is true for the second form, Q_2, producing another hyperboloid with its own pair of linear generators passing through the coordinate origin. Constraint compatibility takes place only if the two hyperboloids *share* a generator which is, of course, an exceptional situation. Apparently then, *incompatibility* is typical, and in fact, can be expected even in cases with a greater number of independent virtual displacements, resulting in compound quasi-variant (type III) systems.

Example 2.6. To explore this case further, consider the system in Fig. 6b, which cannot be type II since it is topologically inadequate:

$$C = 9 < N = 10, \qquad V = 3 > D = 2.$$

While Q_1 given by (14) remains unchanged, the modified form Q_2 contains an additional virtual displacement. The linearized constraint equations yield

$$x_6 = x_2, \qquad x_8 = x_4, \qquad x_1 = x_3 = x_5 = x_7 = x_9 = 0, \qquad (e)$$

so that

$$Q_2 = x_2^2 + (x_2 - x_4)^2 + (x_4 - x_{10})^2 - x_{10}^2 / L_6. \qquad (2.16)$$

Replacing x_4 with the solutions (d) leads, respectively, to the alternative forms

$$Q_2 = 14x_2^2 - 6x_2x_{10} + (1 - 1/L_6) x_{10}^2, \qquad (2.17)$$

$$Q_2 = 2x_2^2 - 2x_2x_{10} + (1 - 1/L_6) x_{10}^2. \qquad (2.18)$$

The first of these forms is found to be definite at $L_6 > 14/5$, and the second at $L_6 > 3/2$. When both of the inequalities are satisfied, Q_1 and Q_2 are incompatible: no set of x_s ($s = 2, 4, 10$) turns them into zero simultaneously. As a result, displacements $x_2 = x_6 = 0$, and the system is quasi-variant. What makes this result remarkable is that at $L_6 > 3$, which is only slightly greater than 14/5, the outcome would be quite trivial. Indeed, at $L_6 > 3$, the original Q_2 in (16) is definite and the second subsystem, being kinematically immobile, alone immobilizes the entire system; in this case, Q_1 becomes irrelevant since it does not contain any of the independent displacements absent from Q_2.

Interestingly, self-stress in both of the systems in Fig. 5 is only virtual and neither of them is prestressable. This follows from the absence of positive-definite combinations of forms (14) and (15) or, respectively, (14) and (16). The existence of such immobile systems was first noticed by Shul'kin (1984). Together with certain higher-order mechanisms (see Section 3), they belong to a distinct class of systems—*unprestressable infinitesimal mechanisms*. As is seen from the preceding examples, these systems can be of either type II or type III.

In general, the issue of infinitesimal mobility in systems with multiple singularity $(D > 1)$ cannot be resolved without an analysis of either D simultaneous reduced quadratic forms or D complete forms subject to the pertinent linearized constraint equations. The first to notice this was, apparently, Kötter (1912), whose condition for kinematic immobility was definiteness of one of the reduced forms, say Q_1, subject to $Q_2 = Q_3 = ... = Q_D = 0$.

Geometrically, each of the equations $Q_d = 0$ with an indefinite form Q_d describes in the coordinate space a convex cone. When all of the cones have just one common point, the apex $x_s = 0$, the compound system is kinematically immobile. It is unlikely that a simple matrix approach similar to the above matrix test can be formulated for determining whether or not the cones intersect, with the exception of one small but notable particular case. Suppose that each of the quadratic forms, except no more than one of them (say, Q_1), can be factored into two linear forms, as was the case in the foregoing example. These linear forms can be included, one at a time, into the test matrix (10) in lieu of the border Jacobian matrices. (The latter are unnecessary since they have been accounted for in evaluating the reduced forms Q_d.) Then, according to Mann's theorem, Q_1 can be checked for definiteness subject to all of the appropriate combinations of linear forms representing the factored quadratic forms.

Analysis of Constraint Compatibility. A proposed approach to the problem emulates one of the classical methods of structural mechanics. Following it, a system is broken into two kinematically mobile subsystems and the two respective loci, prescribed by each subsystem for the node they share, are evaluated. By comparing the loci, the possibility (or otherwise) of compatible displacements of the subsystems is established, leading to a conclusion on the kinematic properties of the assembly as a whole. This analysis requires the r dependent displacements to be presented as power expansions,

$$x_j = a_{js} x_s + (1/2) a_{jst} x_s x_t + \cdots; \quad j=1,\cdots,r; \quad s,t,\cdots = r+1,\cdots,N. \quad (2.19)$$

Example 2.7. For an alternative analysis of the system in Example 5 (Fig. 6a), let the bottom bar 6 be disconnected from the assembly and consider it to be the second subsystem. The nodal loci, in this case, the one-dimensional paths prescribed to the common node by each subsystem, are sought in the form

$$x_7 = a_{78} x_8 + (1/2) a_{788} x_8^2 + \cdots .$$

In this notation, solution (c) of the linearized constraint equations becomes

$$a_{68} = a_{28}, \quad a_{48} = 1, \quad a_{18} = a_{38} = a_{58} = a_{78}^{(1)} = 0. \qquad (f)$$

Evaluation of a_{28} requires the second derivatives of the first three constraint equations (a):

$$a_{18}^2 + X_1 a_{188} + a_{28}^2 + X_2 a_{288} = 0, \qquad a_{38}^2 + X_3 a_{388} + a_{48}^2 + X_4 a_{488} = 0,$$

$$(a_{38} - a_{18})^2 + (X_3 - X_1)(a_{388} - a_{188}) + (a_{48} - a_{28})^2 + (X_4 - X_2)(a_{488} - a_{288}) = 0.$$

Taking into account (b) and (f) these equations yield:

$$a_{28} = a_{68} = 1/3, \ (1), \quad a_{188} = -1/18, \ (-1/2), \quad a_{388} = -1/2. \qquad (g)$$

Here, the parenthesized quantities represent the second of the double-valued solutions, reflecting the existence of two different virtual displacement modes for the first subsystem. Carrying out the second differentiation of the remaining five constraint equations (after a few simple calculations) leads to

$$a_{788}^{(1)} = -1/3, \ (-1), \qquad a_{788}^{(2)} = -1/L_6, \qquad (2.20)$$

describing the respective paths imposed on the common node by the subsystems. As expected, the two paths are found to be compatible to the first order, except that at $L_6 = 3$ and $L_6 = 1$ they are compatible to at least the second order.

The presented constraint compatibility approach might seem somewhat cumbersome since there are many ways of breaking a given system into subsystems and usually it is not known in advance which particular combination of subsystems has the highest order of constraint compatibility. However, only dependent constraints are to be disconnected in the process, and then only any one of them in each singular subsystem; thus, only D subsystem combinations are to be examined. This approach proved indispensable when analyzing both simple and compound systems with higher-order infinitesimal mobility. The analysis is presented in Section 3.

Synthesis of First-Order Mechanisms

Application of the Statical Criterion of Immobility. This criterion is applicable to the synthesis of both simple and complex, but not compound, infinitesimal mechanisms. Consider an underconstrained system IV (e.g., the familiar three-bar chain) and apply an external force at one of its nodes (Fig. 7). Generally, the system geometry changes before it attains an equilibrium configuration that balances the force. Assuming the equilibrium to be stable, add a new link to the system, collinear with the force and connecting its point of application to the foundation. Then either the entire system (Fig. 7a) or the prestressed subsystem (Fig. 7b, c, and d) form a first-order infinitesimal mechanism. Indeed, the constraint reactions produced originally by the external force, together with the internal force in the new link, constitute a pattern of stable self-stress in the system, thus satisfying the statical criterion of immobility. By simultaneously applying several forces to a geometrically variant system, a wide variety of configurations can be obtained and then immobilized by introducing appropriate constraints. Note that in this procedure forces are used only as a convenient means to arrive at prestressable, hence immobile, configurations; such configurations lack kinematic mobility regardless of the actual presence or absence of prestress.

In cases where the prestress is confined to a subsystem, the entire system will still lack mobility if the stress-free subsystem is immobile (Fig. 7b and c). This can be verified by a separate analysis of the stress-free subsystem, assuming

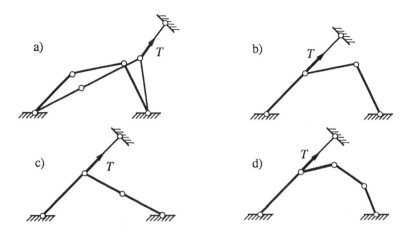

Figure 2.7. Mechanical synthesis of infinitesimal mechanisms: a) and b) first-order mechanisms with comprehensive or partial prestress; c) higher-order mechanism with partial prestress; d) partially immobilized variant system.

the nodes shared with the prestressed subsystem to be fixed. The analysis may indicate prestressability of the subsystem in question (Fig. 7c), but this result must be considered in the context of higher-order infinitesimal mobility. If the stress-free subsystem is mobile, the system as a whole is geometrically variant (Fig. 7d) although it contains an immobile subsystem.

Another version of the approach just described involves introducing internal, rather than external constraints. A variable length bar is inserted between two unconnected points of a variant system, and then shortened or lengthened to produce a prestressable singular configuration. Furthermore, a similar effect can also be achieved by simply varying the length of one of the existing bars until it attains a maximum or a minimum value.

The significance of the statical criterion of immobility goes beyond these elementary examples. Clearly, the best test of prestressability is prestressing: any structural system which is actually prestressed lacks kinematic mobility. Buckminster Fuller recognized and perpetuated this idea in his term *"tensegrity"* structure, emphasizing *tens*ion as a means for attaining structural in*tegrity*. His concept of a tensegrity structure is an infinitesimally mobile system with a closed-loop prestress, i.e., with force-free supports (free-standing, in Fuller's terms). The surprisingly diverse class of tensegrity structures includes, among others, various cable-bar systems, wire nets, and membranes, exemplified, respectively, by a bicycle wheel, a tennis racket, and a musical drum.

An Analytical Approach. A general analytical approach to the synthesis of tensegrity structures is based on generating a singular configuration and verifying its prestressability. This approach is illustrated by the following example of what appears to be the simplest three-dimensional tensegrity system.

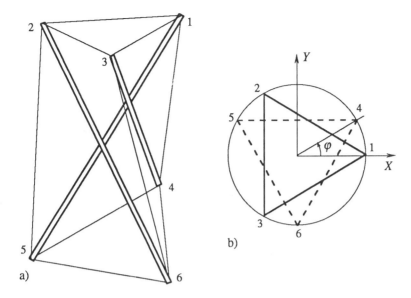

Figure 2.8. A space tensegrity structure: a) three-dimensional view; b) top view.

Example 2.8. Two equilateral pin-bar triangles are located in parallel planes a distance $\Delta Z = H$ apart and connected by six bars or cables. The base triangles are mutually rotated through an angle φ to form a tensegrity structure (Fig. 8). Six external supports needed to restrain free-body motion of the system are assumed to be present at the plane $Z = 0$ so that nodes 4, 5, and 6 cannot have any displacements. In order to capitalize on the cyclic symmetry of the system, it is analyzed as a system of three identical cyclically repeating subassemblies. For the bars connected by node 1, the linearized constraint equations (using a slightly modified notation) are

$$(X_1 - X_2)(x_1 - x_2) + (Y_1 - Y_2)(y_1 - y_2) = 0,$$

$$(X_1 - X_4)x_1 + (Y_1 - Y_4)y_1 + Z_1 z_1 = 0,$$

$$(X_1 - X_5)x_1 + (Y_1 - Y_5)y_1 + Z_1 z_1 = 0.$$

Due to the cyclic symmetry,

$$x_i = r \cos \Theta_i - R\theta \sin \Theta_i ; \quad y_i = r \sin \Theta_i + R\theta \cos \Theta_i ; \quad z_i = z; \quad i = 1, 2, 3,$$

where r, $R\theta$, and z are, respectively, the radial, hoop, and axial displacements common to nodes 1, 2, and 3; Θ_i is the polar angle for the given configuration: $\Theta_1 = 0$, $\Theta_2 = 2\pi/3$, $\Theta_3 = 4\pi/3$.

The necessary coordinates of nodes 4 and 5 of the subassembly in this reference frame are found from the figure and are expressed in terms of the angle of rotation, φ. Denoting $H/R = h$, the constraint function Jacobian matrix can be written:

$$\begin{bmatrix} 3 & 0 & 0 \\ 1 - \cos\varphi & \sin\varphi & h \\ 1 + \cos(\theta_2 + \varphi) & \sin(\theta_2 + \varphi) & h \end{bmatrix}.$$

A singular configuration is characterized by a drop in the rank of this matrix. Equating the determinant to zero gives

$$\sin\varphi - \sin(2\pi/3 + \varphi) = 0,$$

so that the degeneration takes place at $\varphi = \pi/6$. The corresponding self-stress constraint reactions are found by transposing the above Jacobian matrix and solving the obtained equilibrium equations:

$$\Lambda_{1-4} = -\Lambda_{1-5} = \Lambda; \quad \Lambda_{1-2} = \Lambda/\sqrt{3}. \tag{2.21}$$

The resulting bar forces are evaluated according to (1.12) as functions of the respective bar lengths, with an arbitrary N:

$$N_{1-2} = N, \quad N_{1-4} = N\sqrt{2 - \sqrt{3} + h^2}, \quad N_{1-5} = -N\sqrt{2 + \sqrt{3} + h^2}.$$

Obviously, a mirror image of the above system with respect to a vertical plane is also a tensegrity structure. The reason that the obtained solution does not include this alternative is that generating a tensegrity system with the same node numbers but an opposite axial "twist" requires a change in the system connectivity.

After reversing the sign of the member forces in solution (21), it still describes a pattern of virtual self-stress. Since the obtained system has only one independent virtual displacement, the corresponding reduced quadratic form is a monomial. This ensures its positiveness with one of the alternative sign combinations which indicates that the system is prestressable.

Note that the obtained system is topologically adequate $(C = N)$. Selecting its singular, i.e., quasi-invariant, configuration might seem counterproductive, since an adjacent ordinary configuration is invariant and can support an arbitrary load. However, for a configuration formally invariant but nearly degenerate, the stiffness matrix is ill-conditioned. An external load containing a perturbation component of the singular configuration will give rise to large displacements and member forces in the system. For such a case it might be appropriate to choose instead the singular configuration, and to stabilize it by prestress. In doing so, the structural members that under the prestress are in tension (in Fig. 8 they are designated by single lines) can be designed as high-strength wires or cables, efficiently utilizing material strength. In statical-kinematic analysis, such members are represented analytically as unilateral constraints in tension. As shown later, replacement of even one bilateral constraint with a unilateral one generally causes a type II system to become topologically inadequate, i.e., quasi-variant (type III), provided that prestress is still possible. A typical tensegrity structure contains a maximum number of tensile structural members and, therefore, is quasi-variant.

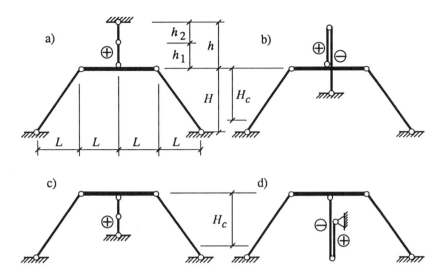

Figure 2.9. Four ways to immobilize a three-bar system with a two-bar chain.

Another Analytical Approach. Examples considered in the preceding section are suggestive of yet another, purely kinematic, analytical approach to the synthesis of systems with infinitesimal mobility. In this method, one or more geometric parameters of a system are kept unspecified. These parameters are then determined so as to satisfy an immobility criterion based on the constraint compatibility condition. The first step is to expand the constraint functions of the system into power series in terms of coordinate increments, i.e., virtual displacements. Next, the arbitrary parameters are evaluated so as to make the expanded constraint equations compatible up to a certain order, but not beyond this order (otherwise, the entire analysis will remain inconclusive). The procedure is illustrated in the following example.

Example 2.9. Consider again the familiar three-bar system in a symmetric configuration. This time the objective is to immobilize it by connecting the midpoint of the second bar to the foundation (frame) with a two-bar chain (Fig. 9a). This will be done by analyzing the motions of the two subsystems separately and then linking them together. First, the midpoint of the second bar is formally identified as a node by using two linear equations,

$$X_1 + X_3 - 2X_5 = 0, \qquad X_2 + X_4 - 2X_6 = 0, \qquad (2.22)$$

to supplement the three constraint equations (5). With displacement x_5 as independent, the nodal path is evaluated as a power expansion (19):

$$x_6 = a_{65} x_5 + (1/2) a_{655} x_5^2 + \dots .$$

Equations for a_{65} and a_{655} are obtained by successively differentiating the constraint equations with respect to X_n (the factor 2 is omitted):

First Differentiation	Second Differentiation
$X_1 a_{15} + X_2 a_{25} = 0,$	$a_{15}^2 + X_1 a_{155} + a_{25}^2 + X_2 a_{255} = 0,$
$(X_3 - X_1)(a_{35} - a_{15})$ $+ (X_4 - X_2)(a_{45} - a_{25}) = 0,$	$(a_{35} - a_{15})^2 + (X_3 - X_1)(a_{355} - a_{155}),$ $+ (a_{45} - a_{25})^2 + (X_4 - X_2)(a_{455} - a_{255}) = 0,$
$-(L - X_3) + X_4 a_{45} = 0,$	$a_{35}^2 - (L - X_3) a_{355} + a_{452} + X_4 a_{455} = 0,$
$a_{15} + a_{35} = 2,$	$a_{155} + a_{355} = 0,$
$a_{25} + a_{45} = 2 a_{65},$	$a_{255} + a_{455} = 2 a_{655}.$

The nodal coordinates in the given configuration are: $X_1 = L$, $X_2 = X_4 = H$, and $X_3 = 3L$. Denoting $H/L = c$, the coefficients of interest are found

$$a_{65} = 0, \qquad a_{655} = -1/H_c, \qquad (2.23)$$

where

$$H_c = H / (1 + 2c^2). \qquad (2.24)$$

Next, the two connecting bars are introduced with their constraint equations

$$(X_7 - X_5)^2 + (X_7 - X_6)^2 = h_1^2, \quad (X_7 - L/2)^2 + (X_8 - H + h)^2 = h_2^2, \qquad (a)$$

where $h = h_1 + h_2$. For this two-bar chain, displacements x_5 and x_7 are chosen as independent. Carrying out the power expansion yields

$$a_{65} = a_{67} = 0, \quad a_{655} = -1/h_1, \quad a_{657} = a_{675} = 1/h_1, \quad a_{677} = -h/h_1 h_2. \qquad (b)$$

According to (23) and (b), the two subsystems independently control one and the same displacement x_6 by prescribing two different nodal loci. The result is a discontinuity which is a function of the independent virtual displacements

$$f(x_s) \equiv x_6^{(1)} - x_6^{(2)} = \Delta x_6 = [(1 - h_1/H_c)x_5^2 - 2x_5 x_7 + (h/h_2)x_7^2]/2h_1. \quad (2.25)$$

The system lacks kinematic mobility if discontinuity is unavoidable under any virtual dislacements, i.e., when form (25) is definite. That is the case when

$$h/h_2 > 0, \quad (h/h_2)(1 - h_1/H_c) - 1 > 0. \qquad (2.26)$$

Four combinations of parameters satisfy both of these inequalities:

$h_1 < 0, \quad h_2 < 0, \quad$ (Fig. 9a), $\qquad\qquad h_1 < 0, \quad h_2 > 0, \quad h > H_c, \quad$ (Fig. 9b),

$h_1 > 0, \quad h_2 > 0, \quad h < H_c, \quad$ (Fig. 9c), $\qquad\quad h_1 > 0, \quad h_2 < 0, \quad h > H_c, \quad$ (Fig. 9d).

The sign of initial forces of stable self-stress patterns is shown in the Fig. 9, with plus denoting tension. Note that the retention of only two arbitrary parameters, h_1 and h_2, produces a wealth of infinitesimally mobile systems. The critical role of parameter H_c given by (24) will be examined in the following section.

Even more variety than demonstrated with this simple mechanism can be attained in the synthesis of compound infinitesimal mechanisms. Here it is necessary to compile an assortment of suitable singular components and to assemble them in a meaningful way. The simplest recipe is to choose a system III as a basis and build upon it a desired superstructure. More challenging is the case when none of the component subsystems is of type III. For example, a closer look at the systems in Fig. 6 reveals that each of them involves two singular subsystems which are finite mechanisms in singular configurations. The vertical virtual displacements of the subsystems are conforming, so that the subsystems can exercise them in concert. However, the respective displacement ratios are, generally, different, hence, unconforming, either accidentally or intentionally. This prevents the displacements from becoming kinematic, and the result is a compound infinitesimal mechanism. Thus, the geometry of a compound system can be manipulated so as to either achieve or, on the contrary, to prevent the conformity of the virtual displacements.

2.3   Higher-Order Infinitesimal Mobility

Higher-Order Virtual Displacements

Constraint Compatibility and Order of Virtual Displacements. The formal description (1) of an infinitesimally mobile system $(V > K = 0)$ only implies that the system possesses virtual mobility, but does not provide any information on its order. Moreover, the very concept of the order of infinitesimal mobility has proved elusive and has defied several known attempts at a concise definition.

A clue for both defining and evaluating the order of infinitesimal mobility is found in expression (25) of the preceding example, which can be viewed as either a constraint discontinuity or a constraint variation necessary to maintain compatibility. It is immediately recognizable as (modulo a constant) the reduced quadratic form of the system (Fig. 9), and this, of course, is not an accident.

If the two subsystems are put together but bar h_1 is not linked (or is cut), the independent displacements produce a gap of the size f in the direction of coordinate x_6. A force Λ, needed to close the gap, induces constraint reactions throughout the system (their pattern corresponds to the system's self-stress) and performs virtual work $Q = \Lambda f$. This work does not depend on the particular distribution of constraint variations: any bar or a group of bars can move, stretch, or contract in order to accommodate the discontinuity. Moreover, since self-stress is determined only as a pattern, the magnitude of the initial forces is just a scale factor which is irrelevant in kinematics and, for this reason, absent in (25). Obviously, if displacements are possible without producing any discontinuity, then constraints are compatible, no constraint variation ensues, and no work is expended. The resolution of this familiar dilemma depends, predictably, on the properties of the quadratic form.

Thus, the entire kinematic analysis (immobility versus infinitesimal or finite mobility, kinematic criteria and definitions of the four types of systems, analysis and synthesis of infinitesimally mobile systems, and so on) stems from a single source—the existence of virtual displacements and the condition of constraint compatibility or incompatibility under these displacements. In this light, it is only natural to turn to the same source for information on the order of infinitesimal mobility.

A complete expression for constraint discontinuity, i.e., one in all virtual displacements, can be presented as the sum of ascending-order forms (linear, quadratic, and so on):

$$f(x_s) = d_s x_s + d_{st} x_s x_t + \dots . \tag{2.27}$$

Note that even though the power expansions of constraint functions may contain only linear and quadratic forms (such is the case with pin-bar assemblies), this does not entail the same property for the resulting expression for constraint discontinuity; generally, it contains higher-order forms. Considered one at a time, these discontinuity forms yield information on the kinematic properties of the system and their order.

To begin with, conventional virtual displacements turn into zero the linear discontinuity form since it is a linear combination of the linearized constraint functions. Those of the displacements that also turn into zero the quadratic discontinuity form are called *second-order* virtual displacements. Proceeding in the same way for forms of higher order leads to the notion of k-th *order virtual displacements* of the system in a given configuration.

Not all (if any) of the virtual displacements of a given order may make it to the next order since each higher-order form imposes additional restrictions on the displacements. A definite form immediately turns all virtual displacements into zero whereas a semidefinite or indefinite form, by relating the displacements involved, reduces the number of independent ones among them and thereby reduces the size of the next-order form. For example, a three-bar system in a dead-center position has two independent virtual displacements of the first order, but only one of the second order. In a stretched rectilinear (quasi-variant) configuration, this system has two first-order virtual displacements but none of the second order. The same is the case with the compound mechanism in Fig. 6. Using prime notation for order designation, this observation can be presented as

$$V^{[k]} \geq V^{[k+1]} \geq K, \tag{2.28}$$

where K and $V^{[k]}$ are, respectively, the degrees of kinematic and k-th order virtual indeterminacy (the respective numbers of kinematic and k-th order virtual degrees of freedom). Thus, the degree of virtual indeterminacy either stays constant or reduces from order to order. Virtual determinacy of any order immediately entails kinematic determinacy (uniqueness of configuration).

Definition. The *order of infinitesimal mobility* (the order of an infinitesimal mechanism) is the order of constraint compatibility; it is the highest order of the constraint discontinuity form that vanishes with nontrivial virtual displacements. The order is k if

$$V^{[k]} > V^{[k+1]} = 0. \tag{2.29}$$

This purely kinematic definition is consistent with the nature of the concept and avoids the conventional reference to the more-sophisticated concept of strain.

Evaluation and Reduction of Higher-Order Discontinuity Forms. A general rule in the analysis of higher-order infinitesimal mobility is that a constraint discontinuity form of any order must be considered in conjunction with all of the vanishing lower-order forms. This was the case with first-order mechanisms where the complete quadratic form was evaluated subject to the linearized constraint equations. One of the alternatives in this analysis was the construction of the reduced quadratic form; the other—matrix test (10). With the second alternative infeasible when dealing with higher-order mobility, the following analysis resorts to the evaluation and reduction of constraint discontinuity forms of successive orders. The resulting analytical procedure is based on the power expansion technique employed in the preceding section.

The first step is an analysis of the linearized constraint equations identifying conventional (first-order) independent virtual displacements. Their number, V, (omitting the prime), is the number of virtual degrees of freedom given by (1.29). Proceeding to the second order, a definite or indefinite quadratic form leads to the conclusions on the system type considered earlier. The case of a semidefinite form, aside from a reduction in the number of independent displacements, is inconclusive and necessitates a higher-order analysis. This is also the situation with an identically vanishing quadratic form, except that in this case all of the virtual displacements are promoted to the next order. The analysis proceeds to the next-order form and does so as long as the sequentially emerging forms keep vanishing, either identically or with a reduction in the number of virtual degrees of freedom. In the process, the hierarchy of gradually diminishing sets of virtual displacements of each order is established. Depending on the character of the lowest-order nonvanishing form, the analysis reveals either infinitesimal mobility of the system, along with its order and number of virtual degrees of freedom, or finite mobility and the number of kinematic degrees of freedom. The first outcome occurs when the reduction process leads to one of the following:

(1) a definite form (simple infinitesimal mechanisms);

(2) a definite linear combination of forms describing singular subsystems (complex mechanisms); and

(3) an incompatible set of simultaneous forms for singular subsystems (compound mechanisms).

The reduction process is illustrated by several examples below.

Example 2.9 (continued). It is convenient to explore higher-order mobility starting with a closer look at the parameter H_C given by (24) in the previous example (Fig. 9). At

$$h = H_C = H/(1 + 2c^2),$$

the quadratic discontinuity form (25) becomes semidefinite, indicating that the displacements of the two subsystems are still compatible, but leaving the question on kinematic mobility of the system unresolved. However, in the very act of turning into zero, the form yields

$$x_7'' = (h_2/h) x_5''. \tag{2.30}$$

Accordingly, of the two first-order independent virtual displacements only one, say, x_5, remains an independent displacement of the second order. Thus, the respective numbers of virtual degrees of freedom of the system are

$$V (\equiv V') = 2, \quad V'' = 1.$$

The coefficients of the linear and quadratic discontinuity forms are given by (23). Proceeding with higher-order terms in the two parallel power expansions, it is found that cubic terms are absent in both of them. The identical disappearance of the third-order discontinuity form indicates constraint compatibility to at least the third order and the possibility of third-order virtual displacements. However, fourth-order displacements of the two subsystems produce a nonzero discontinuity

$$f = \Delta x_6 = [(1+2c^2)^3 - (1 + 22c^2 + 20c^4)] x_5^4/8H^3. \tag{2.31}$$

In this monomial fourth-order form, the quantity in the brackets can be of either sign or equal to zero. Each of these three eventualities leads to a different outcome.

If the form (31) is positive, the fourth-order virtual displacement x_5^{IV} is zero, constraints are compatible only to the third order, and the system lacks kinematic mobility: this case is a type III system of the third order (type III^3). If the discontinuity form vanishes, this leads to an inconclusive outcome, implying merely the possibility of higher-order virtual displacements. When the discontinuity form is negative, curiously, the system is of type IV, in spite of the form being sign definite. The reason is that the condition of being sign definite applies only to forms in all independent variables, while form (31), in its evolution, has "lost" one of its arguments at a lower order. In this case, a stricter requirement of positive definiteness applies since the sign of initial forces cannot be reversed without affecting the stability of self-stress.

The geometric meaning of the analysis is straightforward and illuminating. The origin of parameter H_C shows it to be the local radius of curvature of the nodal path imparted by the first subsystem. Condition (30) indicates that the second subsystem, with its two bars aligned, prescribes a circular path of the same radius. Thus, the two paths not only share the tangent, but have identical curvatures as well. It is quite possible that the paths have an even closer contact

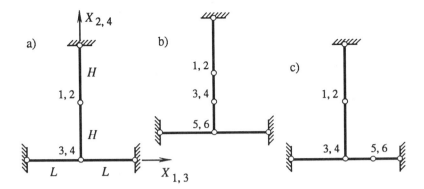

Figure 2.10. Partially prestressable and mixed-order infinitesimal mechanisms: a) type II^3; b) type III^3; c) type $III^{2\wedge3}$.

and share several next infinitesimal vicinities. This is determined by the properties of the corresponding higher-order discontinuity forms.

The implication for the ongoing analysis is straightforward. When, at $c^2 = 2$, the bracketed quantity in (31) becomes zero, further power expansions are needed to refine the paths imposed on the node by the two subsystems, and to evaluate the higher-order discontinuity form. Fifth-order terms are missing in these expansions, but the sixth-order discontinuity form is

$$f = \Delta x_6 = (-0.00317 + 0.00625)x_5^6/H_c^5 > 0. \tag{2.32}$$

Thus, constraints are compatible up to, and including, the fifth order; hence, the system with $c^2 = 2$ and $h = H_c = H/5$ is of type III^5.

Example 2.10. A somewhat different class of singular systems is represented by the infinitesimal mechanism (Fig. 10a) which is only partially prestressable. Here the constraint equations are

$$X_1^2 + (2H - X_2)^2 = H^2, \quad (X_1 - X_3)^2 + (X_2 - X_4)^2 = H^2,$$
$$(X_3 + L)^2 + X_4^2 = L^2, \quad (X_3 - L)^2 + X_4^2 = L^2. \tag{2.33}$$

Let the first three bars be one subsystem and the lower right bar 4 the other. With x_1 as an independent displacement in the first subsystem and x_4 in the second, a power expansion analysis yields

$$x_4 = x_1^2/H, \quad x_3^{(1)} = -x_4^2/2L, \quad x_3^{(2)} = x_4^2/2L. \tag{2.34}$$

The lowest-order nonzero discontinuity form in independent displacements is

$$f = \Delta x_3 = x_1^4/LH^2. \tag{2.35}$$

Since this fourth-order form is monomial, constraint equations (33) are compatible only to the third order and the system is a third-order infinitesimal mechanism; it is quasi-invariant (type II^3) because $D = V (= 1)$.

If a pin is inserted in any bar, the system becomes topologically inadequate, with two independent virtual displacements. It is not difficult to verify that inserting a pin into one of the vertical bars (Fig. 10b) will result in a definite fourth-order discontinuity form in two independent displacements, and the modified system will be type III³. On the other hand, inserting a pin into one of the horizontal bars (Fig. 10c) leads to a different outcome. Although the resulting lowest-order nonzero discontinuity form in two linearly independent displacements is still positive-definite, it is of a mixed order—fourth in x_1 and second in x_6. Such a *mixed-order* infinitesimal mechanism can be considered as type III$^{1\wedge 3}$ or, judging by the higher order, as III³.

Example 2.11. Consider again the compound mechanism in Fig. 6a. When the length of the bar 6 is $L_6 = 1$ or $L_6 = 3$, the second-order analysis of Examples 5 and 7 was inconclusive and the constraint discontinuity form must be extended to a higher order. It turns out that the next step of the analysis requires only the third derivatives of the second triplet of constraint functions (a) on page 49:

$$3a_{58}a_{588} + X_5 a_{5888} + 3a_{68}a_{688} + X_6 a_{6888} = 0, \tag{h}$$

$$(a_{78} - a_{58})(a_{788} - a_{588}) + (X_7 - X_5)(a_{7888} - a_{5888}) - 3(1 - a_{68})a_{688} - (X_8 - X_6)a_{6888} = 0, \tag{i}$$

$$3a_{78}a_{788} + X_7 a_{7888} = 0. \tag{j}$$

Recall that in this analysis the system is divided into two subsystems of which the first one is the original system without bar 6. Multiplying equation (h) by 2 and adding it to (i), while taking into account the coefficient values (b) and (c), gives for the first subsystem

$$2a_{7888}^{(1)} - 3a_{688}(1 - 3a_{68}) = 0.$$

This equation has two solutions for $a_{7888}^{(1)}$ which correspond to the two displacement modes of the subsystem described by coefficients (g). The first solution is zero, but the second (parenthesized) is not:

$$a_{7888}^{(1)} = 0, \quad (a_{7888}^{(1)} \neq 0); \quad a_{7888}^{(2)} = 0. \tag{2.36}$$

The last of the above expressions is the counterpart solution for the second subsystem (the disconnected bar 6); it is obtained from equations (j) and (b).

The obtained solutions are summed up below and the corresponding perturbation modes of the two mechanisms are shown in Fig. 11.

	$L_6 = 1$	Order	$L_6 = 3$	Order
Mode 1	$a_{788}^{(1)} \neq a_{788}^{(2)}$	1	$a_{78888}^{(1)} \neq a_{78888}^{(2)}$	3
Mode 2	$a_{7888}^{(1)} \neq a_{7888}^{(2)}$	2	$a_{788}^{(1)} \neq a_{788}^{(2)}$	1

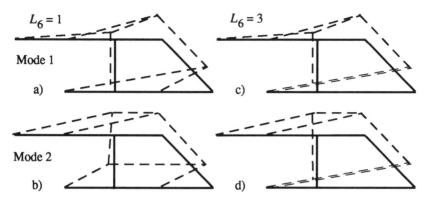

Figure 2.11. Multiple-order infinitesimal mechanisms and their perturbation modes: a) first-order; b) second-order; c) third-order; and d) first-order.

Thus, at $L_6 = 1$ the order of infinitesimal mobility may be either first (Fig. 11a) or second (Fig. 11b), depending on the applied perturbation and imperfections in the system geometry. At $L_6 = 3$ it follows from (20) and (36) that in the first mode (Fig. 11c) the two second-order coefficients are equal and so are the two third-order coefficients. Hence, the paths imposed on the common node by the two subsystems are compatible to at least the third order, leaving the order of infinitesimal mobility in this mode undetermined. Its further analysis requires evaluating the fourth-order derivatives of the constraint functions and solving the resulting simultaneous equations. It has been found that the fourth-order coefficients for the two subsystems are not equal. Therefore, in this mode the constraints are compatible only to the third order and the system is a third-order infinitesimal mechanism. In the second mode (Fig. 11d) the system with $L_6 = 3$, according to (20), possesses first-order mobility.

The foregoing analysis demonstrates the existence of systems having multiple perturbation modes with different orders of mobility; these are *multiple-order* infinitesimal mechanisms. In particular, the system just analyzed can be qualified as type $\mathrm{II}^{1 \vee 2}$ at $L_6 = 1$ and type $\mathrm{II}^{1 \vee 3}$ at $L_6 = 3$. Recall that in mixed-order mechanisms (like the one in Fig. 10c), different orders of mobility with respect to different independent virtual displacements x_s are realized concurrently, in one and the same perturbation mode.

Prestressability and Kinematic Immobility

Even-Order Infinitesimal Mechanisms. Even-order mechanisms have distinct and somewhat unexpected properties. First of all, they are topologically adequate and can only be quasi-invariant (type II). The reason is that a topologically inadequate, geometrically singular system has at least two

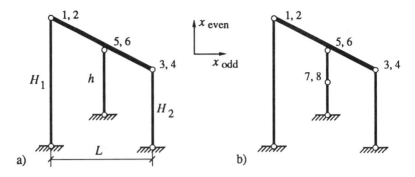

Figure 2.12. Two more varieties of infinitesimal mechanisms: a) second-order type II; b) mixed-order type III.

independent virtual displacements. The corresponding odd-order discontinuity form in more than one variable can always turn into zero with nonzero values of the variables.

In the light of this observation, it is interesting to track the transformations of the topologically inadequate compound system in Fig. 6b in the process of changing the bar length L_6. According to (17) and (18), at $L_6 < 14/5$ the system is quasi-variant (type III) and exibits first-order infinitesimal mobility in each of its two perturbation modes. In the first mode, the middle bars of the two singular horizontal subsystems are tilting, in the second mode these bars remain horizontal. In either mode, the number of independent first-order virtual displacements is three, and there are no nontrivial second-order displacements.

At $L_6 = 14/5$, the quadratic constraint discontinuity form (17) for the first mode becomes semidefinite and the system acquires higher-order mobility in this mode. However, in contrast to its topologically adequate counterpart in Fig. 6a, the system in consideration cannot possess second-order mobilty, so that the order of mobility must be at least the third. This fact is the consequence of the third-order constraint discontinuity form involving more than one (in this case, two) independent virtual displacements. This form being zero cannot annul the two second-order displacements; still, by relating them, it reduces the number of independent displacements of the third order to one, and the latter can be nullified by a higher-order discontinuity form.

At $L_6 > 14/5$, the system mobility in the first mode becomes finite, whereas in the second mode it remains first-order infinitesimal. At $L_6 = 3/2$, the second mode acquires higher-order mobility, and at $L_6 < 3/2$ mobility is finite in both of the modes.

Further noteworthy features of even-order mechanisms are illustrated in the following example.

Example 2.12. The system in Fig. 12a is analyzed by dividing it into two subsystems, with the middle vertical bar being the second one. The constraint equations for the first subsystem are those of a three-bar system, with a few adjustments in the bar lengths and nodal coordinates. Taking the horizontal displacement x_5 as independent, the first-order terms in the power expansions for the two subsystems are found to be identical and, because $V = D$, the system is type II. The respective second-order coefficients for the two subsystems are

$$a_{655}^{(1)} = -(H_1 + H_2)/2H_1 H_2, \qquad a_{655}^{(2)} = -1/h, \qquad (a)$$

and are equal at

$$h = 2H_1 H_2 / (H_1 + H_2). \qquad (2.37)$$

Then the paths of the midpoint of the inclined bar (in the first subsystem) and the counterpart endpoint of the middle vertical bar (in the second) coincide up to the second order. However, the third-order coefficients are different:

$$a_{6555}^{(1)} = 3(H_1 - H_2)^3 / 4L H_1^2 H_2^2, \qquad a_{6555}^{(2)} = 0, \qquad (b)$$

so that the lowest-order nonzero discontinuity form is

$$f = \Delta x_5 = [(H_1 - H_2)^3 / 4L H_1^2 H_2^2] \, x_5^3. \qquad (2.38)$$

The form is cubic, hence, indefinite; nevertheless, being a monomial (and barring the case of $H_1 = H_2$) the form vanishes only with x_5. Furthermore, it also satisfies the above-mentioned qualifier: it is a form in all independent virtual displacements. Thus, the system in Fig. 12a, being topologically adequate, is of type II^2.

Interestingly, inserting a pin in the middle vertical bar of this system (Fig. 12b) makes it topologically inadequate, ostensibly turning it into an even-order type III. Recall, however, that this type has been prohibited; only type II is possible with bidirectional even-order mobility. So instead the system becomes a peculiar mechanism with double-order mobility: finite in the perturbation mode with $x_5 < 0$, and second-order infinitesimal in the mode with $x_5 > 0$. Virtual self-stress of this system, with or without the added pin, is unrealizable. At an attempted prestressing (say, by a thermal action or a turnbuckle), constraint variation occurs bringing the system out of the singular configuration into its corresponding generic class (type IV with the pin, and type I without it). The observations made are summed up in the following conclusion.

A Property of Even-Order Mechanisms. Systems with bidirectional even-order infinitesimal mobility are quasi-invariant (type II) and unprestressable.

This property illuminates the mutual exclusion condition between finite mobility and prestressability and explains why the statical criterion for immobility is only a sufficient one. It has been mentioned that the *absence* of one of the two mutually exclusive features does not necessarily entail the *presence* of the other. As it turns out, systems that are prestressable are kinematically immobile, but not all kinematically immobile systems are prestressable. Specifically, even-order and some mixed-order infinitesimal mechanisms (Fig. 12), as well as certain compound first-order mechanisms (Fig. 6), are *neither mobile nor prestressable*: their displacements and self-stress are both only virtual.

Recall now that even-order mechanisms are topologically adequate and compound mechanisms are multiply singular $(D > 1)$. Without these two types, a condition stronger than the above mutual exclusion condition can be formulated for a narrower class of systems.

Equivalence Condition. For a topologically inadequate system with one geometric singularity $(D = 1)$, prestressability and kinematic immobility are *mutually necessary and sufficient.*

In other words, within this class, a prestressable system is kinematically immobile and a kinematically immobile system is prestressable.

As to the original exclusion condition, it holds for a structural system of any type and any order of mobility. Indeed, prestressability implies that the lowest-order discontinuity form is definite, which entails uniqueness of the configuration. On the other hand, if such a form does not exist or is not positive, prestress is ruled out, but the question on infinitesimal versus finite mobility remains open, with both outcomes possible. Even-order and certain compound infinitesimal mechanisms exemplify one outcome, and singular finite mechanisms the other.

A Geometric Interpretaton and Implications of Stationarity. It is readily seen that the above evaluation of the order of constraint compatibility by dividing the system into two subsystems amounts to establishing the geometric relation between the two respective loci to which the two subsystems confine their common node. For the system as a whole to exist, the loci must have at least one common point. The system is geometrically invariant if this point is an isolated intersection point and the loci do not have a common tangent manifold in it. If (and only if) the loci intersection is not an isolated point, the system is variant and a kinematic displacement is possible over the intersection manifold. A system with infinitesimal mobility is characterized by the fact that the loci share a tangent manifold at their common point.

A geometric illustration of particular situations becomes especially clear when the loci are one-dimensional paths (Fig. 13). If the paths intersect without sharing a tangent (Fig. 13a) the system is invariant, and if the paths coincide, even locally, the system is variant. When the two paths intersect while sharing a tangent (Fig. 13b), the system is an unprestressable infinitesimal mechanism. The order of tangency is the order of infinitesimal mobility and may be either odd or even. Such are, for example, the double-order system in Fig. 6a and the second-order system in Fig. 12a. Finally, when the paths only touch (Fig. 13c and d), the system is a prestressable odd-order infinitesimal mechanism. Such is the system in Fig. 1b; the system in Fig. 1c when $L_3 > L_1 + L_2$; and all of the systems in Figs. 7–9.

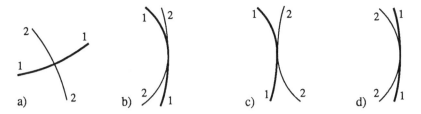

Figure 2.13. A geometric illustration of the order of constraint compatibility: a) intersection without tangency (zeroth order); b) intersection with tangency (any order); c) and d) tangency without intersection (odd order).

As is seen from Fig. 13b, the peculiar nature of unprestressable infinitesimal mechanisms, both compound and even-order ones, is a consequence of a simple analytical fact: instead of a strict maximum or minimum, one of the constraint functions attains only a *stationary* value, subject to the fixed values of the rest of the functions. Stationarity makes the constraints dependent $(D > 0)$, thus rendering the system configuration singular. This suffices for the system to acquire virtual self-stress, but in the absence of an extremum it has no chance of realization. At the same time the system acquires some mobility: virtual when the stationarity is local (the second-order mechanism in Fig. 12a and a simpler system in Fig. 14a), and kinematic when it is global (Fig. 14b and c).

In the case of local stationarity without an extremum, the constraint discontinuity form, which is the difference of the nodal paths (Fig. 13b), describes a curve with an *inflection point*. This is illustrated in the simplest possible way by the one-bar system in Fig. 14a, showing the path to which the end of the bar is confined by a subsystem not shown. (Note in passing that a trajectory imposed on a node by a subsystem can be any algebraic curve [Hilbert and Cohn-Vossen, 1952].) As a result, the system configuration is singular, but both the displacement and the self-stress are only virtual. The system is of type II^{2k}, where k depends on the order of "flatness" of the path at the inflection point. As an even-order mechanism, the system is topologically adequate, and a constraint variation bringing the pin out of the inflection point restores the system to type I.

Although it is not as obvious, the compound infinitesimal mechanisms in Fig. 6 are also characterized by local stationarity. Thus, constrained local stationarity of one of the constraint functions, without an extremum, is an analytical criterion and a source of unprestressable systems with infinitesimal mobility. It complements the classical (Möbius-Maxwell-Mohr) source of infinitesimally mobile systems based on a strict constrained extremum with resulting prestressability.

Perhaps, the simplest example of a system with dependent, yet globally stationary constraints (Kuznetsov, 1988), is a rigid beam supported at three points by three parallel bars of equal length (Fig. 14b). As has been shown already (Chapter 1), the system has $D = V = K = 1$ in an ordinary (globally

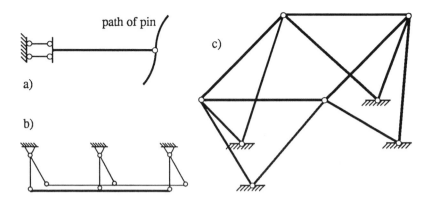

Figure 2.14.   Systems with constrained stationarity: a) local (an even-order infinitesimal mechanism);   b) and   c) global (singular finite mechanisms).

singular) configuration and $K = 1$, but $D = V = 2$, in a doubly singular, horizontally folded, configuration. A distinct feature of this class of systems is the fact that, in spite of being finite mechanisms, they are topologically adequate, and even a slight departure from the nominal member sizes would immobilize such a system. Indeed, by virtue of equation (1.44), at

$$S \geq D = V \tag{2.39}$$

there always exists a constraint variation that makes the system invariant. An interesting example of this class is a space truss (Tarnai, 1980) where the bars are the edges of Archimedes' antiprism (Fig. 14c). The truss represents a statically indeterminate finite mechanism with cyclic and reflection symmetries.

The order of infinitesimal mobility of a system can be interpreted geometrically as the order of "flatness" of the surface $f = f(x_s)$ representing the constraint discontinuity form. In other words, it is the order of contact of the two multidimensional surfaces that represent loci prescribed by the two subsystems to the point they share.

Note in conclusion that although the order of mobility of a system quantifies the rate of stiffening in relation to virtual displacements, it may not characterize the actual physical rigidity of the system, which depends mostly on its geometric parameters and material properties. For example, a real finite mechanism with a small amount of interference is formally a first-order infinitesimal mechanism, but actually can be very flexible. On the other hand, an infinitesimal mechanism of high order can actually be quite rigid if the lowest-order discontinuity form contains a large parameter.

3
Systems Involving Unilateral Constraints

A *unilateral constraint* is an idealized model of a structural component that limits some linear or angular distance to values either no greater than, or no less than, a certain magnitude. Unlike bilateral constraints, unilateral constraints are binding in only one sense of direction. Some structural components modeled as unilateral constraints in tension are wires, cables, flexible bands, membranes, fabrics, and rectilinear pin-bar chains. Situations giving rise to unilateral constraints in compression include contact problems, stacks of building blocks, springs with collapsed coils, components made of brittle, low-tensile strength materials with no reinforcement, and all kinds of liquid and granular media. Typical examples of angular, or bending, unilateral constraints are a door hinge, a spherical hinge with restraints, and a concrete beam with only a bottom reinforcement. The three most common types of unilateral constraints are shown schematically in Fig. 1.

Yet another important class of unilateral constraints are those representing structural members modeled as rigid-plastic. Upon reaching a specified force level, such a constraint becomes unilateral for further loading. This type of model underlies the theory of limit equilibrium and its application to plastic design. Since this approach does not involve conventional (stress–strain) constitutive relations, it falls entirely within the scope of statical-kinematic analysis.

A conceptual difficulty in the statical-kinematic analysis of systems involving unilateral constraints stems from a complicated relation between the structural topology and geometry. Specifically, the overall topology of a system cannot account for a configuration-specific status (engaged or disengaged) of unilateral constraints. For this reason, most of the statical-kinematic properties of a system with unilateral constraints are configuration-specific and, in addition, depend on the status of the unilateral constraints. Only with this status known for a given configuration, can the latter be analyzed and its statical-kinematic properties established, thereby producing the so-called working model of the system. A constraint variation in the current model may alter the engagement pattern and necessitate a modification of the model. The number of such models is combinatorial for a discrete system and infinite for a continuous one.

An adequate tool for an analysis of systems involving unilateral constraints is the theory of linear inequalities. Using this tool, the basic properties of these

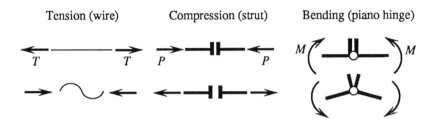

Figure 3.1. Unilateral constraints in tension, compression, and bending.

systems are investigated and the analytical criteria for the four statical-kinematic types are established in the first two sections of this chapter. In the last section the results of bilateral and unilateral constraint imposition and removal are studied. The study reveals, in particular, an unusual role of unilateral constraints in the limit equilibrium analysis due to the possibility of a system conversion into a stationary plastic mechanism.

For the purposes of a statical-kinematic analysis, it is convenient to work with unilateral constraints of the tension and compression types. These generic constraints are sometimes called in the literature a wire and a strut, respectively. Figure 2 contains the simplest possible prototypes of the four statical-kinematic types of structural systems composed solely of wires. The definitions of each type of system, presented earlier, hold. However, the analytical and mechanical criteria to identify each type of system require further elaboration. It is natural to begin with invariant systems.

3.1   Invariant Systems with Unilateral Constraints

Constraint Conditions, Their Linearization and Reduction

Constraint Conditions.   For a system involving unilateral constraints, constraint conditions represent a simultaneous set of equations and *inequalities*

$$F^i (X_1,\cdots, X_n,\cdots, X_N) = 0, \qquad i = 1, 2,\cdots, C_b, \qquad (3.1)$$

$$F^j (X_1,\cdots, X_n,\cdots, X_N) \le 0, \qquad j = 1, 2,\cdots, C_u, \qquad (3.2)$$

where C_b and C_u are the numbers of bilateral constraints (bars) and unilateral constraints (in this case, wires), respectively. It is assumed that the equations and inequalities in (1) and (2) admit at least one real solution corresponding to

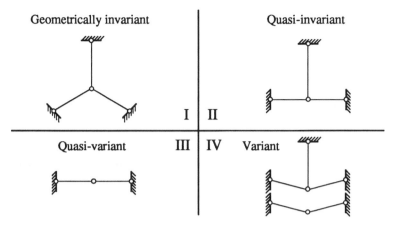

Figure 3.2. The simplest prototypes of the four statical-kinematic types of systems with unilateral constraints in tension (wires).

point $X_n = X_n^0$ in the coordinate space of the system (otherwise, the system could not be assembled). Constraint conditions linearized at the solution point involve simultaneous sets of homogeneous linear equations,

$$f^i = F_n^i x_n = 0, \tag{3.3}$$

and inequalities,

$$f^j = F_n^j x_n \le 0. \tag{3.4}$$

The corresponding equilibrium equations can be obtained by applying the principle of virtual work and are of the form

$$F_n^i \Lambda_i + F_n^j M_j = P_n, \quad M_j \ge 0. \tag{3.5}$$

Here Λ_i are, as before, generalized reactions in bilateral constraints and M_j are generalized reactions in unilateral constraints (wires).

Engaged Versus Disengaged Unilateral Constraints. In contrast with bilateral constraints, a unilateral one can be in one of two states: it is either *engaged* $(f^j = 0)$ or *disengaged* $(f^j < 0)$. For a disengaged constraint, the corresponding conditions (2) and (3) hold as strict inequalities, while the constraint reaction is zero. Because of that, the presence of a disengaged constraint in a system is inconsequential for the given configuration; it is not reflected in the equilibrium equations and the linearized constraint conditions. Accordingly, a disengaged constraint does not affect the force distribution and virtual displacements of the system in the given configuration. On the other hand, when a unilateral constraint is engaged and stressed, the associated constraint condition holds as an equation, and such a constraint is locally (i.e., for the given configuration) equivalent to a bilateral one.

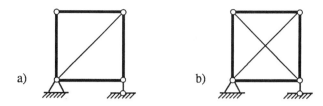

Figure 3.3. Systems involving bars and wires: a) variant; b) invariant.

Obviously, bilateral constraints are always engaged, whereas the state of a unilateral constraint depends on the system configuration. Although detectable by statical means, engagement is a purely geometric property: it only *enables*, but does not cause, the unilateral constraint to develop a reaction. Furthermore, engagement alone does not suffice for the mentioned equivalence with a bilateral constraint. When the status of all unilateral constraints is known, further analysis of the system is straightforward. Thus, the main difficulty in dealing with a system involving unilateral constraints is establishing the current model of the system, that is, the status of each unilateral constraint. This is done using available information on the system composition and external actions.

The foregoing observations can be formalized as a *complementary slackness condition* between the constraint reaction and variation

$$M_j f^j = 0 \quad \text{(no summation).} \tag{3.6}$$

For a wire it follows that, aside from the trivial case of idle (unstressed) engagement, $M_j = f^j = 0$, one of the two situations is possible:

$$M_j > 0, \quad f^j = 0; \quad \text{or} \quad M_j = 0, \quad f^j > 0. \tag{3.7}$$

Each of the two combinations constitutes a dual, statical and kinematic, analytical description of a wire as a unilateral constraint. This approach is also used for describing the properties of rigid-plastic constraints, both bilateral and unilateral. The only difference is that a specified yield force replaces zero as the reference force level determining the status of each constraint. Upon reaching the yield level, the constraint reaction takes on the indicated fixed value; after that, a bilateral constraint becomes unilateral in unloading, and a unilateral constraint becomes kinematically inconsequential.

Reduction of Linearized Constraint Conditions. In the following analysis of systems involving unilateral constraints (Kuznetsov, 1972a), it is assumed that the bilateral constraints alone are insufficient for geometric invariance of the system, i.e., the Jacobian matrix rank of the linearized constraint equations (3) is $r_b < N$. Such are, for example, the two systems shown in Fig. 3, one of them invariant and the other not. In both cases the bilateral constraints alone (they are shown by heavy lines) are not sufficient for geometric invariance: each system has $r_b = 4 < N = 5$.

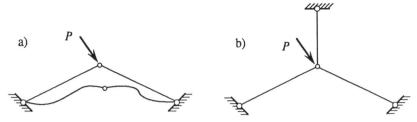

Figure 3.4. Wire systems: a) variant; b) invariant.

For such systems, linearized constraint equations (3) can be solved in terms of $N_u = N - r_b$ virtual displacements, and the obtained solution of the form (1.7) can be used to eliminate the dependent displacements in inequalities (4). After renumbering the virtual displacements from 1 to N_u, the resulting system of inequalities becomes

$$F_n^j \, a_{ns} \, x_s = c_{js} \, x_s \leq 0, \quad s = 1, 2, \cdots, N_u. \tag{3.8}$$

Thereby, the linearized simultaneous constraint equations and inequalities (3) and (4) are reduced to a set of homogeneous inequalities in N_u variables x_s. The variables are the virtual displacements of the system taken without its physical unilateral constraints, and inequalities (8) formally impose these constraints. Thus, determining the statical-kinematic properties of the given configuration of the system calls for an analysis of this set of simultaneous homogeneous linear inequalities.

Analytical Criterion for Invariance

The necessary and sufficient analytical criterion for geometric invariance is the absence of solutions, other than the trivial one, $x_s = 0$, to the set of linear inequalities (8). This would entail both uniqueness of the configuration and, as demonstrated below, the ability of the system to support an arbitrary load in this configuration.

Necessary Conditions for Invariance. To satisfy this requirement, first of all, the rank, r_u, of the matrix c_{js} in (8) must equal the number N_u of virtual displacements. Otherwise, even the set of boundary equations corresponding to inequalities (7) would admit a nonzero solution. At the same time, the number of inequalities, which is the number of engaged unilateral constraints, must be $C_u > r_u$.

These two necessary conditions are illustrated by the systems in Fig. 4. Both systems satisfy the first condition, having $N_u = N = 2$ (two degrees of freedom of the center node) and $r_u = 2$ (the same rank for the three-wire system despite the additional wire). The two-wire system in Fig. 4a does not satisfy the second condition: $C_u = 2$ is not greater than r_u, and, quite obviously, the system is underconstrained. The system in Fig. 4b does satisfy the second condition, since $C_u = 3 > 2$, and therefore it may be invariant, which for this particular configuration happens to be the case.

The two wires of the first system are unable to take away the two degrees of freedom of the center node, while the three wires of the second system can do so. This observation can be generalized. N unilateral constraints can never take away N degrees of freedom; however, with the right arrangement, $N + 1$ unilateral constraints can.

Assuming the above two conditions, $r_u = N_u$ and $C_u > r_u$, to be satisfied for the set (8), let one of its basic subsets, corresponding to one of the principal minors of matrix c_{js}, be

$$c_{st}\, x_s \le 0, \qquad s, t = 1, 2, \cdots , N_u. \tag{3.9}$$

In the space of virtual displacements x_s this subset determines the domain (a narrow polyhedral cone) of admissible values of displacements for a subsystem whose constraints are those represented in the minor c_{st}. Suppose that, among the inequalities (8) not belonging to this subset, there exists (or can be constructed from them as a linear combination with positive coefficients d_k) an inequality

$$c_{os}\, x_s\, (= d_k\, c_{ks}\, x_s) \le 0, \qquad k = N_u + 1, N_u + 2, \cdots , C_u, \tag{3.10}$$

such that the above cone, except for its apex, falls outside the half-space defined by (10). In this case, subset (9) and inequality (10) are mutually exclusive, which eliminates the above cone as a domain of admissible values of the coordinates, *except for a single point* $x_s = 0$. It is not difficult to see that the coexistence in set (8) of a basic subset and an inequality, having this described relation, is necessary and sufficient for the absence of nonzero solutions of the set. The resulting uniqueness of the feasible point in the space of virtual displacements x_s entails the geometric invariance of the system. (Recall that uniqueness of the feasible point in the coordinate space X_n would only amount to kinematic immobility of the system.)

Counteracting Constraints and Inequalities. The conditions under which geometric invariance takes place can be deduced from the Minkowski–Farkas theorem. It reads: all solutions of a set of inequalities of the form (9) are compatible with an inequality $c_{os}\, x_s \le 0$ (called in this case a *consequence inequality*), if and only if nonnegative numbers M_t exist for which the relationship

$$c_{st}\, M_t\, x_s = c_{os}\, x_s \tag{3.11}$$

is satisfied identically for all x_s.

What is needed for invariance is a *counteracting inequality*, which is the opposite of a consequence inequality. By operating in the opposite direction, a counteracting inequality negates the other inequalities of the set, hence, reduces the feasible domain in the space of virtual displacements to a single point. Physically, a counteracting inequality corresponds to a unilateral constraint (wire, strut, or subsystem) that opposes the possible motion of the rest of the system, thereby making the system configuration unique.

From the above theorem, the sought characteristic of a counteracting inequality is obtainable by simple reasoning. For a hyperplane, $c_{oj} x_j = 0$, to intersect the narrow cone engendered by set (9) only at its apex, all M_t in (11) must be positive; otherwise, the hyperplane would pass through at least one of the cone generators. At $M_t > 0$, the change of sign of all coefficients c_{os} in (11) converts this consequence inequality into a counteracting inequality with respect to the same set (9): such an inequality is incompatible with all solutions of the set, except for the trivial (zero) one.

Thus, the necessary and sufficient analytical condition for a counteracting inequality to set (9) is the existence of a positive solution M_u to the system of homogeneous equations

$$c_{us} M_u = 0, \quad u = 1, 2, \cdots, N_u + 1. \tag{3.12}$$

The physical meaning of this condition is not difficult to see. Equations (12) are easily recognizable as the homogeneous equilibrium equations in the unknown generalized reactions, M_u, of the engaged unilateral constraints. The statical possibility of nonzero reactions in the absence of external loads means that the system allows virtual self-stress. The requirement that all reactions are positive guarantees that the unilateral constraints are not only engaged, but stressed. Therefore, for this state of the system (i.e., the geometric configuration and status of constraints), the unilateral constraints are statically and kinematically equivalent to bilateral constraints. For this reason, conditions (7) must be satisfied as equalities and, recalling that the matrix rank $r_u = N_u$, this confirms that all $x_s = 0$.

Proceeding in the opposite direction, it can be shown that for a system satisfying criterion (12), any load is a statically possible (equilibrium) one, which is the statical criterion for invariance. Indeed, under the assumptions made, it is possible to eliminate all of the Λ_i from (5) and to obtain a system of $N_u = N - r_b$ equations of equilibrium in the form

$$c_{js} M_j = P_s - c_{st} \Lambda_t. \tag{3.13}$$

The existence of only a zero solution for the set of inequalities (8) qualifies any homogeneous linear inequality as a consequence of this set. Indeed, in this case it can be said that all solutions of the set are compatible with the inequality. But then, according to the Minkowski–Farkas theorem, the system of equations (13), as a conjugate system, admits a nonnegative solution $M_j \geq 0$ for any right-hand side, that is, for any external load, which was to be proved. Conversely, the existence of a nonnegative solution to system (13) for an arbitrary right-hand side means that any homogeneous inequality in x_s is a consequence of set (8). This is only possible provided the set has no solutions other than the trivial one.

Necessary and Sufficient Condition for Invariance.   To sum up, the necessary and sufficient analytical condition for a system involving unilateral constraints to be geometrically invariant is the absence of nontrivial solutions to the reduced set of homogeneous inequalities (8). This is the case when the set satisfies the following conditions:

(1) the number of inequalities, C_u, exceeds the number $N_u = N - r_b$ of virtual displacements in the absence of unilateral constraints;

(2) the rank r_u of the matrix c_{js} equals N_u; and

(3) among the inequalities not covered by some basic subset, there exists (or is obtainable as a consequence) a counteracting inequality.

Checking the last condition is most laborious and allows a dual approach. On the kinematic side, a counteracting inequality is characterized by the fact that the replacement with its left-hand side of any row in the basic subset results in *sign reversal* of the subset determinant. An equivalent statical condition is the existence of a positive solution to the conjugate system of homogeneous equilibrium equations (12).

In compact form, then, the complete *analytical criterion for invariance* is

$$C_u > r_u = N_u, \quad M_u > 0; \quad \Leftrightarrow \text{ type I.} \tag{3.14}$$

Obviously, for an invariant system, each basic subset of (8) has a counteracting inequality (probably, constructed as a linear combination of inequalities) from this set. If this was not the case, a displacement would be possible into the domain determined by this particular basic subset. The validity of the opposite statement is apparent as well: even if just one of the basic subsets of set (8) does not satisfy the above requirement (i.e., a counteracting inequality cannot be found or constructed for it), none of the basic subsets satisfy this requirement, and the system in consideration is not invariant.

The above analytical criterion for an invariant system with unilateral constraints is computationally tedious, especially in the case of multiple indeterminacy. The reason is that a counteracting inequality must be sought (or, still worse, constructed) when it is unknown in advance whether one exists in the given set. One alternative approach is to seek a general solution for set (8). Another available method of analysis is based on elimination of all consequence inequalities from (8), thereby converting it into a so-called irreducible set. For a system with unilateral constraints to be invariant, every inequality in its irreducible set must be a counteracting one with respect to a basic subset. Recall that a counteracting inequality is recognized by the fact that its substitution for any row in a basic subset reverses the sign of the determinant.

Verifying the geometric invariance of a system involving unilateral constraints has a simple interpretation in terms of linear programming. It means that the feasible domain of displacements, as determined by the simultaneous linearized constraint equations (3) and inequalities (4), consists of a single point.

The computational difficulties of the analytical criterion for invariance can often be avoided by using a statical criterion, as described below.

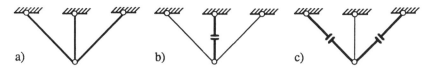

Figure 3.5. Synthesis of invariant systems with unilateral constraints.

Statical Criterion of Invariance

Prestressability of Invariant Systems with Unilateral Constraints.
The most convenient approach to the analysis of geometric invariance employs a
purely statical criterion, which stems from a few additional observations and
interpretations of the foregoing results. The fact is that unilateral constraints,
unlike bilateral ones, reduce mobility of the system by means of mutual
counteraction, which leads to constraint dependence, hence, statical
indeterminacy. As follows from (11), a geometrically invariant system, even if
it involves just one unilateral constraint, is statically indeterminate. Moreover,
virtual self-stress in such a system is always realizable, so that the system is
prestressable. The reason is that an invariant system must resist a force induced
in any of its unilateral constraints (otherwise, displacement in the direction of
the constraint disengagement was possible under some load, contrary to the
assumption on the system invariance).

Prestressability means that the unilateral constraints in this configuration are
equivalent to bilateral, and the corresponding constraint conditions are satisfied as
equations. As before, prestressability in no way implies the actual presence of
prestress, and this is reflected in the formulation of the following criterion.

Statical Criterion of Invariance. A structural system that involves
unilateral constraints is geometrically invariant if, and only if, after replacing all
of the engaged unilateral constraints with bilateral ones, the system:

(1) can support an arbitrary load $(r = N)$;
(2) is statically indeterminate $(r < C)$; and
(3) possesses a pattern of virtual self-stress in which nonzero reactions exist in
all of the replaced constraints, and their signs are appropriate for the original,
unilateral constraints.

This criterion reduces verification of the invariance of a system involving
unilateral constraints to a simpler analysis of a substitute system with bilateral
constraints. Furthermore, the statical criterion makes conventional statically
indeterminate and geometrically invariant systems a source of invariant systems
with unilateral constraints. Specifically, a system with bilateral constraints
which is indeterminate to the first degree gives rise to two invariant systems
with unilateral constraints, as illustrated by the systems in Fig. 5a–c. The two
derived systems mirror each other in terms of the sign reversal of the self-stress
forces and the sense of the corresponding unilateral constraints.

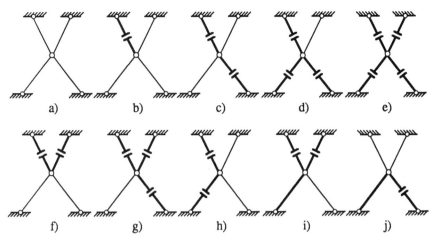

Figure 3.6. Combinatorial variety of systems statically indeterminate to the second degree: a) – e) invariant systems; f) – j) variant systems.

Invariant systems with a higher degree of indeterminacy engender a combinatorial assortment of the sought-for systems, one for each combination of self-stress states with a distinct reaction sign pattern. Each of the resulting systems can either consist exclusively of unilateral members (say, wires and struts (Fig. 6a–e) or involve some bilateral constraints (bars) as well. The obtained variety of invariant systems should not obscure the fact that not every combination of constraints leads to invariance (Fig. 6f–j), even when bilateral constraints are present (Fig. 6i–j).

The foregoing analysis of invariant systems with unilateral constraints leads to a few general observations. First, a certain parity exists between unilateral constraints in tension and compression: mutually exchanging these constraints (while retaining the bilateral ones) preserves geometric invariance of the system. Second, an invariant system with unilateral constraints is adequately constrained (if not overconstrained), hence, is not a tensegrity system. Third, in contrast to systems with bilateral constraints, simultaneous statical and virtual determinacy is precluded for a system involving unilateral constraints. In other words, a virtually determinate system with unilateral constraint is statically indeterminate and, conversely, a statically determinate system (i.e., one not possessing virtual self-stress) cannot be virtually determinate, that is, geometrically invariant.

Another contrasting feature is that an invariant system with unilateral constraints, in spite of being statically indeterminate, may be only adequately constrained and then it has no redundant constraints. The source of this controversy is that the conventional concept of redundancy, related to system reliability, is often understood as synonymous with statical indeterminacy. However, this can be justified only with regard to a statically indeterminate system with bilateral constraints. When the last redundant constraint in such a system fails, the system only loses its redundancy, but does not collapse.

This is not so with systems involving unilateral constraints. Here the loss of statical indeterminacy is the loss not only of redundancy, but also of geometric invariance and, possibly, of structural integrity. The reason is simple: the only mechanism for restraining mobility (both virtual and kinematic) in systems with unilateral constraints is constraint counteraction. This entails constraint dependence, that is, statical indeterminacy, which thereby becomes a prerequisite for kinematic immobility and, the more so, for virtual immobility, i.e., for geometric invariance.

3.2 Systems Other Than Invariant

Prestressability, shown to be the property of a geometrically invariant system with unilateral constraints, happens to be the key concept in analyzing the three noninvariant statical-kinematic types of systems. An analysis of these systems could be expected to be more complicated as compared to their counterparts with bilateral constraints. However, one favorable fact simplifies the situation: in systems with unilateral constraints, stationarity without an extremum of one constraint function is ruled out as a source of kinematic immobility. The reason is that the stationarity by itself does not prevent a displacement in the direction of the constraint disengagement. This rules out the two unprestressable classes of kinematically immobile systems—compound and even-order infinitesimal mechanisms—among systems with unilateral constraints. As a result, all of these systems are covered by the equivalence condition between prestressability and kinematic immobility formulated in Chapter 2; in application to systems with unilateral constraints the condition reads:

Equivalence Condition. For a system involving even one unilateral constraint, prestressability and kinematic immobility are mutually necessary and sufficient.

This condition allows an alternative formulation:

Exclusion Condition. Any system involving unilateral constraints is *either prestressable or variant.*

An obvious necessary condition for prestressability is statical indeterminacy. The fact that it is verifiable by simple matrix means makes very useful the following two, somewhat narrower but stronger, corollaries of the equivalence condition.

Corollary 1. A statically determinate system involving even one unilateral constraint is kinematically mobile (variant, type IV).

Corollary 2. A kinematically immobile system (type I, II, or III) involving even one unilateral constraint is statically indeterminate.

Analytical Criteria for Noninvariant Systems

Among noninvariant systems with unilateral constraints, both quasi-invariant (type II) and quasi-variant (type III) systems are prestressable, while variant systems (type IV) are not. This helps to sort out kinematically mobile systems whereupon other characteristics can be used to distinguish between the two kinematically immobile noninvariant types.

Quasi-Invariant Systems. The two types of structural systems with infinitesimal mobility, that is, types II and III, share many features. Both of them are singular, prestressable, and admit virtual displacements. To distinguish between the two, consider first a system with one degree of singularity ($D = 1$), or a system with $D > 1$ but overlapping virtual self-stress patterns. Assuming the existence of a definite reduced quadratic form, $Q > 0$ (or, respectively, of a definite linear combination of quadratic forms), the sought criterion for a quasi-invariant system (type II) can be obtained from the above analytical criterion for invariance (p. 78) by relaxing and amending it. In fact, only the last condition (3) in this criterion needs to be relaxed so as to allow nonnegative, rather than strictly positive, reactions in the unilateral constraints. The modified criterion reads

$$C_u > r_u = N_u, \quad M_\mu \geq 0, \quad Q > 0; \quad \Leftrightarrow \quad \text{type II.} \tag{3.15}$$

Note, first, that at least one of M_μ must be zero, otherwise the system is invariant, and second, the construction and investigation of the reduced quadratic form is the same as for a system with bilateral constraints.

A system satisfying (15) is topologically adequate: its independent virtual displacements, although disengaging one or more unilateral constraints, bring the system out of degeneration. The resulting adjacent configuration admits constraint variations re-engaging the constraints and making them prestressable.

The system in Fig. 7a illustrates the situation. In the configuration shown, prestress involves only the three horizontal wires, while the vertical members (a wire and a strut), although engaged, cannot be prestressed: $M_4 = M_5 = 0$. The virtual displacements of the system are in the directions of disengagement of these idle constraints. After a suitable variation of all constraints (Fig. 7b), both the wire and the strut are engaged and prestressable, so that the resulting system satisfies the statical criterion of invariance. Accordingly, the original system, as one allowing a constraint variation rendering it invariant, qualifies as quasi-invariant.

In contrast to systems with bilateral constraints, not every variation of unilateral constraints will bring a quasi-invariant system out of degeneration in such a way as to restore invariance. For example, constraint variations producing configurations shown in Fig. 7c and d, were not coordinated with the virtual displacements of the original system. As a result, the feature of prestressability is lost, and the obtained configuration, although still adequately constrained and statically indeterminate, is variant.

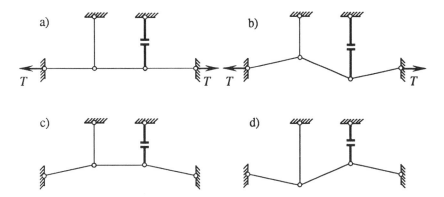

Figure 3.7. Transformations of a quasi-invariant system with unilateral constraints: a) type II; b) type I; c) and d) type IV.

Quasi-Variant Systems. When comparing quasi-variant systems (type III) to quasi-invariant ones (type II), the key difference is that not every in ndent virtual displacement disengages a unilateral constraint. The system shown in Fig. 8a has only one virtual displacement, but it is bidirectional and does not disengage any constraints. Because of this, a constraint variation bringing the system out of degeneration (Fig. 8b), restores kinematic mobility—an indication that the original system is topologically inadequate, hence, quasi-variant. Such a system is kinematically characterized by

$$V \geq D, \quad K = 0, \tag{3.16}$$

where D is the degree of singularity. Note that in the counterpart description (2.4) of systems of type III with bilateral constraints, equality $V = D$ is precluded, since in this case exiting from a degenerate configuration fully restores the rank r , leading to invariance $(V = 0)$. However, in the case of unilateral constraints, invariance requires not only full rank, but also the presence of a counteracting constraint. In the absence of at least potentially counteracting constraints, no constraint variation can lead to invariance, so that the system is topologically inadequate. An example is the system in Fig. 8 and, in general, any system that contains a material point restrained by only three unilateral constraints in space or only by two such constraints in a plane.

In light of this reasoning, a comprehensive analytical criterion of a quasi-variant system with unilateral constraints is

$$r_u < N_u, \quad r_u < C_u, \quad M_u \geq 0, \quad Q > 0; \Leftrightarrow \text{type III.} \tag{3.17}$$

It is clear from this criterion that tensegrity structures of the type generated in the previous chapter (Fig. 2.8) are quasi-variant. Indeed, the desired prestressable singular configuration has been obtained by adjusting the geometric parameters of an invariant, statically determinate $(r = C = N)$ system with bilateral constraints. Replacing only one (or more so, several) of its bilateral constraints

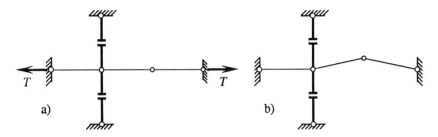

Figure 3.8. Transformations of a quasi-variant system with unilateral constraints:
a) type III; b) type IV.

with unilateral constraints immediately makes the system underconstrained; hence, assuming prestressability, it is a quasi-variant system.

Variant Systems. This is the only type among systems with unilateral constraints that is not prestressable. This leads to a simple statical criterion for a variant system with unilateral constraints: an *unprestressable* system (the more so, a statically determinate system) with even one unilateral constraint is *geometrically variant* (type IV).

Three distinct and escalating requirements must be met for prestressability. First, the system must be statically indeterminate, otherwise it does not possess even virtual self-stress. Second, one of the virtual self-stress patterns (or their linear combination) must have the member force signs appropriate for the unilateral constraints. Third, this pattern of virtual self-stress must be stable, thereby eliminating the case of a variant system in a singular configuration. Note that in verifying each of the three conditions, the analysis is the same as for systems with bilateral constraints. This is due to the fact that an engaged and stressed unilateral constraint is locally (i.e., for the given configuration) indistinguishable from a bilateral one.

Failure at any step of this three-level test condemns the system to the lowest type (variant) and constitutes an analytical criterion for type IV.

Structural Properties of Systems with Unilateral Constraints

The presence of unilateral constraints causes some important and, at times, counterintuitive differences in the structural properties of systems of all four types as compared to their respective counterparts with bilateral constraints. The following paragraphs discuss a few of these unique characteristics.

Structural Indeterminacies. The three structural indeterminacies in systems involving unilateral constraints are interrelated closely but in a different way than they are in systems with bilateral constraints. As shown earlier, a geometrically invariant system with unilateral constraints is virtually determinate and statically indeterminate. A stronger and more general condition can be formulated for all systems involving unilateral constraints, with kinematic determinacy replacing the virtual one.

Exclusion Condition Between Determinacies. For a system involving even one engaged unilateral constraint, statical and kinematic determinacies are mutually exclusive.

Thus, for all structural systems involving unilateral constraints, the concept of redundancy must be divorced from that of reliability: a structural member failure resulting in the loss of statical indeterminacy is not just the loss of redundancy, but the loss of kinematic immobility and, perhaps, of structural integrity.

As follows from the preceding analysis, the degree of statical indeterminacy S, in a given configuration of a system involving unilateral constraints, is

$$S = C_b + C_u - (r_b + r_u), \tag{3.18}$$

where C_b and C_u are the respective numbers of bilateral and engaged unilateral constraints; r_b and r_u are the corresponding ranks of matrices (3) and (4).

The degree of virtual indeterminacy V of a given configuration equals the dimension of the solution cone of the set of inequalities (8). Establishing it is a computationally tedious task even for simple systems. For a compound system, the domain of virtual displacements must be sought as the intersection of the solution cones corresponding to each of the singular subsystems.

A variant system, by definition, is kinematically indeterminate, with the degree of indeterminacy K being the dimensionality of the feasible domain in the configuration space of the system.

Constraint Variations and Disengagement. An important difference between systems involving unilateral constraints and those with only bilateral constraints is the possibility of constraint disengagement, which has some far-reaching ramifications. For example, in a type II system with purely bilateral constraints, a change in configuration caused by changing the member sizes will generally convert the system from type II to type I, geometrically invariant, although a conversion into a singular type IV is not ruled out. For systems with unilateral constraints, however, a usual transformation is among types I, II, and IV. This may happen, for example, as a result of the thermal expansion or contraction of structural members. Such transitions are shown in Fig. 7, where the constraint variations transformed a partially prestressable configuration, in one case, into a comprehensively prestressable and invariant configuration, and in the other case, into an unprestressable, hence, variant one.

Constraint disengagement has an effect on equilibrium loads and (what is more important) vice versa. In fact, manipulating the external loads and, especially, prestress, is a conventional method of stabilizing systems with unilateral constraints and preventing the constraint disengagement.

Disparity Between Wires and Struts. A parity, in the sense of simultaneous mutual exchangeability, between unilateral constraints in tension and compression exists in *invariant* systems (type I). It can be tracked back to the analysis of the set of linear inequalities and the resulting analytical criterion of invariance. Physically, a comprehensive reversal of unilateral constraints only reverses the sign of constraint counteraction, but does not affect the pattern. This parity also exists in *nonsingular* variant systems: no constraint reversal can make them prestressable, which would be the only possibility for changing the system type. These observations can be generalized as follows.

Preservation of System Type. Simultaneous reversal of the sense of all unilateral constraints in a *nonsingular* system does not affect its statical-kinematic type.

The parity breaks down in infinitesimally mobile systems (types II and III) because of the additional relevance of prestressability. For example, whereas systems II and III may involve only wires (see Fig. 2), they cannot consist entirely of struts. The reason is that, to be prestressable, each of the struts would have to be of a *maximum* length compatible with the fixed lengths of the remaining struts. In analytical terms, all of the constraint reactions—that is, all of the Lagrange multipliers—must be negative. However, in the basic theorem on a constrained extremum, a multiplier corresponding to the function to be maximized is conventionally stipulated as positive (and often taken as 1). Simultaneous reversal of all of the multiplier signs does not help since the quadratic form in the theorem must be negative. This makes the constrained maximum in question impossible, leading to the following conclusion.

Restriction on Strut Systems. A system comprised solely of struts can belong only to one of the two main types: invariant (type I) or variant (type II).

This breakdown of parity entails, for instance, the possibility of an all-wire, but not all-strut, quasi-variant system.

The following example illustrates the above observations and analytical results for systems involving unilateral constraints.

Example 3.1. The system in Fig. 9 has two bars and three wires. The constraint conditions and their linearized versions are:

$$X_1^2 + X_2^2 = L^2, \qquad\qquad X_1 x_1 + X_2^2 x_2 = 0,$$

$$X_3^2 + X_4^2 = L^2, \qquad\qquad X_3 x_3 + X_4^2 x_4 = 0,$$

$$X_1^2 + (L - X_2)^2 \le 2L^2, \qquad\qquad X_1 x_1 - (L - X_2) x_2 \le 0,$$

$$X_3^2 + (L - X_4)^2 \le 2L^2, \qquad\qquad X_3 x_3 - (L - X_4) x_4 \le 0,$$

$$(3L - X_1 - X_3)^2 \qquad\qquad (X_1 + X_3 - 3L)(x_1 + x_3)$$
$$+ (X_4 - X_2) \le L^2, \qquad\qquad - (X_4 - X_2)(x_4 - x_2) \le 0.$$

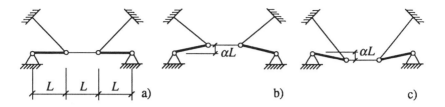

Figure 3.9. Transformations of a quasi-invariant system with unilateral constraints resulting from constraint variation: a) type II; b) type I; c) type IV.

In writing down these conditions, advantage has been taken of the symmetry of the given configuration by using two separate coordinate systems. One has the origin at the left support and serves as a reference for coordinates X_1 and X_2, while the other, being a mirror image of the first one, originates at the right support and forms a reference for X_3 and X_4. In the given configuration, $X_1 = X_3 = L$, $X_2 = X_4 = 0$, and the linearized constraint conditions take the form (a) below:

$$
\begin{aligned}
x_1 &= 0, & x_1 + \alpha\, x_2 &= 0, \\
x_3 &= 0, & x_3 + \alpha\, x_4 &= 0, \\
x_1 - x_2 &\le 0, \quad (a) & x_1 - x_2\,((1 - \alpha) &\le 0, \quad (b) \\
x_3 - x_4 &\le 0, & x_3 - x_4\,(1 - \alpha) &\le 0, \\
-x_1 - x_3 &\le 0, & -x_1 - x_3 &\le 0.
\end{aligned}
$$

(Equations (b) above and (d) below are to be disregarded until later.) Eliminating the dependent displacements x_1 and x_3 leads to the set of constraint inequalities with coefficient matrix (c) below:

$$
\begin{bmatrix} -1 & 0 \\ 0 & -1 \\ 0 & 0 \end{bmatrix}, \qquad (c) \qquad
\begin{bmatrix} -1 & 0 \\ 0 & -1 \\ \alpha & \alpha \end{bmatrix}. \qquad (d)
$$

Kinematic analysis is performed first. The first two rows in (c) form a basic subset. For the system to be invariant, the last inequality must be counteracting the inequalities of the basic subset. As is easily seen, this does not happen: instead of reversing the sign of the basic minor, the replacement of the first or the second row with the third one turns the determinant into zero. The system is not invariant.

Turning now to the statical analysis, the homogeneous equilibrium equations (12) are formed by transposing the coefficient matrix (c):

$$
\begin{bmatrix} -1 & 0 & 0 \\ 0 & -1 & 0 \end{bmatrix}
\begin{bmatrix} M_1 \\ M_2 \\ M_3 \end{bmatrix} =
\begin{bmatrix} 0 \\ 0 \end{bmatrix}. \qquad (e)
$$

The nontrivial solution is

$$M_1 = M_3 = 0, \qquad M_2 = M.$$

The solution is not positive, which confirms that the system is not geometrically invariant. It is however, quasi-invariant (type II), since it is adequately constrained. Indeed, the system allows a constraint variation leading to invariance which is impossible for topologically inadequate systems. A changed configuration of the system shown in Fig. 9b is the result of a constraint variation involving virtual displacements x_2 and x_4. This would happen if the inclined wires were shortened and the middle one lengthened, while keeping all of them engaged.

There is no need for the evaluation of updated bar and wire lengths and writing down the modified constraint equations. In the new configuration the nodal coordinates are $X_1 = X_3 = L$, $X_2 = X_4 = \alpha L$, and the new system of linearized constraint conditions can be written down immediately; it has the form (b), given earlier. In the kinematic analysis, the coefficient matrix (d) replaces the original (c). Now each of the three 2×2 subsets is basic and satisfies the condition of determinant sign reversal following a row replacement. The alternative, statical, check yields an expected result—positive constraint reactions

$$M_1 = M_3 = \alpha M_2 > 0.$$

Hence, the system in Fig. 9b is invariant. Furthermore, it possesses a certain redundancy. Treating each of the two bars as a wire–strut assembly shows that upon removal of the two struts the resulting all-wire system is still geometrically invariant. It also remains type I if all five structural members are struts—an example of a system I composed entirely of struts.

The configuration in Fig. 9c, corresponding to $\alpha < 0$, fails both the kinematic and statical tests and is variant. Thus, the original quasi-invariant (type II) configuration is a transition from an invariant to a variant one—a typical transformation for a system involving unilateral constraints. As expected, for both nonsingular systems (Fig. 9b and c), the replacement of all wires with struts does not change their respective system types. This is not so for a partial replacement: upon replacing the horizontal wire with a strut, the first system becomes variant—and the second invariant. The reason is that a selective constraint reversal, unlike a comprehensive one, affects the pattern of constraint counteraction.

3.3 Constraint Imposition and Removal

Imposition and removal of constraints is an important operation in analytical modeling of structural systems. Aside from being one of the classical techniques in the structural analysis of statically indeterminate systems, this operation is exercised as a mental or numerical experiment in the process of statical-kinematic analysis and synthesis of structural systems. It also reflects actual transformations of a system resulting from engagement, disengagement, yielding, or failure of structural members. Finally, imposition and removal of constraints may be a physical action, such as reinforcement, modification, or destruction of a structure.

Constraint imposition and removal, as an operation affecting the system topology, but not geometry is, in a sense, complementary to the previously considered variation of constraints. Constraint variations are analytically implemented by varying the system coordinates, which may bring the system in or out of degeneration by making independent constraints dependent or vice versa. This leads to a change in the Jacobian matrix rank, the related degrees of statical, virtual and kinematic indeterminacies, and the statical-kinematic type of the system. The effect of unilateral constraint variations is even more far-reaching, due to the possibility of attaining or losing constraint engagement. This, in turn, may affect the pattern of counteraction and the associated feature of prestressability.

In contrast to constraint variations, constraint imposition or removal does not affect the configuration geometry but changes the system connectivity. An increase or reduction in the number of constraints affects the degrees of structural indeterminacies and the statical-kinematic type of the system in its own way. The operation outcome depends on the properties of both the original system and the constraint in question, especially on whether the constraint is independent or dependent.

In considering the consquences of constraint imposition or removal it is natural to begin with bilateral constraints.

Imposition and Removal of Bilateral Constraints

Independent Versus Dependent Constraints. Imposition of an *independent* constraint increases by one the number of constraints and the Jacobian matrix rank r, whereas the removal of an independent constraint has the exact opposite effect on the two parameters. As seen from (1.28) and (1.29), the degree of statical indeterminacy is preserved in both cases but the degree of virtual indeterminacy changes. This change is, in fact, the definitive sign of an independent constraint: if imposition or removal of a constraint affects the number V of virtual degrees of freedom of the system, the constraint is independent.

Imposition of an independent constraint on an invariant system (type I) is impossible since its rank r is already full. Removal of an independent constraint from such a system converts it into either type II or type IV, but never into type III. The reason is that, according to (2.4), a system of type III (quasi-variant) has at least two virtual degrees of freedom. Therefore, obtaining a

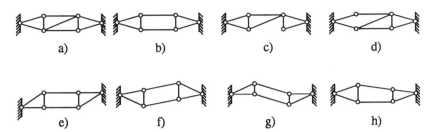

Figure 3.10. Transformations of a system with bilateral constraints: a) the original geometrically invariant (type I), statically indeterminate system (S = 1); b) quasi-invariant (type II) system (S = 1); c) – h) geometrically invariant, statically determinate systems.

system III, if at all feasible, requires removal of no less than two independent constraints from a statically indeterminate system I. Conversely, imposition of one independent constraint on a quasi-variant system may at best elevate it to type II, and then it will take another independent constraint to attain invariance.

In general, sequential imposition of V independent constraints eliminates all virtual degrees of freedom, thereby unavoidably converting any system into an invariant one. If the original system is singular but not prestressable, then in the course of constraint imposition it remains singular, eventually acquires prestressability, and becomes a system II prior to the imposition of the last independent constraint. Thereupon, it converts into a statically indeterminate invariant system.

Dependent constraints are different in that their imposition or removal does not affect the rank r. Accordingly, the degree of statical indeterminacy changes, while the degree of virtual indeterminacy is left intact, although the kinematic mobility of the system and its type may be affected. Once again, parameter V serves as a check: if imposition (or removal) of a constraint does not affect the number of virtual degrees of freedom, the constraint is dependent. In the particular case of an invariant system, any *imposed* constraint is dependent, whereas a *removed* constraint may be either independent or dependent, with very different consequences for the system, respectively.

Example 3.2. Some of the described transformations are illustrated in Fig. 10. As is readily seen, the original system (Fig. 10a) is invariant, statically indeterminate, and described by

$$N = 8, \quad C = 9, \quad r = 8, \quad S = 1, \quad V = K = 0; \quad \Rightarrow \text{ type I.}$$

The system allows self-stress in which the diagonal bar is the only force-free member, indicating that it represents the only independent constraint. Because of that, the system in this configuration cannot be prestressed by varying the length of the diagonal bar. Removal of the bar (Fig. 10b) results in

$$C = 8, \quad r = 7, \quad S = 1, \quad V = 1, \quad K = 0; \quad \Rightarrow \text{ type II.}$$

Note that the expected drop in the rank does not affect the degree of statical indeterminacy, and it is obvious that the system is still prestressable. At the same time, the system acquired virtual, but not kinematic, indeterminacy and, therefore, is quasi-invariant. Except for the diagonal bar, all other constraints in the original system are dependent. Removal of any of them reduces the degree of statical indeterminacy but does not affect the degree of virtual indeterminacy or the system type. Accordingly, the systems in Fig. 10c and d are invariant and statically determinate.

All constraints of the system in Fig. 10b are dependent; removal of any of them would produce a topologically inadequate and kinematically mobile system (type IV). However, the constraint dependence in this system is the result of its singular geometry. In any nonsingular configuration (Fig. 10e–h) the constraints become independent, so that the system is invariant and statically determinate:

$$C = 8, \quad r = 8, \quad S = V = K = 0; \quad \Rightarrow \text{ type I}$$

which is easily verifiable by inspection.

The last possible category of constraint is a *locally dependent* one. The very notion of such a constraint is meaningful only for systems whose geometric configuration is nonunique (type IV, $K > 0$). Imposition of a locally dependent constraint on a system IV leads to singularity and a pattern of virtual self-stress (or, perhaps, it increases by one the number of both). As a result, the degree of statical indeterminacy increases by one and the system may become prestressable, hence, kinematically immobile. This outcome underlies one of the previously described techniques employed in the synthesis of systems with infinitesimal mobility.

Imposition and Removal of Unilateral Constraints

Independent Versus Dependent Constraints. When exploring the consequences of imposition or removal of unilateral constraints, the latter are assumed engaged. Somewhat surprisingly, imposition or removal of an *independent* unilateral constraint changes neither the degree of statical indeterminacy nor the degrees of virtual and kinematic indeterminacies. As a result, it does not affect the statical-kinematic type of the system either. The reason for all that lies in the fact that an independent unilateral constraint has no counteracting constraint or subsystem in the current configuration of the system. Hence, an independent constraint cannot impart virtual self-stress, the more so, prestressability, which would be the only way of restraining the system mobility. The only kinematic consequence of imposition or removal of an independent unilateral constraint is that it alters the feasible domains in the space of virtual displacements and, in the case of type IV, in the coordinate space of the system.

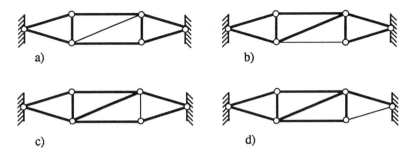

Figure 3.11. Transformations of a system resulting from imposition and removal of unilateral constraints: a) type II; b) – d) type I.

On the statical side, imposition or removal of an independent unilateral constraint imparts or, respectively, detracts the system ability to support an external load acting in the direction of the constraint engagement. Thereby, the operation affects the composition of the equilibrium loads of the system. It is not difficult to see that an invariant system contains no independent unilateral constraints, and any constraint imposed on it is dependent. Conversely, a system containing even only one independent unilateral constraint cannot be geometrically invariant because of the absence of a counteracting constraint or subsystem.

Imposition of a *dependent* unilateral constraint results in an additional consequence inequality or counteracting inequality. The influence of either one on the statical properties of the system is the same—the degree of statical indeterminacy is increased by one. However, the kinematic properties are affected by the two types of dependent unilateral constraint in different ways. A consequence constraint does not affect them at all and may only alter the feasible domain in the coordinate space of a system IV. A counteracting constraint deprives the system of one or more kinematic (and, perhaps, virtual degrees of freedom), depending on the subsystem that the constraint counteracts. This may lead to prestressability and kinematic immobility, hence, the possibility of elevating the statical-kinematic type of the system. Removal of a globally or locally dependent unilateral constraint reverses the described results of their imposition.

The familiar system in Fig. 11 illustrates most of the transformations caused by imposition or removal of unilateral constraints. The diagonal wire (Fig. 11a) is the only independent constraint in the system. Its removal or imposition does not affect either the system type (it is quasi-invariant) or any of its degrees of indeterminacy which are $S = 1$, $V = 1$, $K = 0$. Introduction of the second diagonal wire, which becomes dependent and counteracting, would elevate the system to type I. The systems in Fig. 11b–d are invariant with or without the shown wires, which are in this case dependent and only affect the degree of statical indeterminacy. The same is the situation if diagonal wires are introduced into the invariant systems in Fig. 10e–h. Since the parity between unilateral

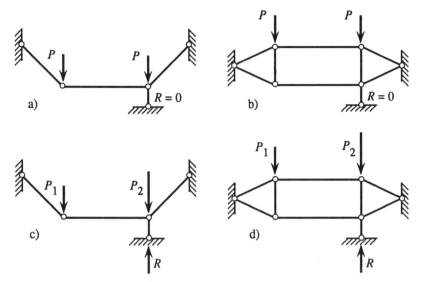

Figure 3.12. An illustration of the statical property of an independent constraint.

constraints in tension and compression holds for invariant systems, in each case the wire can be replaced with a strut without causing any change in the system properties.

One Statical Property of Independent Constraints.   The foregoing observations on the effects of constraint imposition or removal lead to some interesting implications for structural behavior like, for instance, the following proposition (Kuznetsov, 1979).

A Statical Property of Independent Constraints. If a system is in equilibrium, imposition of an independent constraint has no effect on the force distribution under the given load, and the introduced constraint is force-free. For the same system, removal of an independent constraint which is *not* force-free makes equilibrium in the given configuration impossible.

For proof, note first of all that the first part of the proposition applies only to systems II – IV, since introduction of an independent constraint into an invariant system is impossible. In applicable cases, the operation does not change the number of equilibrium equations while increasing by one the number of unknown forces. This amounts to adding an independent column to the equilibrium matrix, which does not affect the solution (unique or otherwise) of the equilibrium equations. As a result, reactions in the introduced constraints must be absent. A similar analysis of the effect of constraint removal on the equilibrium matrix proves the second part of the proposition.

The systems in Fig. 12a and b illustrate the effects of constraint imposition on a statically determinate and indeterminate system. In both cases, the vertical support bar is a newly introduced independent constraint. Accordingly, under any equilibrium load of the original system, this support is force-free. One such load, shown in the figure, involves a force applied to the pin above the support itself, and yet the support is force-free.

On the other hand, under the load shown in Fig. 12c and d, the support bar has a nonzero reaction. According to the second part of the proposition, removal of the bar from either system makes equilibrium in the original configuration impossible.

Limit Equilibrium

Limit Equilibrium and the Limit Load. The above statical property of an independent constraint is relevant in the theory of limit equilibrium and the associated plastic limit analysis and design methods. As has been mentioned, since this theory employs the constitutive model of an ideal rigid-plastic material, it falls within the scope of statical-kinematic analysis. Furthermore, all of the attributes of the limit equilibrium theory are of a local character, i.e., they characterize only a fixed configuration of a system rather than the system as such. It is known, for example, that in the case of a geometrically nonlinear system, the limit load found for its initial configuration may be either greater or smaller than the actual load-carrying capacity of the system. The first, nonconservative, outcome is typical of convex arches and shells, the second, conservative, with tensile systems.

Although the traditional object of the limit analysis is an invariant system, the analysis can be extended to a general system carrying an equilibrium load. If an equilibrium load increases proportionally to a parameter, any displacements of the system must be due exclusively to the flexibility of the structural members. Given their undeformability (within the material yield limit), the system configuration does not change, regardless of whether or not the system is kinematically mobile and no matter what its statical-kinematic type is. Accordingly, the existing theorems and methods (both kinematic and statical) of limit analysis should be applicable to systems of any statical-kinematic type. Nevertheless, extension of the theory to an arbitrary system calls for certain adjustments.

In the conventional approach to limit analysis, the state of limit equilibrium is assumed to set in at the moment of formation of a plastic mechanism. The drawback of this criterion is more than reminiscent of the situation with the conventional concept of statical indeterminacy discussed in Chapter 1: the purely *statical* concept of limit *equilibrium* is defined in *kinematic* terms. Not surprisingly, the key to refinement is once again found in recovering and consistently implementing the statical origin of the underlying concept. Indeed, in statical terms, the occurrence of limit equilibrium means that, in the course of loading, the structural members of the system have reached a state at which the external load can no longer be supported in the given geometric configuration. Since the only alternative to equilibrium is motion, the following statical definitions come naturally:

The (one-parameter) *limit load* is an equilibrium load of such a magnitude that it renders equilibrium in a given configuration impossible, thus leading to the onset of motion.

Limit equilibrium is the state of a system under a limit load.

These definitions apply to all types of systems and are different from conventional ones in that they address two shortcomings of the latter. First, a system other than invariant is *already* a mechanism (either infinitesimal or finite), and the presense of a yielding constraint does not add much new information.   Second, when applied to invariant systems with unilateral constraints, the existing criteria are not always meaningful. The reason for that is described below.

Formation of a Stationary Plastic Mechanism. The unusual features of geometrically invariant systems with unilateral constraints are reflected in their structural behavior. In particular, such a system can convert into a stationary (nonworking) plastic mechanism, i.e., one for which the acting load continues to be equilibrated. This situation promptly raises a number of problems under the conventional approach.

First of all, it is not justified to regard a state with a nonworking plastic mechanism as a limit state, because under certain conditions the load may be further increased, until the equilibrium becomes impossible and the system is converted into a working plastic mechanism. This depends on the property— stable or otherwise—of the equilibrium of the first, nonworking, mechanism. If it is stable, further loading is feasible and it represents the reserve load capacity of the system.

If the equilibrium of the nonworking mechanism is unstable, there is a risk of overestimating the load capacity: since the plastic mechanism is balanced and therefore stationary, the corresponding load parameter *cannot be obtained by equating the virtual work of internal and external forces.* As a result, the conventional procedure, based on the examination of all kinematically possible displacement patterns from the point of view of plastic work, will not detect this value of the load parameter. Instead, the analysis will go on to a higher (unconservative) value corresponding to some other failure mode—one with a working plastic mechanism.

Although the above definitions do not obviate these difficulties, they serve to recognize them and to avert the possibility of a nonconservative design. As for the problem itself, it can be demonstrated that the described situation is confined to systems involving unilateral constraints. Indeed, conversion of a system with bilateral constraints into a plastic mechanism can only result from the yielding of an independent constraint. However, according to the second part of the above proposition, such a mechanism cannot be in equilibrium. Thus, *conversion of a system with only bilateral constraints into a stationary plastic mechanism is impossible.*

As regards a system with unilateral constraints, a plastic mechanism (in particular, a balanced one) may also form through the yielding of a dependent constraint. Moreover, the constraint itself can even be bilateral; the only important condition is that it be counteracting. Then the system can acquire one or more virtual degrees of freedom, depending on the subsystem that the constraint was counteracting.

The described difference in structural behavior under constraint yielding is rooted in the respective generic statical-kinematic properties of the two systems. In a system with only bilateral constraints, each constraint reaching the yield stress and becoming plastic reduces the degree of statical indeterminacy by one. Just prior to its conversion into a plastic mechanism, the system is statically determinate and geometrically invariant and then, with the onset of yielding in the next constraint, it acquires kinematic mobility. Such a situation is impossible in a system involving unilateral constraints. This system is statically indeterminate at least to the first degree as long as it is invariant, including the stage immediately preceding the conversion into a plastic mechanism. The reason, once again, is that unilateral constraints, as a matter of principle, restrain the system mobility only by means of their counteraction which implies statical indeterminacy.

The following example illustrates the consequences of constraint yielding and the resulting peculiarities of the limit equilibrium analysis of systems with unilateral constraints.

Example 3.3. The bottom-reinforced concrete beam of the system shown in Fig. 13a is capable of resisting a bending moment of only one direction. The system is invariant and statically indeterminate. Let the stiffness ratio of its members be such that the beam yields first over one of the posts, say, the right one (Fig. 13b). This suffices for the system to convert into a plastic mechanism. However, the given load is an equilibrium load for the formed mechanism which, therefore, is stationary: the work of the external forces over the antisymmetric virtual displacement (Fig. 13c) is zero. A computer program based on the kinematic theorem of the theory of limit equilibrium will attribute to the resulting limit load a very large (theoretically, infinite) magnitude and will continue its search for another plastic mechanism as a potential failure mode. Eventually, it will be found among symmetric mechanisms, like the one depicted in Fig. 13d. Here the equation of virtual work is

$$2M \varphi + T \Delta L = 2P d, \tag{a}$$

where M and T are, respectively, the plastic moment of the beam and the yield tension force of the bottom bar. The corresponding rotation at the hinge and bar elongation are expressed in terms of deflection d by simple kinematic relations

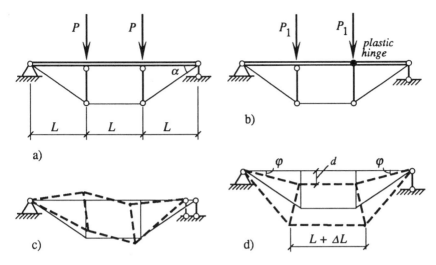

Figure 3.13. Formation of a stationary plastic mechanism in a system with unilateral constraints and its consequences: a) system geometry and external load; b) plastic hinge forms above one of the posts; c) displacement mode of the formed stationary plastic mechanism; d) symmetric failure mode.

$$\varphi = d/L, \quad \Delta L = 2d \tan \alpha. \qquad (b)$$

The resulting limit load is

$$P = M/L + T \tan \alpha. \qquad (c)$$

Depending on T, this limit load may be arbitrarily large, leading to an overestimated load capacity.

The differences in the behavior of the described system with a bottom reinforced beam (a system with unilateral constraints) and a similar system with a doubly reinforced beam (a system with bilateral constraints) can be summed up as follows. With the formation of the plastic hinge, the first system turns into a stationary, nonworking, plastic mechanism, while the second system only loses its statical indeterminacy (and redundancy) but remains geometrically invariant. In both cases, the resulting state is not that of limit equilibrium, and the corresponding load cannot be found by conventional methods. Further loading is certainly possible for the second system, and its limit load can be evaluated in a usual way. As to the first system, it may happen that its load capacity is also not exhausted, and further loading may be possible until a working plastic mechanism forms (Fig. 13d).

It is this last situation that represents a potential danger, since the equilibrium of an undetected stationary plastic mechanism may be unstable. Such, probably, is the case with the system in Fig. 14 where most of the structural members are in compression, making the system susceptible to an abrupt structural failure. This is in contrast with conventional frame structures, where a typical process of

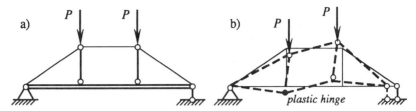

Figure 3.14. A system involving a bottom-reinforced beam and the displacement mode of the ostensibly unstable stationary plastic mechanism.

failure is relatively benign and is accompanied by gradually increasing plastic deformations. Thus, care must be taken to detect stationary plastic mechanisms and to evaluate their stability. This can be done by taking into account elastic unloading of yielding structural members using, for example, the Shanley approach.

4
Kinematic and Elastic Mobility

Mobility, either virtual or kinematic, is a definitive feature of underconstrained structural systems. The two types of mobility are intricately intertwined. For instance, when a quasi-variant system (type III) made of a real, flexible material is subjected to a perturbation load, it undergoes displacements and constraint variations bringing it out of its singular configuration. As a result, the system acquires kinematic mobility, i.e., the possibility of finite displacements without any additional constraint variations. Under the same conditions, a quasi-invariant system (type II) behaves differently: in a new configuration it loses even its virtual mobility and becomes capable of supporting an arbitrary load.

In this chapter, kinematic mobility is first explored in its own right, and later on, in conjunction with elastic mobility. The first section is concerned with a general description in matrix terms of the linearized problem. The resulting concept of the statical-kinematic stiffness matrix becomes a basis for describing the response of an underconstained structural system to small perturbations. This concept is later adapted to continuous underconstrained systems as the modal statical-kinematic stiffness matrix. The next section illustrates the application of the statical-kinematic stiffness matrix to an analysis of displacements in variant and quasi-variant systems. This is done using the simplest possible prototypes (usually, the familiar three-bar systems) possessing the intended features.

An interplay of kinematic and elastic displacements in underconstrained structural systems is studied in the last section. A certain formal demarcation line is crossed when elasticity becomes involved, one separating analytical mechanics from structural mechanics. Analytical mechanics, in particular statical-kinematic analysis, is a hypothetical-deductive discipline aspiring to mathematical rigor in statements and proofs. On the other hand, structural mechanics employs additional quantitative and phenomenological assumptions. In linear structural mechanics, one of the common (although sometimes tacit) assumptions is the extrapolation of, strictly speaking, infinitesimal displacements to small but finite magnitudes. When displacements are accompanied by nontrivial variations of constraints, experimentally established constitutive relations are employed in the analysis. The implication is that, when dealing with elastic displacements, the subject matter is no longer confined to analytical mechanics and, in fact, will frequently traverse into the domain of structural and solid mechanics.

4.1   The Statical-Kinematic Stiffness Matrix

Virtual and, the more so, kinematic mobility of underconstrained systems complicate their analysis by making it geometrically nonlinear. The first problem to be addressed here is determining the equilibrium configuration of a variant system under a given load. This is done by evaluating the system displacements in an incremental analysis employing a special tangent stiffness matrix of statical-kinematic origin.

Load Resolution and Incremental Analysis

Orthogonal Load Resolution and Projection Matrix.   When an underconstrained system is subjected to a general load containing a perturbation load component, the system must change its configuration in order to balance the load. This makes the problem nonlinear and, moreover, nonlinearizable by conventional load incrementing techniques: the resulting linear system of equations is overdeterminate and, generally, incompatible. Physically, it means that the system is incapable of supporting even an arbitrarily small perturbation load without undergoing some displacements. Their evaluation requires quantifying the resistance of the system to perturbation loads.

An underconstrained system in a state of stable equilibrium possesses a certain stiffness of a statical-kinematic (as opposed to elastic) nature. Its source is the presence of member forces induced by the acting equilibrium load, including prestress. In this case the problem can be linearized and solved, either in increments or, perhaps, in one step, provided that the perturbation load component is sufficiently small.

An analysis capitalizing on the resolution of a given general load P_n into the orthogonal (equilibrium and perturbation) components is based on the following considerations. An incompatible system of equilibrium equations in unknown constraint reactions Λ_i can be solved only approximately. An approximate solution does not satisfy the equations, producing instead an error

$$E_n = P_n - F_n^i \, \Lambda_i. \qquad (4.1)$$

The best solution in the sense of the minimum mean square error is one with the error vector E_n orthogonal to the column space of the equilibrium matrix, i.e., one satisfying

$$F_n^j \, E_n = F_n^j \, (F_n^i \, \Lambda_i - P_n) = 0. \qquad (4.2)$$

Since the matrix rank is $r \leq C$ (recall that C is the number of constraints), r independent reactions can be found from (2) by retaining in it any set of columns of rank r:

$$\Lambda_p = (F_m^p \, F_m^q)^{-1} \, P_n^q \, P_n = G_n^p \, P_n, \quad p, q = 1, 2, ..., r . \qquad (4.3)$$

These reactions equilibrate a load

$$P_n^* = F_n^p \, G_m^p \, P_m \equiv R_{mn} \, P_m, \tag{4.4}$$

which is in the range (column space) of the equilibrium matrix. This is the complete equilibrium component of the load P_n. The remaining part,

$$p_n = P_n - P_n^* = E_n, \tag{4.5}$$

is the pure perturbation component which, together with the error vector, is outside of the column space, that is, in its orthogonal complement.

The $N \times N$ matrix R_{mn} evolved in the foregoing analysis and defined in (4) appears in the theory of generalized inverse and is called a projection matrix (Strang, 1988). It is symmetric, has rank r, and projects an arbitrary vector onto the column space of the equilibrium matrix (the space of constraint reactions). The projection matrix is idempotent ($R_{lm} \, R_{ln} = R_{mn}$) which ensures that an equilibrium load is unaffected by the projection; obviously, a perturbation load has a zero projection:

$$R_{mn} \, P_m^* = P_n^*, \qquad R_{mn} \, p_m = 0. \tag{4.6}$$

Note that constraint reactions Λ_i can be found immediately from (3), without the load resolution. However, the resolution is convenient when evaluating the perturbation load component, which serves as a measure of error of the linearized solution and is necessary for an incremental displacement analysis. Regardless of the method for their evaluation, the unknown constraint reactions can be found only for a nonsingular, hence, statically determinate configuration $(r = C)$; otherwise, their evaluation is impossible without the constitutive relations.

A projection matrix can be constructed in a similar way for resolving a given constraint variation vector, f_i, into the compatible and anticompatible components. In this case, r independent displacements

$$x_p = (F_m^p \, F_m^q)^{-1} \, P_i^q \, f_i = G_i^p \, f_i, \tag{4.7}$$

correspond to a compatible constraint variation

$$f_i^* = F_i^p \, G_j^p \, f_j \equiv R_{ij} \, f_j. \tag{4.8}$$

This is the complete compatible component residing in the column space of the constraint Jacobian matrix; the remaining part of the constraint variation is the pure anticompatible component.

Incremental Solution. Upon determining the constraint reactions, the system displacements are to be evaluated. If they are sufficiently small, both the obtained reactions and the deformed configuration of the system can be considered final; otherwise, a further solution refinement is necessary. To evaluate the displacements, equilibrium equations for the given configuration must be *varied*. This operation, used, for instance, in an analysis verifying the stability of a given state of a system, involves a simultaneous incrementation of both sides of the equations; it yields

$$\delta(F_n^i \Lambda_i) \equiv F_{mn}^i \Lambda_i x_m + F_n^i \lambda_i = \delta P_n. \tag{4.9}$$

Here λ_i are small increments of the respective constraint reactions Λ_i, and the right-hand side represents a variation of the present equilibrium load. The variation symbol δ applied to an array implies an independent incrementation of the array components.

Equations (9) must be used in conjunction with the linearized constraint equations (1.21) reproduced below:

$$F_n^i x_n = f^i.$$

Constraint variations f^i in the right-hand side are assumed known and represent small changes in the geometric parameters of the system due to elastic, plastic, rheological, thermal, and other possible deformations of the system. The resulting combined system of equations can be presented in a block-matrix form:

$$\begin{bmatrix} K_{mn} & F_m^p \\ F_m^p & 0 \end{bmatrix} \begin{bmatrix} x_m \\ \lambda_p \end{bmatrix} = \begin{bmatrix} \delta P_n \\ f^p \end{bmatrix}. \tag{4.10}$$

The square, symmetric, $(N+r) \times (N+r)$ matrix in the left-hand side is the tangent statical-kinematic matrix for a given state (configuration and force distribution) of an underconstrained system. The $N \times N$ upper left block,

$$K_{mn} = F_{mn}^i \Lambda_i, \tag{4.11}$$

is the familiar linear combination of the constraint function Hessian matrices. The off-diagonal blocks in (10) are the constraint Jacobian matrix and its transpose. This matrix is represented in (10) by a set of any r independent rows. The reason is that a singular Jacobian matrix $(r < C)$ is a manifestation of statical indeterminacy, and such a problem cannot be solved by purely statical-kinematic means, that is, without constitutive relations. When the degree of statical indeterminacy $S = C - r = 1$, the tangent matrix in (10) coincides with the test matrix (2.10) used for detection of infinitesimal mobility; this, of course, is no accident.

Statical-Kinematic Stiffness Matrix

Derivation of the Matrix. The simultaneous equations (10) form a closed system with N unknown displacements x_n and r unknown increments λ_p of constraint reactions. It is computationally advantageous to separate the equilibrium load component in advance, evaluate the corresponding constraint reactions, and incorporate them into K_{mn} according to (11). This is unavoidable when $P_n = \delta P_n$ is the only load applied to the system. In the absence of constraint variations, the equations of the lower echelon in (10) are homogeneous. Their solution, $x_m = a_{ms} x_s$, can be used to eliminate the dependent displacements from the remaining equations:

$$K_{mn} a_{ms} x_s + F_n^p \lambda_p = \delta P_n. \tag{4.12}$$

As has been demonstrated in Chapter 1, the unknown $(N - r)$ independent displacements x_s and r incremental reactions λ_p belong to two orthogonal complement subspaces. Accordingly, the equilibrium component of the load P_n produces λ_p but no displacements, while the perturbation component gives rise to x_s (and x_n) without incremental reactions. More precisely, the effect of a perturbation load on the constraint reactions is nonlinear and must be sufficiently small to justify the incremental linearization.

With this in mind, the kinematic and statical aspects of the problem (the evaluation of the system displacements and forces) at each incremental linear step of the analysis can be uncoupled. To this end, both sides of equations (12) are multiplied by $x_n = a_{nt} x_t$. Since the resulting equation of virtual work must be satisfied identically with respect to each of the independent virtual displacements x_s, it yields V simultaneous equilibrium equations in unknown displacements

$$K_{mn} a_{ms} a_{nt} x_t = a_{ns} P_n. \tag{4.13}$$

Introducing a $V \times V$ matrix,

$$K_{st} = K_{mn} a_{ms} a_{nt}, \tag{4.14}$$

allows the system of equilibrium equations to be rewritten as

$$K_{st} x_t = a_{ns} P_n. \tag{4.15}$$

where K_{st} is the *statical-kinematic stiffness matrix*. As seen from (15), this matrix acts on displacements that are all independent.

The stiffness expressed by the matrix K_{st} is peculiar in two ways. First, it is engendered by, and depends on, the *member forces*, which are induced by prestress and the equilibrium component of the applied load. Second, this stiffness is meaningful only in the context of a *perturbation load*, to which it relates the system displacements. Aside from these unusual features, the statical-kinematic stiffness matrix (14) possesses the properties of conventional, elastic tangent stiffness matrices used in linear or iterative analyses. Moreover, since this matrix quantifies the response of an underconstrained system in a conventional matrix form, its role is not limited to an analysis of such systems alone. If a structural assembly involves an underconstrained component, the latter is readily incorporated into the overall analytical model, much in the same way as is done when using substructuring or finite element techniques.

Positiveness of the Statical-Kinematic Stiffness Matrix. The most important property of a statical-kinematic stiffness matrix (and, in fact, of any stiffness matrix) is positiveness. It entails stability of deformation under a perturbation loading: positive work is performed, and when the load is removed the system returns to its unperturbed state. This implies that the reference equilibrium state of the system is stable, which takes place under certain, but not all, equilibrium loads and virtual self-stress patterns.

In the case of self-stress, the conditions of positiveness of the stiffness matrix (14) and of the reduced quadratic form Q, introduced in Chapter 2, are identical. This follows from the very origin of the statical-kinematic stiffness matrix showing that K_{st} *is the matrix of the reduced quadratic form* Q:

$$Q(x_s, x_t) = K_{st} \, x_s \, x_t. \tag{4.16}$$

This relation links the properties of the incremental solution matrix in (10), which also happens to be the test matrix (2.10), to the properties of the statical-kinematic stiffness matrix (14). It also reveals the physical meaning of the formal mathematical condition of the matrix test for infinitesimal mobility. In fact, the relation amounts to the following more comprehensive statement on kinematic immobility.

Alternative Criteria of Kinematic Immobility.   If an underconstrained structural system satisfies one of the following four conditions:

(1) the reduced quadratic form Q is positive-definite;
(2) the statical-kinematic stiffness matrix K_{st} is positive-definite;
(3) the matrix test (2.10) is passed successfully;
(4) the system is prestressable;

then it satisfies all four of them and is kinematically immobile.

Modal Discretization and the Statical-Kinematic Stiffness Matrix for Continuous Underconstrained Systems.   The concept of the statical-kinematic stiffness matrix and the orthogonal load resolution is the key to the analysis of discrete underconstrained systems. This prompts a question whether a generalization of the concept to continuous systems, in the form of a continuous statical-kinematic stiffness operator and a load projection operator, is feasible. This prospect, however, does not appear promising. The reason is that the construction of the stiffness and projection matrices involves the operation of matrix inversion. Since very few boundary value problems with continuous operators allow an explicit inverse, a projection operator could be constructed only in a symbolic form or formulated as an algorithm.

Nevertheless, there exists an attractive practical alternative to such an operator. The fact is that continuous systems are usually analyzed using some kind of discretization (finite differences, finite elements, or modal discretization in variational methods). This provides a natural framework for employing the above techniques. For example, in the Galerkin method a problem is discretized by approximating the unknown field functions by linear combinations of base functions (modes). The latter are obtained by truncating a suitable complete set of functions, which assures the solution convergence. The unknown modal amplitudes are evaluated from the condition that the approximation error, often called residual, is orthogonal to a finite set of linearly independent functions, usually (but not necessarily) coinciding with the base functions.

Note that the very idea of the residual orthogonalization underlying the Galerkin method is conceptually nothing but projection. Indeed, any discretization, by reducing the problem dimension, generally rules out an exact solution; instead, it settles for the solution projection onto the lower dimension space of the chosen (and then usually truncated) set of base functions. Orthogonalization makes this solution space error-free, thereby confining the error to its orthogonal complement, the error space. Only when an external action happens to be in the solution space, can the solution be exact. In particular, if an entire complete set of base functions can be retained, an exact infinite series solution is obtained.

In a more general sense, the projection idea permeates not only the computational aspect of analysis but the very process of analytical modeling. In this sense, every assumption made represents an implicit or, perhaps, even explicit reduction of the solution space dimension. In yet another sense, projection may be interpreted as an imposition of certain formal and, sometimes, physically meaningful constraints onto the system. Thereby, the number of degrees of freedom and, with it, the size of the problem are reduced.

When dealing with continuous underconstrained systems, it is only natural to utilize the projection approach twice: first, as a means of discretization, and second, for analyzing the discretized system which is still underconstrained. Note that a nonlinear constraint equation can be linearized either before or after discretization. In both cases, discretization requires an introduction of approximating sets of base functions for displacements; the number of these functions is the number of degrees of freedom of the discretized model. This analysis is illustrated in the next chapter, in application to a cable.

4.2 Applications of the Matrix

The main application of the statical-kinematic stiffness matrix is a determination of the equilibrium configuration and the resulting stress state of a system under a given load and constraint variation. This problem is considered below, starting with variant systems.

Variant Systems

An incremental analysis determining the equilibrium configuration of an underconstrained system supporting a given load involves the following steps: (a) resolving the applied load into the equilibrium and perturbation components; (b) evaluating the reactions Λ_i produced by the equilibrium load component and

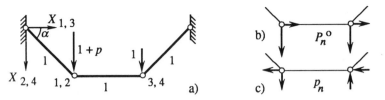

Figure 4.1.   Variant system under a general load: a) reference configuration; b) orthogonal load resolution.

adding them to the prestress (if it is present);   (c) forming the statical-kinematic stiffness matrix K_{st};   and finally,   (d) solving for the displacements x_s produced by the perturbation load component.   In a nonlinear incremental analysis, the next step starts with an update of the system configuration.

This incremental procedure, as well as an alternative one, not using the orthogonal load resolution, are illustrated by the following example of a variant system (type IV).

Example 4.1.   The system in Fig. 1a consists of three identical bars of unit length. The external load is not an equilibrium one and cannot be equilibrated in the original symmetric configuration of the system. The objective is to evaluate the final state of the system in a linearized analysis employing the statical-kinematic stiffness matrix.

The necessary input for constructing the stiffness matrix involves the constraint function Jacobian and Hessian matrices.   As has been shown, for a pin-bar system, this does not require knowing the constraint functions explicitly: the Jacobian matrix is determined by the bar-end coordinate differences, and the Hessian matrices by the system connectivity.   For any in-series three-bar chain, the respective generic matrices are given by $(2.6)_1$ and (2.12).   The system configuration is specified by the angle α.   Denoting, as before,

$$\sin \alpha = s, \quad \cos \alpha = c,$$

the constraint Jacobian matrix can be written (omitting, as usual, the factor 2)

$$F^i_n = \begin{bmatrix} c & s & 0 & 0 \\ -1 & 0 & 1 & 0 \\ 0 & 0 & -c & s \end{bmatrix}. \tag{4.17}$$

Note that it is of full rank, $r = 3$.

Constructing the load resolution matrix R_{mn} requires the matrices

$$F^i_n F^j_n = \begin{bmatrix} 1 & -c & 0 \\ -c & 2 & -c \\ 0 & -c & 1 \end{bmatrix}, \quad (F^i_n F^j_n)^{-1} = (1/2s^2) \begin{bmatrix} 1+s^2 & c & c^2 \\ c & 1 & c \\ c^2 & c & 1+s^2 \end{bmatrix}.$$

According to equation (6)

$$R_{mn} = F_m^i \left(F_l^i F_l^j\right)^{-1} F_n^j = \frac{1}{2} \begin{bmatrix} 1+c^2 & sc & -s^2 & -sc \\ sc & 1+s^2 & sc & c^2 \\ -s^2 & sc & 1+c^2 & -sc \\ -sc & c^2 & -sc & 1+s^2 \end{bmatrix}. \qquad (a)$$

With the aid of this matrix, the applied load,

$$P_n = [0, \quad 1+p, \quad 0, \quad 1]^T,$$

is resolved into the equilibrium and perturbation components

$$P_n^* = (1/2) [psc, \quad 2 + p(1+s^2), \quad psc, \quad 2 + pc^2]^T,$$

$$p_n = (1/2) [-psc, \quad pc^2, \quad -psc, \quad -pc^2]^T,$$

shown in Fig. 1b and c. As it should be, the perturbation load component mimics the pattern of virtual displacements.

Constraint reactions Λ_i induced by the equilibrium load P_n^* are found from an overdetermined but compatible system of equilibrium equations obtained by transposing the Jacobian matrix (17). The omitted factor of 2 is compensated for by halving the loads on the right-hand side:

$$F_n^i \Lambda_i = P_n^*/2.$$

Solving this system yields the same solution as the streamlined alternative calculation employing expression (5)

$$\Lambda_1 = [2 + p(1 + s^2)]/4s, \quad \Lambda_2 = c(2 + p)/4s, \quad \Lambda_3 = (2 + pc^2)/4s. \qquad (b)$$

With these Λ_i and Hessian matrices (2.12), their linear combination, the matrix K_{mn} can be formed

$$K_{mn} = \begin{bmatrix} \Lambda_1 + \Lambda_2 & 0 & -\Lambda_2 & 0 \\ 0 & \Lambda_1 + \Lambda_2 & 0 & -\Lambda_2 \\ -\Lambda_2 & 0 & \Lambda_2 + \Lambda_3 & 0 \\ 0 & -\Lambda_2 & 0 & \Lambda_2 + \Lambda_3 \end{bmatrix}. \qquad (4.18)$$

Constructing the statical-kinematic stiffness matrix (14) requires the matrix a_{ns} that relates all displacements to the independent ones. Designating x_2 as the only independent displacement, the linearized constraint equations with matrix (17) in the left-hand side are solved:

$$x_n = a_{n2} x_2, \quad a_{n2} = [-t, \quad 1, \quad -t, \quad -1]^T, \quad (t = \tan \alpha). \qquad (c)$$

With matrix a_{n2} in hand, both sides of equation (15) can be evaluated:

$$K_{st} = K_{mn} a_{m2} a_{n2} = K_{22} = (1 + t^2) (\Lambda_1 + \Lambda_3) + 4\Lambda_2 = d(2 + p)/2sc \approx d/sc^2,$$

$$(1/2) a_{n2} P_n = p/2 \quad (d = 1 + 2c^3). \tag{d}$$

The approximate equality sign in (d) reflects the fact that the perturbation load parameter p is treated in these calculations as negligible compared to unity, which is consistent with the accuracy of the linearized incremental analysis. The problem reduces to one equation in one unknown, x_2, and the solution is

$$x_2 = psc^2/2d. \tag{e}$$

The remaining, dependent, displacements are found with the aid of (c). Both the displacements and the constraint reactions (b) constitute the outcome of the one-step linearized solution. The next step would start with an update in the system configuration based on the exact constraint equations and the evaluation of the new Jacobian matrix. It is possible, however, to make a simple "half-step" forward by incrementing the constraint reactions prior to the configuration update. The reaction increments λ_p are obtained from system (12) which is constructed with the aid of (18) and (c) and has the form

$$\begin{bmatrix} -1/2c & c & -1 & 0 \\ (1+2c)/2s & s & 0 & 0 \\ -1/2c & 0 & 1 & -c \\ -(1+2c)/2s & 0 & 0 & s \end{bmatrix} \begin{bmatrix} x_2 \\ \lambda_1 \\ \lambda_2 \\ \lambda_3 \end{bmatrix} = \begin{bmatrix} -psc/4 \\ pc^2/4 \\ -psc/4 \\ -pc^2/4 \end{bmatrix}. \tag{f}$$

With x_2 already known, the increments λ_p are found

$$\lambda_{1,3} = \mp psc^3/2d, \quad \lambda_2 = 0,$$

and the final values of the constraint reactions are

$$\Lambda_1 = [2 + p(1 + s^2/d)]/4s, \quad \Lambda_2 = c(2 + p)/4s, \quad \Lambda_3 = [2 + p(1 - s^2/d)]/4s.$$

The bar forces corresponding to these generalized reactions are found using formula (1.12). Since in this example all three bars are of unit length, the bar forces are simply the above Λ_i doubled.

Alternative Load Decomposition and Comparison. It is interesting to compare the above solution with an alternative one, base on the statical-kinematic stiffness matrix but *not* employing the orthogonal load resolution. Obviously, for a system not as simple as the above three-bar system, it is not a trivial task to isolate the equilibrium and perturbation load components. But, at least in theory, any equilibrium component of the applied load can serve for the analysis. With some insight, such a component may be identified and used in the construction of the statical-kinematic stiffness matrix. However, the resulting solution is likely to be inferior to the previous one, based on the complete equilibrium and pure perturbation components. Specifically, a load in the right-hand side of equation (15), instead of being a perturbation load, will contain an equilibrium component. Although the latter will be reflected in the incremental reactions, it does not contribute to the system stiffness, thereby resulting in exaggerated displacements.

Example 4.1 (continued). In the alternative approach, the first task is to extract from a given load an equilibrium load component suitable for the construction of the stiffness matrix. To make the point, suppose that the above load P_n is decomposed as follows:

$$P_n^* = [0, \ 1/2, \ 0, \ 1/2]^T, \quad \delta P_n = [0, \ 1/2 + p, \ 0, \ 1/2]^T. \tag{g}$$

These components are not orthogonal, since δP_n is not a pure perturbation load. The constraint reactions corresponding to the equilibrium load in (g) are easily found. They are exactly half of their respective previous values (b), after disregarding the small parameter p:

$$\Lambda_i = (1/4) \, [1/s, \ c/s, \ 1/s]^T.$$

The system of equations (12) has the form (f), except that the first column in the matrix is smaller by a factor 2, reflecting the reduced values of Λ_i, and the right-hand side is replaced by $\delta P_n/2$, compensating, as usual, for the omitted factor 2. The solution is

$$x_2 = psc^2/d, \quad \lambda_{1,3} = [1 + p(1 \pm s^2/d)]/4s, \quad \lambda_2 = (1 + p)c/4s. \tag{h}$$

It is no surprise that the obtained displacement is double the displacement (e) in the previous procedure, one based on the complete equilibrium load and the corresponding stiffness matrix. On the other hand, the respective reactions Λ_i are identical; the originally unsupported portion of the equilibrium load in the second method is recovered in the incremental reactions.

Although the preceding solution illustrates that it is possible to perform a nonlinear statical-kinematic analysis without the orthogonal load resolution (and in some cases, even to reduce the amount of computation), there are several compensating advantages to the rigorous load decomposition. First of all, it is a consistent systematic procedure that identifies unique, mutually orthogonal, components of the applied load. Moreover, certain features of the structural behavior of underconstrained systems depend directly on these load components. For example, the stabilizing effect of an equilibrium load, introduced into the linearized analysis through the statical-kinematic stiffness matrix, is proportional to the constraint reactions induced by this load. Since the load resolution approach extracts no more and no less than the complete equilibrium component of the applied load, the result is the maximum attainable accuracy in evaluating both forces and displacements at each incremental step of analysis.

In contrast, any alternative load resolution unavoidably leads to exaggerated displacements. This is a disadvantage, since even the displacements obtained with the proper load resolution are, generally, overestimated, due to the tangent linearization used in conjunction with a stiffening load-displacement diagram.

Other Uses for the Linearized Analysis. The linearized system of equations (10) is quite versatile and can be employed in solving a variety of problems. One of them is finding the perturbation load necessary to produce a required displacement. This is done by substituting the assigned independent displacements into the left-hand side of the equations and evaluating the resulting load. Of course, this load will depend on the existing prestress level and the present equilibrium load. Some other possible applications are illustrated in the following example.

Example 4.2. Suppose the system in Fig. 1 supports two equal vertical loads of unit magnitude, as before, and the first bar now undergoes a thermal expansion $\delta L_1 = l_1$. The corresponding constraint variation

$$f^1 = 2L_1 \, \delta L_1 = 2 l_1$$

is introduced into the right hand side of the lower echelon in equations (10), and the resulting system of seven simultaneous equations is easily assembled from the previously obtained submatrices. It is presented in the following table form with the right-hand side designated RHS1:

x_1	x_2	x_3	x_4	λ_1	λ_2	λ_3	RHS1	RHS2
$(1+c)/2s$	0	$-c/2s$	0	c	-1	0	0	0
0	$(1+c)/2s$	0	$-c/2s$	s	0	0	0	0
$-c/2s$	0	$(1+c)/2s$	0	0	1	$-c$	0	$l\lambda_3$
0	$-c/2s$	0	$(1+c)/2s$	0	0	s	0	0
c	s	0	0	0	0	0	l_1	0
-1	0	1	0	0	0	0	0	0
0	0	$-c$	s	0	0	0	0	$-cl$

The solution is

$$x_1 = x_3 = c\,(1+2c)\,l_1/2d, \qquad x_2 = (s^2+d)\,l_1/2s\,d, \qquad x_4 = c^2(1+2c)\,l_1/2s\,d,$$

$$\lambda_1 = -\,[d+s^2(1+2c)]\,l_1/4s^3 d, \qquad \lambda_2 = -\,c\,l_1/4s^3, \qquad \lambda_3 = [d-s^2(1+2c)]\,l_1/4s^3 d.$$

Yet another type of problem deals with variations in any other geometric parameters of the system present in the constraint equations. Let the overall span of the system increase by $\delta L = l$. Parameter L figures in the third constraint function and its derivatives [see matrix (2.6)]. After tracking and incrementing this parameter in equations (10), the corresponding terms are transferred into the right-hand side and appear as column RHS2 in the above system of equations. As could be expected, the obtained solution is symmetric:

$$x_1 = x_3 = l/2, \qquad x_2 = x_4 = -lc/2s, \qquad \lambda_1 = \lambda_3 = lc/4s^3, \qquad \lambda_2 = l/4s^3.$$

Figure 4.2. Quasi-variant system under a general load: a) reference configuration; b) orthogonal load resolution.

The above techniques are used and elaborated on in the next chapter, in application to cables and cable systems.

Quasi-Variant Systems

When employing the statical-kinematic stiffness matrix in an analysis of quasi-variant systems (type III), it should be borne in mind that here displacements are possible only at the expense of constraint variations of second order, which are ignored in a linear analysis. The following examples illustrate the analysis of quasi-variant systems and their structural behavior.

Example 4.3. The quasi-variant system in Fig. 2a represents a singular configuration of the system just considered; yet, the load resolution matrix cannot be derived by specifying the appropriate geometric parameters in the previously obtained R_{mn}. Because of the singularity of the constraint Jacobian matrix used in the construction of R_{mn}, a new calculation is required. Since the degree of singularity is $D = 1$, any two rows of the singular Jacobian matrix are independent and can be used in constructing the projection matrix following equations (4)–(6):

$$F_n^p = \begin{bmatrix} 1 & 0 & 0 & 0 \\ 0 & 0 & -1 & 0 \end{bmatrix}, \qquad F_m^p F_n^p = \begin{bmatrix} 1 & 0 \\ 0 & 1 \end{bmatrix} = \left(F_m^p F_n^p \right)^{-1},$$

$$R_{mn} = \begin{bmatrix} 1 & 0 & 0 & 0 \\ 0 & 0 & 0 & 0 \\ 0 & 0 & 1 & 0 \\ 0 & 0 & 0 & 0 \end{bmatrix}, \qquad K_{mn} = \begin{bmatrix} T+1/2 & 0 & -T/2 & 0 \\ 0 & T+1/2 & 0 & -T/2 \\ -T/2 & 0 & T & 0 \\ 0 & -T/2 & 0 & T \end{bmatrix}.$$

The obtained matrix R_{mn} facilitates the resolution of the load P_n (Fig. 2b):

$$P_n = [1, \ p, \ 0, \ 0]^T, \qquad P_n^* = [1, \ 0, \ 0, \ 0]^T, \qquad p_n = [0, \ p, \ 0, \ 0]^T.$$

The matrix K_{mn} expressed in terms of constraint reactions is given by (18) in the preceding example. It remains to evaluate these reactions which in this case come from two sources, the equilibrium load P_n^* and prestress.

Suppose the system is prestressed to a tension level $N_i = T$. Recalling relation (1.12) between generalized constraint reactions and member forces, the corresponding reactions are easily evaluated as $\Lambda_i = T/2$. However, the member forces and reactions induced by the equilibrium load cannot be found without the constitutive relations because of statical indeterminacy. In one arbitrarily chosen statically possible state, only one bar develops a reaction, $\Lambda_1 = 1/2$, $\Lambda_{2,3} = 0$. Combining these values with the above Λ_i from prestress, and introducing the results into (18), produces the above written matrix K_{mn}.

The matrix a_{ns} needed for the evaluation of the stiffness matrix is obtained from the matrix F_n^p. With x_2 and x_4 as independent,

$$a_{ns} = \begin{bmatrix} 0 & 1 & 0 & 0 \\ 0 & 0 & 0 & 1 \end{bmatrix}^T, \qquad K_{st} = K_{mn}\, a_{ms}\, a_{nt} = \begin{bmatrix} T+1/2 & -T/2 \\ -T/2 & T \end{bmatrix}.$$

The right-hand side of equation (15) is

$$(1/2)\, a_{ns}\, P_n = [p/2,\ 0]^T$$

and the sought independent displacements are

$$x_2 = 2p/(3T+2), \qquad x_4 = p/(3T+2).$$

Although constraint reactions induced by prestress and by an equilibrium load combine in matrix (11), their respective roles in producing statical-kinematic stiffness are not identical. The difference stems from the fact that the prestress pattern in a system is a function of the system configuration and no prestress at all can be induced in an independent constraint. In contrast, an equilibrium load can always be fashioned so as to obtain a desired force in any constraint, dependent or independent, and regardless of the system geometry. In fact, all of the member forces can be assigned arbitrarily, whereupon the corresponding equilibrium load is easily found. If the resulting statically possible state is stable, it it physically realizable.

The difference between equilibrium with prestress and that under an equilibrium load is best observed in systems with a higher-order infinitesimal mobility. The following example illustrates.

Example 4.4. For the system in Fig. 3, the Jacobian matrix is

$$F_n^i = \begin{bmatrix} 0 & -h & 0 & 0 & 0 & 0 \\ 0 & h & 0 & -h & 0 & 0 \\ 0 & 0 & 0 & h & 0 & 0 \\ 0 & 0 & 0 & 0 & l & 0 \\ 0 & 0 & 0 & 0 & -l & 0 \end{bmatrix}, \qquad (a)$$

and the matrix K_{mn} is easily written down:

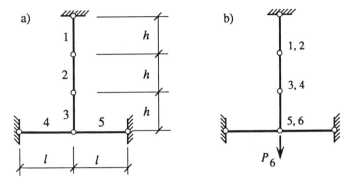

Figure 4.3. A complex quasi-variant system: a) prestress does not impart first-order stiffness; b) an appropriate equilibrium load does.

$$K_{mn} = \begin{bmatrix} \Lambda_1+\Lambda_2 & 0 & -\Lambda_2 & 0 & 0 & 0 \\ 0 & \Lambda_1+\Lambda_2 & 0 & -\Lambda_2 & 0 & 0 \\ -\Lambda_2 & 0 & \Lambda_2+\Lambda_3 & 0 & -\Lambda_3 & 0 \\ 0 & -\Lambda_2 & 0 & \Lambda_2+\Lambda_3 & 0 & -\Lambda_3 \\ 0 & 0 & -\Lambda_3 & 0 & \Lambda_3+\Lambda_4+\Lambda_5 & 0 \\ 0 & 0 & 0 & -\Lambda_3 & 0 & \Lambda_3+\Lambda_4+\Lambda_5 \end{bmatrix}. \quad (b)$$

The rank of the Jacobian matrix (a) is 4 and, with x_1 and x_3 independent,

$$a_{ns} = \begin{bmatrix} 1 & 0 & 0 & 0 & 0 & 0 \\ 0 & 0 & 1 & 0 & 0 & 0 \end{bmatrix}^T, \qquad K_{st} = \begin{bmatrix} \Lambda_1+\Lambda_2 & -\Lambda_2 \\ -\Lambda_2 & \Lambda_2+\Lambda_3 \end{bmatrix}^T. \quad (c)$$

Solving the homogeneous equilibrium equations given by the transposed matrix (a) yields a nontrivial solution

$$\Lambda_1 = \Lambda_2 = \Lambda_3 = 0, \qquad \Lambda_4 = \Lambda_5 = \Lambda.$$

It turns out that the nonzero constraint reactions are not present in the stiffness matrix (c) which, therefore, is a zero matrix. Thus, prestress does not impart first-order stiffness to the system.

On the other hand, under the action of an equilibrium load P_6,

$$\Lambda_1 = \Lambda_2 = \Lambda_3 \neq 0,$$

and the statical-kinematic stiffness matrix in (c) is positive-definite. Thus, *an appropriate equilibrium load induces first-order stiffness*, regardless of the order of infinitesimal mobility of the system. The order of mobility is relevant only for the stiffness induced by prestress. This distinction is the above-mentioned difference between the two sources of stiffness in underconstrained structural systems.

The presented incremental analysis based on the statical-kinematic stiffness matrix is computationally efficient and helpful in understanding the structural behavior of underconstrained systems. At the same time, it has a "brute-force" alternative using conventional means for solving geometrically nonlinear elastic problems. Such an analysis becomes feasible after turning an underconstrained system into a formally invariant one by introducing V auxiliary, very flexible constraints. While making a conventional incremental solution possible, this may lead to a substantial increase in the problem size, which may be unwarranted or even unacceptable as, for instance, in the case of a real-time shape control of a variant system.

4.3 Interplay of Kinematic and Elastic Mobility

Structural members of a system made of a real material undergo deformations of various nature (elastic, plastic, rheological, thermal, etc.) Among these, elastic deformations are always present and sometimes must be taken into account together with kinematic deformations. When studying the interplay of kinematic and elastic displacements, two distinct cases will be addressed. The first one deals with the elastic deformations of structural members in underconstrained systems of different types. The second case, which seems more interesting and, probably, more important, is that of a system with both undeformable and elastic components. This combination is typical of many conventional analytical models in structural mechanics. In particular, a structural system or component may be considered perfectly rigid with respect to certain selected modes of resistance but elastic in other modes. Perhaps the best example of such a model is any version of the bending theory of plates and shells which assumes an inextensible middle surface.

Elastic Displacements in Underconstrained Systems

Elastic Flexibility of a Constraint. When constraint variations are known in advance (like, for instance, unrestrained thermal deformations), they can be accounted for by introducing them into the right-hand side of equations (10). However, in the case of elastic and some other types of deformations, the constraint variations depend on their respective reactions. In this case, a two-stage approach is necessary. First, the purely kinematic displacements and the resulting member forces are found for the system with idealized constraints using, if necessary, a multistep incremental analysis. Then second, with the constraint reactions Λ_i for the kinematically deformed configuration available, the corresponding constraint variations can be evaluated:

$$\delta F^i = f^i = e_i \Lambda_i \qquad \text{(no summation)}, \qquad (4.19)$$

where e_i is the elastic flexibility parameter of the i-th constraint. If the structural member is a bar, then, according to (1.10)–(1.12),

$$e_i = 4(L^3/EA)_i, \qquad (4.20)$$

with the conventional notation EA for the axial stiffness of the bar.

Example 4.5. The familiar three-bar system in Fig. 1 supports two equal vertical loads and is in a symmetric equilibrium configuration with

$$P_n = [0,\ 1,\ 0,\ 1]^{\mathrm{T}}, \qquad \Lambda_1 = \Lambda_3 = 1/2s, \qquad \Lambda_2 = c/2s.$$

After kinematic displacements are accounted for, it is appropriate to consider the elastic properties of the bars. This can be done with the aid of equations (10). For the given system, the resulting 7×7 matrix is the one of Example 2, but the right-hand side in this case is

$$[0,\ 0,\ 0,\ 0,\ e_1/4s,\ e_2c/4s,\ e_3/4s]^{\mathrm{T}},$$

where, in the usual way, the loads have been halved to compensate for the omitted factor 2 in the equilibrium equations. If the three bars of the system are identical, the solution is

$$x_1 = -x_3 = -c/2sEA, \quad x_2 = x_4 = (2 + c^2)/2s^2EA,$$

$$\lambda_1 = \lambda_3 = -(2 + c^2)/4s^4EA, \quad \lambda_2 = -[1 + (2 + cs^2)/2s^4EA.$$

Usually, elastic displacements are much smaller than kinematic ones, and their evaluation takes just one incremental step. Otherwise, the elastic aspect of the problem is itself geometrically nonlinear and then it requires either a multistep incrementation or iteration.

Direct Design. The examples considered so far in this chapter utilize one and the same incremental solution matrix (10) with various modifications in the right-hand side. They represent different forms of a computational algorithm known as the parameter incrementation method. In this method, a parameter is selected in (or purposely introduced into) a nonlinear problem. The state of the system for one value of the parameter must be known and is taken as a reference state. Next, the parameter is given a small increment, and the resulting change in the system state is evaluated. The increment must be sufficiently small to justify the use of linearized equations, thereby solving the original nonlinear problem in a recurring sequence of linear steps.

A particular choice of an incrementation parameter, while generally not affecting the computational effort at each step, gives the analysis a certain physical meaning. Therefore, it makes sense to choose (or to introduce) the parameter so as to generate some useful information at each step, and to complete the calculation in a minimal number of steps. For example, a usual design objective is evaluating or, better still, optimizing the required sizes of structural components, rather than checking intuitively selected member sizes. Therefore, a procedure based on the incrementation of the yet unspecified design parameters makes more sense than the common trial-and-error analysis based on the load incrementation.

A closer look at expression (20) for the elastic flexibility of a constraint shows that it may be used to account for a change in either the elastic properties or the cross-sectional area of the structural member. The second alternative underlies a direct design procedure that is especially suitable for structural systems requiring a nonlinear analysis.

The procedure involves the following sequence of steps. First, a statically possible state of the system with a given load is assigned or obtained in a statical-kinematic analysis. At this stage, all of the member flexibility parameters (20), being inversely proportionate to the member stiffnesses, have zero values. Next, the flexibility parameters are given increments of a magnitude appropriate for a linearized step of the analysis: sufficiently small to maintain the intended accuracy, yet not too small as to slow down the calculation. (If too large or small a step is taken, it is easy to correct the results by scaling them up or down, since the incremental analysis is linear.) The increase in flexibility, which amounts to a reduction in the member cross-sectional areas, produces some displacements and increases the member stresses. Upon updating the system state, the process continues until eventually either the stresses or the displacements, or, ideally, both of them simultaneously, reach an assigned limit. The flexibility parameter values attained at this stage determine the required member areas. Thus, the described process combines the nonlinear analysis and design optimization in one direct design procedure.

Systems Involving Undeformable and Flexible Components

Elastic Reinforcement of an Underconstrained System. A structural system involving both flexible (elastic) and very rigid (practically undeformable) components may be either geometrically invariant or underconstrained. Yet, in both cases it may be advantageous to model such a system as a combination of underconstrained and conventional elastic subsystems. Since the respective statical-kinematic and elastic stiffness matrices are of the same form, the two of them can be easily merged.

Example 4.6. The faithful three-bar system of Example 1 is furnished with two elastic springs and supports the load shown in Fig. 4a. The equilibrium load component and the resulting constraint reactions Λ_i in matrix (18) preserve, and so does the right-hand side in (15). The given spring constants, k_2 and k_4, can be added directly to the respective matrix elements, K_{22} and K_{44}, of the matrix (11), whereupon the combined stiffness matrix of the form (14) is assembled. Once again, it is a 1×1 matrix (cf. (*d*) in Example 1):

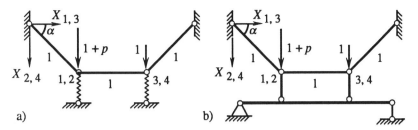

Figure 4.4. An underconstrained system interacting with: a) elastic springs; b) an elastic beam.

$$K_{22} = d/sc^2 + k_2 + k_4. \tag{a}$$

In verifying the dimensions in this equation, recall that the generalized reactions Λ_i and, with them, the first term in (a), have the same dimension (force/length) as the spring constants. The solution is

$$x_2 = p/2K_{22}, \quad N_1 = -k_2 x_2, \quad N_2 = k_4 x_2,$$

where the sign of the spring forces is reversed to obey the sign convention adopted earlier (positive for tension). The bar reactions are still given by (b) in Example 1 and can be refined with the aid of the system of equations (f), taking into account the evaluated displacement x_2. Note that this displacement is smaller than its counterpart (e) in Example 1, since the introduced elastic supports share the perturbation load with the original system. As is the case with any statically indeterminate system, the load sharing is proportionate to the respective subsystem stiffnesses, their dissimilar (statical-kinematic versus elastic) origins being irrelevant.

In an alternative arrangement shown in Fig. 4b, the springs are replaced by a flexible beam, giving rise to a slightly different interaction problem. The contribution of the beam is characterized by

$$k_{22} = \frac{3EI(1+2c)}{c^2(1+c)^2} = k_{44}, \qquad k_{24} = k_{42} = \frac{6EI(1+2c)}{c^2(1+4c+2c^2)}.$$

Adding these elastic coefficients to the respective elements of matrix (11), and forming the combined stiffness matrix, yields

$$K_{22} = d/sc^2 + k_{22} + k_{44} - 2k_{24}. \tag{b}$$

This 1×1 matrix replaces the one in (a) and the rest of the solution is a repetition of the previous one.

Kinematic–Elastic Interaction and Stability. Note that both systems in Fig. 4 are geometrically invariant and could be treated by the conventional means of nonlinear analysis. Still, it often makes sense to represent such a system as an underconstrained system with rigid elements interacting with an elastic subsystem. For example, the combined stiffness matrix K_{st} for such a system provides an immediate solution to the stability problem under a one-parametric equilibrium load. Setting the determinant of K_{st} to zero yields an equation in the unknown critical load parameter.

To illustrate, introduce a load parameter P as a scale factor for the given unit load of the system in Fig. 4a. The 1×1 stiffness matrix (a), above, becomes

$$K_{22} = P\,d/s\,c^2 + k_2 + k_4, \qquad\qquad (c)$$

and its determinant turns into zero at

$$P_{cr} = -(k_2 + k_4)\,s\,c^2/d. \qquad\qquad (d)$$

Similarly, for the system in Fig. 4b with the stiffness matrix (b),

$$P_{cr} = -(k_2 + k_4 - 2k_{24})\,s\,c^2/d. \qquad\qquad (e)$$

As is readily seen, the critical load is proportional to the stiffness of the elastic members. If the latter approaches zero, so does the critical load, indicating *geometric instability*, a characteristic of any system of type IV. Reversing the load sign not only stabilizes the system but induces stiffness of a statical-kinematic nature.

A much more meaningful and important problem of kinematic-elastic interaction, and its effect on stability, is considered in detail in Chapter 10, when analyzing the phenomenon of bending instability of elastic plates and shells.

Geometric Hardening. When an underconstrained structural system, subjected to a perturbation load, changes its geometric configuration, it does so in a very interesting way. The system mostly deforms by adapting to the applied load kinematically, while the elastic deformations reflect the accompanying changes in the member forces. Both of the deformations, and especially the kinematic one, modify the geometric configuration of the system so as to enable it to support the applied load in the best possible way, that is, with the lowest possible member forces. This type of geometrically nonlinear behavior, known as geometric hardening, is reflected in the incremental analysis utilizing the statical-kinematic stiffness matrix. The analytical process shows how an ever-increasing portion of the perturbation load is gradually absorbed by the system, and in the end the entire load is incorporated into an equilibrium load of the final deformed configuration. In the process, the statical-kinematic stiffness of the system, proportionate to the magnitude of the member forces, steadily increases, together with these forces.

The feature of geometric hardening manifests itself in the stiffening load-displacement diagram. The span variation in the symmetrically loaded three-bar system considered in Example 2 provides a ready illustration. The horizontal force increment required to produce a small displacement, $\delta L = l$, in the original system is given by the expression for λ_2 in solution (a) of Example 2. As is easily seen, this force depends on the system configuration; specifically, it is an increasing function of the span, approaching infinity in the limit (the analysis is purely kinematic). The introduction of elastic reinforcing elements into the system modifies this response. In this case, the incremental reactions are evaluated by incorporating the stiffness parameters of the elastic members, $k_{22} = k_{44}$ and $k_{24} = k_{42}$, into the respective elements of the 7×7 master matrix of Example 2. The solution of the resulting system of equations with the modified matrix and the same column RHS2 in the right-hand side is

$$\lambda_1 = \lambda_3 = cl\,(1 + 2sk)/4s^3, \quad \lambda_2 = l\,(1 + 2c^2sk)/4s^3 \quad (k = k_{22} + k_{24}).$$

The obtained expressions clearly indicate the role of the elastic elements in the system response to a variation of its span. As is readily seen, the response of the reinforced system is a blend of its kinematic and elastic properties, current geometric configuration, and the magnitude of the member forces induced by the equilibrium load.

Globally Statically Indeterminate Finite Mechanisms

Conceptual Description. A somewhat different pattern of interaction between kinematic and elastic mobility is exhibited by a special subclass of variant systems—statically indeterminate finite mechanisms. The simplest system of this unusual kind, presented in Fig. 1.7, is reproduced in Fig. 5. It represents an adequately constrained system that has degenerated into a singular variant system with global statical indeterminacy.

One of the peculiar features of this system is that it possesses virtual self-stress in any configuration, but the stability analysis cannot be accomplished without the constitutive relations. Indeed, the forces of self-stress are easily found, but the resulting linear combination (2.3) of the constraint functions vanishes identically. Thus, the statical-kinematic analysis only shows that the system is kinematically mobile and that self-stress is statically possible.

In investigating this system, two distinct situations must be addressed. First, prestressing may change the natural (strain-free) member sizes such that the assembly ceases to be a finite mechanism. For example, if the horizontal bar bends, the system loses purely kinematic mobility and becomes a mechanism with elastic interference. The second, apparently different and more interesting, situation arises when the assembly appears to acquire kinematic mobility in its prestressed configuration, that is, it acquires the appropriate member sizes after the prestress. This is the case if the horizontal bar either originally has the right camber, or is straight and perfectly rigid in bending, but not in tension and compression. Assuming the second alternative, hereafter the bar is modeled as a spring sliding along a rigid rod.

Regarding this type of finite mechanism, note, first of all, that stable self-stress must be ruled out. Indeed, since kinematic motion preserves the member sizes and forces, it requires no increase in the elastic strain energy which, in the absence of external loads, could be the only source of a restoring force. Hence, a

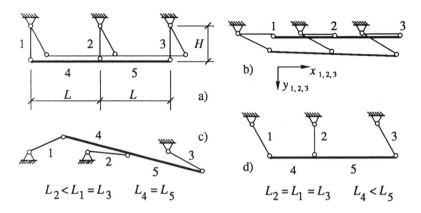

Figure 4.5. A statically indeterminate finite mechanism: a) reference configuration; b) folded configuration, singular to the second degree; c) and d) nonsingular, geometrically invariant configurations.

finite mechanism, if at all prestressable, may be only in neutral equilibrium. However, even this is feasible only in exceptional trivial cases of rigid motion (e.g., a tensioned chain with one end hinged and the other sliding along a circular guide). In general, a kinematic perturbation upsets the equilibrium of the existing member forces, since their nodal resultants will no longer add up to zero. This means that adjacent equilibrium states are absent and virtual self-stress is unstable (cf. the exclusion condition between prestressability and kinematic mobility in Chapter 2).

Example 4.7.   To explore the possible situations and to illustrate the preceding statements, consider first the configuration in Fig. 5a. Let the statically possible self-stress state, perhaps, unstable,

$$N_1 = N_3 = N_0, \quad N_2 = -N_0, \quad N_4 = N_5 = 0, \tag{a}$$

be induced while restraining the system. It turns out that upon its release, the system will fold up (Fig. 5b) and become a finite mechanism with elastic interference. To prove it, note that in the folded configuration the elastic nodal displacements in the horizontal direction are

$$x_1 = x_3 = -x_2/2 = N_0(3k_L + k_H) = x. \tag{4.21}$$

As a result, the member forces and lengths become

$$N_1 = N_3 = -N_2/2 = N_4 = -N_5 = 3k_L N_0(3k_L + k_H) = N,$$

$$L_1 = L_3 = H(1 - \xi), \quad L_2 = H(1 + 2\xi), \quad \xi = x/H, \tag{4.22}$$

$$L_4 = L(1 + 3\eta), \quad L_5 = L(1 - 3\eta), \quad \eta = x/L.$$

Here $k_L = L/EA_L$ and $k_H = H/EA_H$ are the stiffness parameters for the horizontal bar and support bars, respectively. The normalized displacements ξ and η, however small they may be, are finite.

To investigate stability of the state described by (22), the system is given a small perturbation and the elastic strain energy increment is evaluated to the second-order terms:

$$\Delta U = N_i \, (\varepsilon L)_i + (1/2k_i) \, (\varepsilon L)_i^2. \tag{4.23}$$

Within the required accuracy the elastic strains are

$$(\varepsilon L)_i = (x_i + y_i^2/2H), \quad i = 1, 2, 3;$$
$$(\varepsilon L)_4 = x_2 - x_1 + L\alpha^2/2, \quad (\varepsilon L)_5 = x_3 - x_2 + L\alpha^2/2. \tag{b}$$

Since the horizontal bar does not bend,

$$y_1 = y_2 - \alpha L, \quad y_3 = y_2 + \alpha L, \tag{c}$$

so that only five out of the six nodal displacements are independent. After the necessary substitutions and rearrangements with consistently maintained accuracy, and upon the introduction of $\delta = y_2/H$, the strain energy increment becomes

$$\Delta U = 3NH\,\xi(\delta - \alpha)^2 + N(\alpha L)^2/H$$
$$+ k_H (x_1^2 + x_2^2 + x_3^2)/2 + k_L [(x_2 - x_1)^2 + (x_3 - x_2)^2]/2. \tag{4.24}$$

At $N > 0$ or, which is the same, $N_0 > 0$, form (24) is positive-definite. Thus, with the middle bar in compression, the folded configuration is stable and the system is a prestressed infinitesimal mechanism. Purely kinematic displacements are impossible, but first-order elastic displacements require only second-order strains.

At $N < 0$, form (24) is indefinite and the folded configuration is unstable. In this case, the combination of the natural bar lengths

$$L_1^0 < L_2^0 = L_3^0, \quad L_4^0 = L_5^0, \tag{d}$$

is compatible, and the system acquires a configuration (Fig. 5c) which is geometrically invariant and statically determinate (hence, stress-free). Although the system can be forced (elastically) into the self-stressed configurations shown in Fig. 5a and b, it will not stay in either of them.

Similarly, a set of bars with natural lengths

$$L_1^0 = L_2^0 = L_3^0, \quad L_4^0 < L_5^0, \tag{e}$$

can be assembled (Fig. 5d) to form an invariant statically determinate system. The system can be forced into two distinct folded self-stressed configurations, left and right, but both of them are unstable. Thus, different combinations of bar lengths can render the considered system invariant and statically determinate (nonsingular configurations); variant and indeterminate to the first degree (singular configurations); and quasi-invariant and indeterminate to the second degree (doubly singular configurations).

With the aid of the expression (24) for the elastic strain energy, the response of the prestressed assembly in Fig. 5b to a vertical load, V, can be evaluated

$$\partial \Delta U / \partial \alpha = 2NL^2 \alpha / H - 6NH\xi(\delta - \alpha) = 0,$$

$$\partial \Delta U / \partial \delta = 6NH\xi(\delta - \alpha) = V.$$

(f)

From here,

$$\alpha = VH^2 / 2NL^2, \quad \delta = V/6N\xi.$$

(g)

Expressing N and ξ in terms of the original prestressing force with the aid of equations (21) and (22) gives

$$V = \frac{18N_0^2 \delta}{Hk_L(3 + k_H/k_L)^2}.$$

(4.25)

Resistance to Displacement for Various System Types. As is seen from the foregoing analysis, a singular system IV (a globally statically indeterminate finite mechanism) possesses kinematic mobility only as an idealization. Any component deformation or the slightest departure from the nominal member sizes leads to an elastic interference or self-stress. In the latter case, the system will spontaneously move into a configuration where the strain energy is minimized, thereby rendering the self-stress stable and acquiring elastic resistance to any displacements. This response, exemplified by solution (24), is relatively weak, but otherwise quite remarkable: the resistance of the system is proportional to the *square* of the prestressing force N_0.

This response contrasts with the behavior of conventional structural systems and first-order infinitesimal mechanisms having a similar prestress level and typical member stiffness, EA. Table 1 compares the three types of systems in terms of the respective external loads producing the same nondimensional displacement δ.

Table 4.1. Resistance to Displacement for Various System Types.

System description and load type	Resistance to δ due to	
	Prestress	Member stiffness
Any system, equilibrium load	$N_0 \delta$	$EA \delta$
First-order infinitesimal mechanism, perturbation load	$N_0 \delta$	$EA \delta^3$
Singular finite mechanism with interference, perturbation load	$N_0^2 \delta / EA$	

As is seen from the first entry in the table, the resistance of any system to an equilibrium load is the same and is practically unaffected by prestress because of $N_0 \ll EA$. In particular, independence of stiffness from prestress is well known for invariant systems, where perturbation loads simply do not exist. In contrast, resistance of an infinitesimal mechanism to a perturbation load is proportionate to the prestress level, whereas the role of the member stiffness is negligible even in first-order mechanisms and further reduces in higher-order mechanisms. However, in the absence of prestress and for unprestressable infinitesimal mechanisms, the member stiffness is the only source of resistance.

The peculiar, mixed-origin, resistance to perturbation loads in singular finite mechanisms with interference can be explained by noting that

$$N_0^2 \delta / EA = N_0 \varepsilon_0 \delta, \tag{4.26}$$

where an elastic strain ε_0 induced by prestress is a measure of the interference. Obviously, the resistance should increase with both the amount of interference and the prestress level.

Part II
Applications
to Particular Systems

Imagination is more important than knowledge.

Albert Einstein

God is really only another artist. He invented the giraffe, the elephant and the cat. He just goes on trying other things.

Pablo Picasso

5
Cables and Cable Systems

A cable is the simplest continuous underconstrained structural system. Its analytical model is a flexible wire, or a fiber, i.e., a perfectly flexible structural element capable of resisting only one type of force, axial tension. A cable is a variant system, with the exception of a rectilinear configuration with both ends fixed, when it is quasi-variant. The role of the cable as a structural component in engineering applications is diverse and growing. One of the reasons is the very logic of technical progress, resulting in an ever-increasing materials strength. As a matter of principle, tension is the only type of stress fully capitalizing on this trend. The cable also serves as an analytical model for a membrane in a uniaxial stress state. This can be either a wrinkling membrane or a membrane in a state of plane strain, like a membrane trough under a hydrostatic load considered in Section 1.

Generally, the presence of cables in a structural assembly complicates its analysis in several ways. To begin with, the analytical model is likely to be geometrically nonlinear due to relatively large deformations resulting from the interplay of kinematic and elastic mobility. Another potential source of nonlinearity is a possible disengagement of wires as unilateral constraints, an effect nonlinearizable in a conventional way.

Since an extensive technical literature exists on cables and cable systems, this chapter includes only selected generic problems, mostly new. The first two sections are concerned with a single cable (inextensible or elastic), first, as a structural system and then as a system component. The second task is accomplished by constructing the tangent stiffness matrix for a sagging cable supporting a given load. In the next section, one- and two-layer radial cable assemblies are investigated, exhibiting some interesting and favorable features of structural behavior. These are associated with the force-leveling effect of kinematic mobility inherent in sagging cables.

The concluding section is devoted to a support structure, in the form of a closed contour ring, for a radial cable system of an arbitrary planform. It is shown that a variety of radial cable systems of a given planform can be designed such as to make the contour ring momentless. On the other hand, the momentless shapes of the contour ring, including some unexpected ones, are established for a few particular types of cable systems. Finally, a statical analysis of the contour ring interacting with the cable system is presented. It is reduced to a contact problem where, upon linearization, the cables are represented by a special type of elastic foundation.

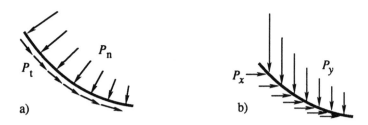

Figure 5.1. Alternative representations of a distributed load on a cable: a) normal and tangential components; b) vertical and horizontal components.

5.1 Some Problems of Statics of Elastic Cables

A Cable or a Membrane Trough Under Hydrostatic Pressure

Basic Assumptions and Types of Loads. An external load acting on a cable is usually given by its tangential and normal components. Except for field-induced loads (gravity, centrifugal, magnetic, etc.), transmitting a tangential load on a cable requires some kind of a mechanical fixture. For example, a joint or a grip can transmit a concentrated force, whereas friction can be a means for transmitting a distributed load, with the resulting uncertainty in the actual load. distribution.

Equilibrium equations for an infinitesimal element of a cable (Fig. 1a) in the tangential and normal directions are

$$dT + P_t \, ds = 0, \quad \sigma T + P_n = 0, \tag{5.1}$$

where $T = T(s)$ is the cable tension as a function of arclength, P_t and P_n are, respectively, tangential and normal loads per unit length, and σ is the cable curvature.

In practical analyses, two types of loading are encountered most often: loads normal to the cable and loads representing a system of parallel forces, say, vertical, usually field-induced. In the first case, $P_t = 0$ and, as follows from (1), the cable tension does not vary along the length: $T = $ const. In the second case (Fig. 1b), if the horizontal load component is absent, it makes sense to use the equilibrium condition of a cable element in this direction

$$d \, (T \cos \beta_c)/dx = 0. \tag{5.2}$$

From here

$$T \cos \beta_c = H = \text{const.} \tag{5.3}$$

where β_c is the cable slope. Thus, in these two most important cases of cable loading, either the cable force, T, or its horizontal projection, H, is constant along the length.

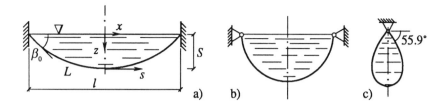

Figure 5.2. A cable (or a membrane trough cross section) under hydrostatic pressure: a) a general configuration; b) configuration with vertical slopes at the edges; and c) zero-span configuration.

Inextensible Cable under Hydrostatic Pressure. A typical normal load is a pneumatic or hydrostatic pressure. Under a uniform pneumatic pressure, a cable has the form of a circular arc and the tension force is $T = P_n R$. A much more interesting load is hydrostatic pressure:

$$P_n = \gamma z, \tag{5.4}$$

where γ is the specific weight of the liquid. Shown in Fig. 2a is a membrane trough whose transverse strip of a unit width can be considered (after switching from a plane strain to plane stress formulation) as a cable under a normal pressure proportional to the depth. The inadequacy of a linear formulation for this problem has been demonstrated by Kerr and Coffin (1990).

Under a hydrostatic load (4), the equilibrium condition in the normal direction of the cable,

$$\sigma T = \gamma z, \tag{5.5}$$

happens to coincide with the equation of column buckling

$$\sigma EI = P z.$$

In this equation, the column compression force, P, replaces specific weight, γ, and the bending stiffness, EI, comes in place of cable tension, T. The identity of the two equations is not purely coincidental. Indeed, both P and γ represent external actions, and tension T, being the source of the statical-kinematic stiffness of the cable, is the counterpart of the bending stiffness of the column. The analogy immediately provides a comprehensive closed-form solution to the cable problem by rephrasing the corresponding results from Euler's elastica theory.

After expressing the cable curvature as a function of slope, β,

$$\sigma = d\beta/ds = d(\sin \beta)/dx,$$

and introducing a new variable by

$$k \sin \varphi = \sin \beta/2, \qquad k = \sin \beta_0/2, \tag{5.6}$$

the differentials of arclength, s, abscissa, x, and ordinate, z, of the cable are obtained as

$$ds = d\beta/\sigma, \quad dx = ds \sin \beta, \quad dz = ds \cos \beta.$$

Taking into account relations (5) and (6), integration is carried out, producing a solution in terms of elliptic integrals and functions:

$$as = F(\varphi, k), \quad ax = 2E(\varphi, k) - as, \quad az = 2k \operatorname{cn} as, \quad a^2 = \gamma/T. \quad (5.7)$$

Here

$$F(\varphi, k) = \int_0^\varphi (1/\operatorname{dn} as)\, d\varphi, \quad E(\varphi, k) = \int_0^\varphi (\operatorname{dn} as)\, d\varphi, \quad (5.8)$$

are, respectively, incomplete elliptic integrals of the first and second kind with amplitude φ and modulus k, and

$$\operatorname{sn} as = \sin \varphi, \quad \operatorname{cn} as = \cos \varphi, \quad \operatorname{dn} as = \sqrt{1 - k^2 \sin^2 \varphi},$$

are the Jacobian elliptic functions which depend on both the arclength s and, implicitly, the modulus k.

For a cable of a given length L and span l supporting a hydrostatic load (4), the maximum sag, S, and the tension force are evaluated by observing that at the cable end

$$s = L/2, \quad x = l/2, \quad \sin \varphi_0 = 1, \quad \varphi_0 = \pi/2.$$

With that, equations (7) yield

$$aL = 2K, \quad al = 2(2E - K), \quad aS = 2k, \quad T = \gamma/a^2, \quad (5.9)$$

where $K(k)$ and $E(k)$ are complete elliptic integrals of the first and second kind. The unknown quantities in (9) are a, k, S, and T. Eliminating a between the first two equations gives

$$2LE(k) - (L + l)\, K(k) = 0.$$

From this transcendental equation, the modulus k is found, and the rest of the solution is straightforward.

The analysis simplifies when the end slope β_0 is assigned as one of the two input parameters, making k available from (7). For example, at $\beta_0 = \pi/2$ (vertical slope, Fig. 2b), $k = \sqrt{2}/2$ and

$$K = 1.86, \quad E = 1.35, \quad al = 1.68, \quad L = 2.21l, \quad S = 0.84l, \quad T = 0.354\gamma l^2.$$

Incidentally, the tension force is related to the weight of the liquid, W, and the cross-sectional area of the trough profile, A, by

$$2T \sin \beta_0 = W = \gamma A.$$

If the span l is set to zero, the trough acquires a drop-shaped profile (Fig. 2c). According to (9) this occurs at $2E - K = 0$, leading to

$$k = 0.828, \quad S = 0.357L, \quad T = 0.0465\gamma L^2, \quad \beta_0 = 68.2°, \quad A = 0.0864L^2.$$

The utter simplicity of the above solution makes feasible its generalization in two ways: accounting for the material flexibility and considering a more general load pattern.

More Comprehensive Solutions

Extensible Cable. When the cable material is extensible, the simplest problem statement is one stipulating the geometry of the deformed cable profile, for example, by assigning the final sag S. Upon evaluating the tension force and the stretched length of the cable using equations (9), the initial length, L_i, is easily found with the aid of constitutive equations. This length can be greater, equal to, or smaller than the cable span.

If the initial, rather than final, length is assigned, the first of the simultaneous equations (9) is replaced by

$$aL_i \, [1 + \varepsilon \, (T)] = 2K$$

and the resulting system is solved iteratively. However, the computational process is unstable if the deformed cable profile is shallow. This difficulty is overcome (and its cause is exposed) in an asymptotic solution based on power expansions.

Consider first a cable with a rectilinear initial shape, $L_i = l$, and a shallow deformed profile such that

$$\sin^2 \beta_0 / 2 = k^2 \ll 1.$$

Only two consecutive terms are needed to be retained in the power expansions of complete elliptic integrals

$$K(k) \approx (1 + k^2/4) \, \pi/2, \quad E(k) \approx (1 - k^2/4) \, \pi/2 . \tag{5.10}$$

Within this accuracy, the length-to-span ratio of the cable, found from (9), is

$$L/l = K/(2E - K) \approx 1 + k^2. \tag{5.11}$$

Since this ratio equals $(1 + \varepsilon)$, the cable strain must be $\varepsilon = k^2$, and it only remains to relate this strain to the cable tension. For example, for a linearly elastic cable, $\varepsilon = T/EA$ (EA is the extensional stiffness of the cable) and, taking into account (9),

$$k = \sqrt{T/EA} = \sqrt{\gamma / EAa^2}.$$

From here, with the aid of (9) and (10), an explicit relation between the external load and the normalized elastic deflection, in this case, the cable sag, is found:

$$(S/l)^2 = (2k/al)^2 = 4\gamma l^2 / \pi^4 EA. \tag{5.12}$$

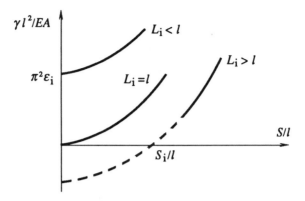

Figure 5.3. Normalized load-sag diagrams for elastic cables with prestressed, rectilinear stress-free, or sagging initial configurations.

The corresponding load-deflection diagram is a quadratic parabola with a horizontal tangent at the origin, so that the problem is nonlinearizable at the onset of loading. This comes as no surprise; after all, this is a property of any quasi-variant system subjected to a perturbation load, and a cable with $L_i = l$ is such a system. However, the particular hydrostatic load (4) makes the problem peculiar in yet another way. This transpires in the analysis of a shallow cable with $L_i < l$. Such a cable must be prestressed to a certain tension level, T_i, with the corresponding elastic strain ε_i. In terms of the span l, the final length of the cable is

$$L = L_i(1 + \varepsilon) = l\,(1 - \varepsilon_i)\,(1 + \varepsilon).$$

Equating the resulting L/l ratio to that given by the geometric expression (11),

$$(1 - \varepsilon_i)\,(1 + \varepsilon) = 1 + k^2,$$

produces the sought load-deflection relation for a prestressed cable:

$$(S/l)^2 = 4\varepsilon_i\,(1 - \varepsilon_i)\,(\gamma l^2/\pi^2 T_i - 1)/\pi^2. \tag{5.13}$$

Analysis of this solution leads to several interesting observations:

(1) The lowest-order stiffness of the cable is imparted by the prestress and is independent of the material properties and the cross section area of the cable.

(2) Since $\varepsilon_i/T_i = 1/EA$, it is readily seen that at $T_i \to 0$ (no prestress), expression (12) is recovered, whereby the second-order stiffness is related to the extensional stiffness of the cable.

(3) The condition $(S/l)^2 > 0$ entails

$$\gamma > \pi^2 T_i/l^2 = \gamma_{cr}, \tag{5.14}$$

meaning that equilibrium in a deflected configuration is possible only if the specific weight of the liquid exceeds the critical value (14) which is a function of

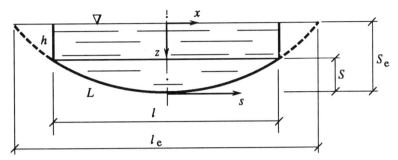

Figure 5.4. A membrane trough with vertical walls.

the cable prestress level. In other words, the γ versus S diagram, quite like the Euler solution, is approximated by a quadratic parabola with a horizontal tangent at the point $(0, \gamma_{cr})$.

(4) Even with $\gamma > \gamma_{cr}$, the solution for a prestressed membrane trough is meaningful only in the context of displacement loading. If the membrane is deflected say, with a templet, and then filled with a liquid of the right specific weight, the equilibrium will be maintained after the templet is removed. Under a different γ, the membrane will either change the equilibrium configuration or, at $\gamma < \gamma_{cr}$, spring back completely.

The normalized load-sag diagrams presented in Fig. 3 display the structural behavior at the onset of loading of systems with the initial length L_i smaller, equal to, or greater than the span. For a prestressed membrane, the sag is zero as long as $\gamma l^2 / EA < \pi^2 \varepsilon_i$, and the diagram initially has a horizontal tangent. On the other hand, the diagram for a membrane with $L_i > l$, i.e., one with an initial sag, has a nonzero slope at the onset of loading. In all three cases the system behavior is characterized by a stiffening load-sag diagram (the familiar phenomenon of geometric hardening).

General Hydrostatic Load. Shown in Fig. 4 is a membrane trough with rigid vertical walls. In terms of the analogy with elastica, this problem corresponds to a compressed column with two end moments. Their magnitudes are Ph_1 and Ph_2, where h_1 and h_2 are the respective pressure heads at the two, perhaps not equally elevated, ends of the trough. The problem can be reduced to a familiar analysis of a statically equivalent trough with the as-yet unknown span l_e terminating at the surface of the liquid. This requires incorporating the known geometric parameters of the trough (length L, span l, and wall height h) into equations (7):

$$2F(k, \varphi) = aL, \quad 4E(k, \varphi) - aL = al, \quad S_e \, cn \, (aL/2) = h. \quad (5.15)$$

This system of simultaneous transcendental equations in the unknown φ, a, and S_e can be solved in a simple iterative process as shown in the following example.

Example 5.1. Let the geometric parameters of the trough be $L = 5$, $l = 4$, and $h = 1$ (Fig. 4). The chosen initial approximations of the sought quantities are $a = 0.20$ and $S_e = 3.0$. The corresponding values of modulus k and amplitude φ are found, respectively, from (9) and $(15)_1$:

$$k = aS_e/2, \quad F(\varphi, k) = aL/2 = u, \quad \varphi = \sin^{-1}(\operatorname{sn} u). \tag{a}$$

These values are introduced into the second of equations (15) which must be solved simultaneously with the third one. In doing that, the modified Newton method with a simplified derivative matrix will be used, which avoids differentiation of $E(\varphi, k)$ and at the same time uncouples the equations. With this in mind, the second equation is incremented only with respect to a, and the third one in both a and S_e:

$$(L + l)\Delta a = 4E(\varphi, k) - (L + l)a. \tag{b}$$

$$\Delta S_e \operatorname{cn} u - \Delta a \, S_e \, (\operatorname{sn} u \, \operatorname{dn} u) \, L/2 = h - S_e \operatorname{cn} u. \tag{c}$$

Implementing formulas (a), (b), and (c) produces an iterative chain of calculations with the results tabulated below:

Itrn	a	S_e	k	u	sn u	cn u	dn u	φ	E	Δa	ΔS_e
0	.200	3.000	.300	.500	.836	.551	.988	.990	.980	.017	−.803
1	.217	2.197	.238	.543	.882	.471	1.080	.993	.982	.001	−.096
2	.218	2.101									

Considering the process converged, the outcome is

$$S = S_e - h = 2.101 - 1.000 = 1.101, \quad T = \gamma/a^2 = 21.0\gamma. \tag{d}$$

At this stage, it may be appropriate to take into account elastic deformations.

5.2   Shallow Cables

In many applications cables are shallow and their analysis simplifies. A cable is considered shallow if the maximum slope β_c is such that $\beta_c^2 \ll 1$, leading to

$$z' = \tan \beta_c \approx \beta_c, \quad \cos \beta_c = 1/(1 + z'^2) \approx 1, \quad \sigma = z''/(1 + z'^2)^{3/2} \approx z'',$$

and, as a consequence,

$$P_n \approx P_z, \quad H \approx T.$$

According to these relations, for a shallow cable, tension T and its horizontal projection, thrust H, are identical in magnitude; more exactly, the difference between the two is beyond the accuracy of the analysis based on the above assumptions. Hence, in this analysis there is no need (or, indeed, sense) to discriminate between a vertical load and a load normal to the cable. Furthermore, in both cases the cable can be considered as uniformly stressed (isotensoid).

Modal Discretization and Load Resolution

A typical problem in cable analysis calls for evaluating the shape and the tension force in a cable of a given length under a given load. One possible approach is to apply the modal discretization and projection matrix techniques described in Chapter 4. A shallow inextensible cable, being the simplest continuous underconstrained system, is an opportune object for illustrating this approach.

Differential Constraint Equations. The analysis starts with the development of a kinematic relation expressing the cable strain as a function of horizontal, U, and vertical, W, displacements. The original and final lengths of an infinitesimal element of the cable are

$$ds^2 = (dx^2 + dz^2) = dx^2(1 + z'^2),$$

$$ds_1^2 = dx^2[(1 + U')^2 + (z' + W')^2].$$

The strain is evaluated assuming small extensions U' but finite rotations W':

$$\varepsilon \approx (ds_1^2 - ds^2)/2ds^2 \approx U' + z'W' + W'^2/2.$$

The condition of cable inextensibility, $\varepsilon = 0$, is the nonlinear differential constraint equation. Upon linearization at $U = W = 0$, the equation becomes

$$u' + z'w' = 0,$$

where u and w are small (strictly speaking, infinitesimal) displacements.

Discretization and Modal Jacobian Matrix. A cable will be modeled as a system with two degrees of freedom, confining its deformation to a combination of one symmetric and one antisymmetric mode. In this model, the displacements can be approximated as

$$u = u_k \sin a_k x, \quad w = w_k \sin a_k x, \quad a_k = k\pi/l, \quad k = 1, 2. \quad (5.16)$$

A convenient reference configuration for the cable with even supports is

$$z(x) = S \sin a_1 x, \quad (5.17)$$

where S is the cable sag. The operation of discretization involves substituting the above expressions into the left-hand side of the linearized constraint equation and orthogonalizing the result to the two base functions:

$$\int_0^l (u' + z'w') \sin a_k x \, dx = 0, \qquad k = 1, 2.$$

This leads to two algebraic constraint equations relating four displacement amplitudes u_1, w_1, u_2, w_2. The equation matrix is the modal Jacobian matrix of a cable as a system with two degrees of freedom introduced by (16)

$$F_n^k = \frac{\pi}{2B} \begin{bmatrix} 0 & 1 & 0 & 0 \\ B & 0 & 0 & 1 \end{bmatrix} \qquad (B = l/\pi S), \qquad (5.18)$$

where the dimensionless ratio B is determined by the cable geometry.

The transpose of (18) is the matrix of equilibrium equations in the unknown generalized constraint reactions; their relation to the cable tension is yet to be established. In doing that, note that for a continuous boundary value problem, the counterpart of a transposed matrix is the conjugate operator. Its origin and physical meaning become clear in the process of its construction, based on the principle of virtual work. The force that performs work over the cable elongation is the cable tension, $T \approx H$. Adding to it the work done by external loads, P_x and P_z, and integrating by parts, the total work is evaluated

$$\int_0^l [H(u' + z'w') + P_x u + P_z w] \, dx$$

$$= H(u + z'w) \big|_0^l - \int_0^l \{(H' - P_x) u + [(Hz')' - P_z]w\} \, dx = 0.$$

Due to the boundary conditions (zero displacements at both ends), the first term in the obtained expression vanishes. For the resulting equation to be satisfied with any choice of kinematically admissible u and w, the following two conditions must hold:

$$H' = P_x, \qquad (Hz')' = P_z. \qquad (5.19)$$

These are the equilibrium equations of a shallow inextensible cable under a general load.

Modal Loads. To be consistent with the two-degree-of-freedom model, the discrete version of equilibrium equations is obtained from the above equation of virtual work by introducing displacements (16)

$$\int_0^l [Ha_k \cos a_k x \, (u_k + z'w_k) + (P_x u_k + P_z w_k) \sin a_k x] \, dx = 0.$$

This equation must be satisfied identically with respect to u_k and w_k. Upon incorporating (17) and carrying out the integration, a system of four equilibrium equations in two unknown parameters, H_0 and H_1, is obtained. Its matrix is the transpose of the modal Jacobian matrix (18), which was expected. More importantly, this calculation reveals the nature of the modal constraint reactions: they turn out to be the coefficients of the cosine Fourier series of function $H(x)$,

$$H(x) = H_i \cos a_i x , \quad i = 0, 1. \tag{5.20}$$

The modal load parameters P_m figuring in the right-hand side of the discretized equilibrium equations are

$$P_m = \begin{cases} \int_0^l P_x \sin a_k x \, dx, & m = 2k - 1, \\ \int_0^l P_z \sin a_k x \, dx, & m = 2k. \end{cases} \tag{5.21}$$

With this, the modal discretization of the problem is complete; it has a familiar format, with an underdetermined system of linearized constraint equations and an overdetermined system of equilibrium equations sharing the modal Jacobian matrix (18).

The cable tension is evaluated with the aid of the matrix G_n^i introduced in (4.7) as a function of constraint Jacobian matrix:

$$G_n^i = (F_m^j F_m^i)^{-1} F_n^j = \frac{2B}{\pi} \begin{bmatrix} 0 & 1 & 0 & 0 \\ B/B_1 & 0 & 0 & 1/B_1 \end{bmatrix}, \qquad \begin{matrix} B = l/\pi S, \\ B_1 = B^2 + 1. \end{matrix}$$

The resulting modal constraint reactions are

$$H_i = G_m^i P_m = (2B/\pi) \, [P_2, \ (BP_1 + P_4)/B_1]^{\mathrm{T}}. \tag{5.22}$$

Note the absence of the symmetric horizontal load component represented by P_3 and related to longitudinal displacement $x_3 = u_2$. Accounting for this load would require at least one additional degree of freedom (the one involving $w_3 \sin a_3 x$).

The same result is obtained with the load resolution (projection) matrix

$$R_{mn} = F_m^i G_n^i = (1/B_1) \begin{bmatrix} B^2 & 0 & 0 & B \\ 0 & B_1 & 0 & 0 \\ 0 & 0 & 0 & 0 \\ B & 0 & 0 & 1 \end{bmatrix}.$$

The modal parameters of the equilibrium load are

$$P_n^0 = R_{mn} P_m = (1/B_1) \, [B(BP_1 + P_4), \ B_1 P_2, \ 0, \ BP_1 + P_4]^{\mathrm{T}}.$$

With this load, the discretized equilibrium equations, whose matrix is the transpose of (18), are compatible and yield the above modal constraint reactions.

Example 5.2. A shallow inextensible cable is supporting a lopsided vertical load shown in Fig. 5a. According to (21), the amplitudes of the modal load components are

$$P_1 = P_3 = 0, \qquad P_2 = (2 + p) \, l/\pi, \qquad P_4 = -pl/\pi. \tag{a}$$

Substituting these values into (22) gives

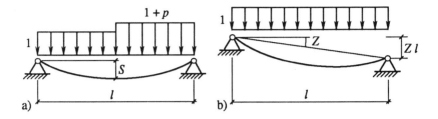

Figure 5.5. Examples of cables analyzed as systems with two degrees of freedom: a) cable supporting a lopsided load; b) cable with uneven level supports.

$$H_0 = 2B(2 + p)/\pi^2, \qquad H_1 = -2Bpl/\pi^2 B_1. \qquad (b)$$

The cable tension obtained from here for the case of a uniform load $(p = 0)$ is

$$H_0 = 4l^2/\pi^3 S \approx l^2/7.75 S. \qquad (c)$$

This is very close to the exact solution, $H = l^2/8S$, for the actual (parabolic) equilibrium profile, corresponding to a uniform vertical load.

For a shallow inclined cable, i.e., one with uneven supports (Fig. 5b), a uniform vertical load P produces components parallel and perpendicular to the chord. The corresponding modal load parameters are

$$P_1 = 2PZl/\pi, \qquad P_2 = 2Pl/\pi, \qquad P_3 = P_4 = 0. \qquad (d)$$

From (25),

$$H_0 = 4PBl/\pi^2, \qquad H_1 = 4PZB^2l/\pi^2 B_1. \qquad (e)$$

Obviously, the above solutions are meaningful only when the perturbation load component is small; otherwise, the equilibrium profile of the cable must be refined by evaluating the displacements.

Evaluation of Displacements. Evaluation of displacements caused by a perturbation load calls for the construction of the statical-kinematic stiffness matrix which, in turn, requires variation of equilibrium equations. For a continuous system, these are differential equations, and their variation is carried out prior to discretization. In this analysis, convected coordinates are used, implying a fixed association with material points of the cable. Therefore, displacements u and w are treated as the respective variations of the independent, x, and dependent, z, coordinates:

$$u = \delta x, \qquad w = \delta z.$$

The operations of variation and differentiation are commutative, $\delta d = d\delta$; hence

$$\delta\, dx = u'\, dx, \qquad \delta\, dz = w'\, dx, \qquad \delta z' = w' - z'u',$$

$$\delta(dz') = d(\delta z') = (w'' - z''u' - z'u'')\, dx.$$

These relations are used in varying the equilibrium equations (19):

$$h' - H'u' = p_x,$$

$$(hz')' + H'(w' - 2z'u') + H(w'' - 2z''u' - z'u'') = p_z.$$

After replacing the displacements and forces with their expansions (16) and (20), this system is discretized by orthogonalizing it to the two base functions. The matrix of the resulting equations is supplemented with the modal Jacobian matrix (18) to form the closed system of tangent statical-kinematic equations presented in the following table:

u_1	w_1	u_2	w_2	h_0	h_1	RHS
$\dfrac{2\pi H_1}{3l}$	0	0	0	0	$-\dfrac{\pi}{2}$	p_1
0	0	$\dfrac{8\pi H_1}{15l}$	0	0	0	p_2
$\dfrac{5\pi^3 H_1 S}{8l^2}$	$-\dfrac{\pi^2 H_0}{2l}$	0	$-\dfrac{\pi^2 H_1}{2l}$	$-\dfrac{\pi^2 S}{2l}$	0	p_3
$\dfrac{3\pi^3 H_0 S}{4l^2}$	$\dfrac{\pi^2 H_1}{2l}$	0	$-\dfrac{2\pi^2 H_0}{l}$	0	$-\dfrac{\pi^2 S}{2l}$	p_4
0	$\dfrac{\pi^2 S}{2l}$	0	0	0	0	0
$\dfrac{\pi}{2}$	0	0	$\dfrac{\pi^2 S}{2l}$	0	0	0

The elimination of dependent displacements from these equations yields a system in two independent displacements with a singular 2×2 matrix. This is another consequence of the modal displacement u_2 and load P_3 being related to a higher-order symmetric deformation, not represented in the current two-degree-of-freedom model. In fact, both u_2 and P_3 should have been excluded from the analysis. They were retained, first, to show that this inconsistency, perhaps not obvious in advance, is not detrimental to the model, and second, because a more accurate model would involve the obtained matrices.

With only one unknown independent displacement, $x_4 = w_2$, one row and one column must be omitted from the obtained singular matrix, producing the sought 1×1 reduced stiffness matrix

$$K_{44} = 2\pi H_1/3lB^2 - \pi^2 H_0(3/4B^2 + 2)/l. \tag{5.23}$$

Evaluating the right-hand side of the resulting equilibrium equation requires the matrix (1.7) expressing all displacements in terms of the independent one, x_4:

$$a_{n4} = [-1/B, \quad 0, \quad 0, \quad 1]^{\mathrm{T}}.$$

From here, in accordance with (4.15),

$$a_{n4} P_n = - P_1/B + P_4. \tag{5.24}$$

Note that both of the load components present are antisymmetric, meaning that the symmetric loads do not produce any displacements. This is consistent with the above observation on the retained number of the degrees of freedom and their pattern.

Example 5.2 (continued). For the cable in Fig. 5a, assuming $p/2 \ll 1$, the substitution of (b) into (23) yields

$$K_{44} = - (3 + 8B) = - (3 + 8l/\pi S). \tag{f}$$

The right-hand side is found from (24) and (a)

$$a_{n4} P_n = - pl/\pi. \tag{g}$$

From here, the unknown independent displacement w_2 is found and then the dependent displacements are evaluated

$$w_2 = pl/\pi(3 + 8l/\pi S), \quad u_1 = - \pi S w_2/l, \quad w_1 = 0. \tag{h}$$

As expected, modal displacement u_2 remains undetermined.

For the slightly tilted cable with $Z \ll 1$ (Fig. 5b), K_{44} coincides with (f). Inserting (d) into (24) gives

$$a_{n4} P_n = - 2PZS, \tag{i}$$

and the resulting displacements are obtained from (h) following an obvious replacement of (g) with (i).

Stiffness Matrix for a Shallow Cable

As a geometrically nonlinear element of a structural system, a cable can be characterized by a tangent stiffness matrix quantifying the cable response to small displacements of its ends. For an inextensible cable, the resulting stiffness matrix is statical-kinematic, otherwise it accounts for the elastic properties as well. Here the stiffness matrix will be evaluated for a shallow elastic cable supporting an arbitrary vertical load. This matrix, describing the cable response to unit displacements of its ends in the coordinate directions, can be transformed to account for the end displacements in any direction.

As with any underconstrained system, statical analysis of a cable can be separated into two stages. The first is shape finding, i.e., a statical-kinematic analysis whereby the cable is treated as inextensible and its equilibrium configuration under a given load is established. When the cable can be assumed inextensible, evaluation of the tension force in the found configuration completes the analysis. Otherwise, this solution provides a reference state for the second stage of the analysis, accounting for the cable extensibility.

Differential Equations of a Shallow Elastic Cable. The first, statical-kinematic, stage of the analysis proceeds from the condition of equilibrium of a cable element in the vertical direction:

$$z''H + P(x) = 0, \tag{5.25}$$

where the prime designates a derivative with respect to x and $P(x) = P_z \approx P_n$ since the cable is shallow. Note that for a shallow cable differentiation with respect to the cable arclength is identical to differentiation over the length of the chord connecting the ends of the cable (coordinate x). Integration yields

$$z' = -\frac{1}{H}\int_0^x P\,dx + C_1, \qquad z = -\frac{1}{H}\int_0^x dx \int_0^x P\,dx + C_1 x + C_2.$$

In the chosen coordinate system, $z(0) = 0$ and $z(l) = Zl$, so that

$$C_2 = 0, \qquad C_1 = \beta = V_0/H + Z.$$

Here $\beta = \beta_c(0)$ is the cable slope at $x = 0$ and V_0 is the support reaction which would be produced in a simply supported beam of the span l by the load $P(x)$. As a result,

$$z' = V/H + Z, \qquad z = M/H + Zx,$$

where $V = V(x)$ and $M = M(x)$ are, respectively, the beam shearing force and bending moment, which are assumed known.

This solution will be complete provided either the thrust H or one of the geometric parameters of the equilibrium configuration is known. The most convenient geometric parameter is

$$\Theta = \beta - Z = V_0/H. \tag{5.26}$$

Then

$$T \approx H = V_0/\Theta, \tag{5.27}$$

and assigning either H or Θ completes the first (shape-finding) stage of the analysis for a given load. Note that the cable length does not appear in this analysis and can be evaluated only after the shape has been established.

The second stage of the analysis, accounting for the elastic flexibility of the cable, employs a linearized formulation of the problem

$$T(z + w)'' + P(x) = 0, \tag{5.28}$$

$$u' - z''w = FT. \tag{5.29}$$

In these equations, the equilibrium configuration of the inextensible cable found at the first stage serves as a geometric reference for the small tangential, u, and normal, w, elastic displacements; $T \approx H$ is the cable tension which does not vary along the length; z'' approximates the cable curvature; and $F = 1/EA$ is the elastic flexibility of the cable. Equation (28) is the equilibrium condition in the vertical direction, while (29) equates the cable strains obtained, respectively, from the kinematic and constitutive relations.

Mutual displacements of the cable ends in the coordinate directions are implemented as boundary conditions for the above differential equations. When integrating equation (28), the displacement at the near end of the cable is taken as $w(0) = z_0 - \beta x_0 \approx z_0$, while the far end is fixed, $w(l) = 0$. With these boundary conditions the solution is

$$w' = \vartheta V/V_0 - z_0/l, \quad w = \vartheta M/V_0 + z_0(1 - x/l),$$

where

$$\vartheta = V_0/T - \Theta. \tag{5.30}$$

The first-order equation (29) is solved with an initial condition $u(0) = x_0 + \beta z_0$. The integration is carried out taking into account (30), as well as the following identity from statics of a simple beam:

$$\int_0^l P\, M\, dx = -\int_0^l V^2\, dx.$$

Upon integration, the condition $u(l) = 0$ for the cable far end is implemented:

$$u(l)/l = FT - C\Theta\vartheta + x_0/l + Zz_0/l = 0. \tag{5.31}$$

The nondimensional constant C depends on the beam shearing force, $V(x)$:

$$C = \int_0^l V^2 dx / V_0^2 l.$$

As is readily seen, C is determined by the load pattern but not magnitude (intensity), thus being a geometric characterization of the load. The values of C for several common load patterns are presented in Table 5.1.

A closed system of equations for a linearized analysis of shallow elastic cables is obtained by incrementing equations (27) and (31). Its matrix form is

$$\begin{bmatrix} F & -C\Theta \\ \Theta & T \end{bmatrix} \begin{bmatrix} t \\ \vartheta \end{bmatrix} = \begin{bmatrix} -fT - (x_0 + Zz_0)/l \\ 0 \end{bmatrix}. \tag{5.32}$$

Here, as before, the lowercase letters designate increments of the quantities denoted by the respective capital letters. The horizontal and vertical components of force increment t, due to the unit displacements $x_0 = -1$ and $z_0 = -1$ of the near end of the cable, are the elements of the sought tangent stiffness matrix. The unit displacement signs are chosen so as to produce positive t (tension). This matrix, obtained by solving equations (32), is

$$K = \frac{T}{l(C\Theta^2 + \varepsilon)} \begin{bmatrix} 1 & Z \\ Z & \beta Z \end{bmatrix}, \tag{5.33}$$

where $\varepsilon = FT$ is the current elastic strain of the cable.

Table 5.1. Nondimensional Parameter C for Several Types of Load.

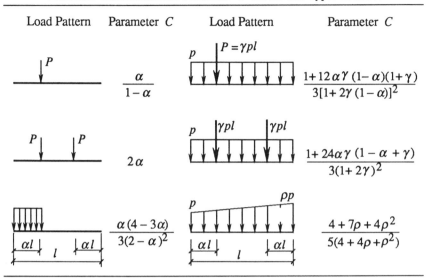

Load Pattern	Parameter C	Load Pattern	Parameter C
P	$\dfrac{\alpha}{1-\alpha}$	p, $P=\gamma pl$	$\dfrac{1+12\alpha\gamma(1-\alpha)(1+\gamma)}{3[1+2\gamma(1-\alpha)]^2}$
P P	2α	p, γpl γpl	$\dfrac{1+24\alpha\gamma(1-\alpha+\gamma)}{3(1+2\gamma)^2}$
αl, l, αl	$\dfrac{\alpha(4-3\alpha)}{3(2-\alpha)^2}$	p, ρp, αl, l, αl	$\dfrac{4+7\rho+4\rho^2}{5(4+4\rho+\rho^2)}$

For an inextensible cable, $\varepsilon = 0$, and the stiffness matrix is statical-kinematic. For a rectilinear cable, the transverse load must be absent, resulting in $\beta = Z$ and $\Theta = 0$. Then the common factor in (33) becomes EA/l, producing the familiar elastic stiffness matrix for a truss bar. Comparison of $C\Theta^2$ with ε provides an immediate assessment of the relative roles of the statical-kinematic and elastic aspects of the cable stiffness.

Example 5.3. When analyzing the system in Fig. 6, the simplest and least accurate approach is to assume all of the system components inextensible. The system is statically determinate and, given the geometry of the equilibrium configuration, all the forces are obtained from elementary statics. Under a uniform load w, the central cable profile is a quadratic parabola with

$$Z = 0, \quad \Theta = \beta = V_0/H = wl/2H. \tag{a}$$

Either the end slope, Θ, or the thrust, H, in equation (a) can be assigned arbitrarily (within the assumption of shallowness), and then the forces in the column and the guy cable are easily found:

$$N = H \tan\varphi, \quad T_g = H/\cos\varphi. \tag{b}$$

The first refinement of this analysis accounts for the elastic extension of the central cable. This is done assuming both Θ and $T \approx H$ known and setting the displacements $x_0 = z_0 = 0$. After taking the initial cable flexibility as $F = 0$ (the cable has been considered inextensible) and the incremental flexibility as $f = 1/EA$, the system of equations (32) yields

$$t = -fT^2/C\Theta^2 = -\varepsilon T/C\Theta^2, \quad \vartheta = -\Theta t/T = \varepsilon/C\Theta. \tag{c}$$

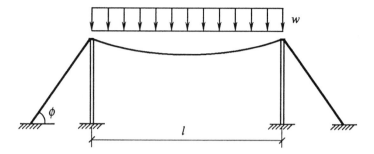

Figure 5.6. A cable interacting with other structural components.

From Table 1, for a uniform load, $C = 1/3$. The refined solution is obtained by incrementing the tension force and the geometric parameter Θ:

$$T^1 = T + t, \quad \Theta^1 = \Theta + \vartheta, \tag{d}$$

with the ensuing changes in the column and guy-cable forces. The increments in (d) must be sufficiently small to justify the one-step linearization.

In another incremental refinement, the flexibility of the guy cables is accounted for, while the columns, for simplicity, are still assumed rigid. The resulting horizontal stiffness of supports, K_s, and the stiffness of the main cable, K_c, obtained from (32) are

$$K_s = EA_g \cos^2 \varphi, \quad K_c = T^1/l \, [C(\Theta^1)^2 + \varepsilon^1], \tag{e}$$

where $\varepsilon^1 = T^1/EA$. The incremental displacement x_0 of each support is assumed sufficiently small; it is found from the equilibrium equation

$$(K_s + K_c) \, x_0 = T^1. \tag{f}$$

The resulting tension level in the central cable is

$$T = T^1 + t = T^1[1 - K_c/(K_s + K_c)]. \tag{g}$$

5.3 Radial Cable Systems

With the tangent stiffness matrix for a cable available, global stiffness matrices for a variety of cable systems can be assembled. As a result, a variant or a quasi-variant cable system (not necessarily prestressed) can be analyzed by means of multistep incremental linearization, like any other geometrically nonlinear structure. This section is concerned with three-dimensional radial cable systems, involving one or two arrays of cables radiating from a central node (in reality, a small central ring or hub).

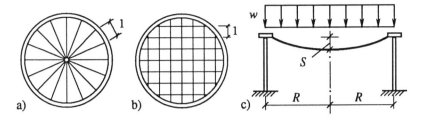

Figure 5.7. A comparison of a radial cable system with an orthogonal cable net:
a) radial system; b) orthogonal net; c) geometric parameters of both systems.

Comparison with a Cable Net. Structural behavior of radial cable systems of various types turned out to be surprisingly interesting and favorable. This can be illustrated by comparing a radial cable system and a cable net of a circular planform of the same radius R and sag S (Fig. 7).

In both systems the cables are attached to a circular contour ring continuously supported in the vertical direction. Under a uniform vertical load per unit horizontal area, the cables of the radial system form a surface of revolution with a cubic parabola $z_1 = Sr^3/R^3$ as meridian. The cable net forms a paraboloid of revolution of second order, which is at the same time a surface of translation with identical generators, $z_2 = Sx^2/R^2$. The respective cable-end slopes for the two systems are

$$\beta_1 = \Theta_1 = z_1'(R) = 3S/R, \quad \beta_2 = \Theta_2 = z_2'(R) = 2S/R.$$

Given a unit spacing of radial cables along the contour ring, and a unit spacing of parallel cables in the two arrays of the cable net (Fig. 7), the corresponding beam shearing forces are

$$V_0^{(1)} = wR/2, \quad V_0^{(2)} = wR/4.$$

Employing formula (27) for shallow cables gives the respective cable forces in the two systems

$$T_1 = wR^2/6S, \quad T_2 = wR^2/4S. \tag{5.34}$$

Although the force T_2 has been evaluated for the middle (longest) cable of the net, it can be easily verified that the ratio V_0/β_2, and hence the cable force, is the same for all of the cables. For the purpose of comparison, the lengths of shallow cables can be identified with their spans; then the required total cable length for each system is easily evaluated and is found to be the same. Thus, judging by (34), other things being equal, the radial cable system is 50% more efficient than the orthogonal net; in other words, for a given load, the radial system is that much lighter, and for a given weight, it supports that much more load. Furthermore, this advantage spills over to the support structures of the two systems: the horizontal forces transmitted onto the respective two contour rings also differ by the same factor.

One-Layer Radial Cable Systems

Basic Equations. Consider a shallow radial cable system of an arbitrary planform (Fig. 8) referred to a cylindrical coordinate system, z, r, φ. With the coordinate origin at the central node, a unit sector $d\varphi = 1$ contains a certain constant number of cables. Due to the system shallowness, normal and vertical surface loads are indistinguishable and, in the absence of tangential loads, the cable forces in such a system do not vary along the length: $T = T(\varphi)$. Because of that, equations (28) and (29) are valid and can be solved as ordinary differential equations, with the polar angle φ as the only independent variable. Among the boundary conditions, the one at the near end of a typical cable undergoes an obvious modification

$$u(0)/R = x \cos \varphi + y \sin \varphi + \beta z.$$

Here and below, the parameters x, y, and z, which are no longer necessary as Cartesian coordinates, designate the normalized displacements of the central node: $x = x_0/R$, $y = y_0/R$, and $z = z_0/R$. The obtained solution, after being linearized, differs only slightly from (32). It must be complemented by three integral conditions of equilibrium of the central node:

$$I[T(\varphi)\cos\varphi] = 0, \quad I[T(\varphi)\sin\varphi] = 0, \quad P_0 + I[V_0 - (Z + z)T] = 0,$$

where P_0 is an external load at the node and

$$I() = \int_0^{2\pi} ( )\, d\varphi, \tag{5.35}$$

The incremental version of the resulting system of five simultaneous equations can be presented in table form:

t	ϑ	z	x	y	RHS
F	$-C\Theta$	Z	$\cos\varphi$	$\sin\varphi$	$-fT$
$C\Theta$	CT	0	0	0	Cv_0
IZ	0	IT	0	0	q_0
$-I\cos\varphi$	0	0	0	0	0
$-I\sin\varphi$	0	0	0	0	0

$$\tag{5.36}$$

Here RHS designates the right-hand side column and $q_0 = p_0 + I(v_0)$ is the incremental load at the central node. Only the quantities figuring in the first two equations, including the cumulative magnitudes of parameters Θ and Z, are functions of φ, while the rest of the matrix elements are numbers. The second row, representing the incremental version of equation (30), is multiplied by C

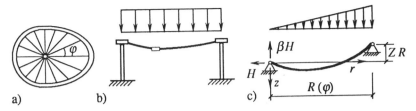

Figure 5.8. A shallow, one-layer radial cable system: a) and b) the system geometry; c) a typical cable and its load.

to reveal the matrix symmetry, which is a property of any linear or linearized system of equations describing a conservative structural problem.

Since the vector of unknowns involves both forces and displacements, the matrix exhibits the kind of symmetry found in the matrices of the mixed method of structural mechanics (i.e., antisymmetry between elements relating to forces and displacements). Somewhat unusual is the symmetry between the matrix elements of the first two rows and the symmetrically located integrands of the first two columns. This reflects the fact that the first two unknowns, t and ϑ, are functions of φ, whereas the last two, x and y, are parameters. Accordingly, in solving the system (36), the increments t and ϑ are expressed in terms of displacements using the first two equations and then introduced into the last three equations, from which the displacements are evaluated.

In many particular cases, equations (36) simplify considerably. For example, in the presence of one or two vertical planes of symmetry, the number of degrees of freedom of the central node reduces respectively and so does the number of simultaneous equations to be solved. For an axisymmetric circular system, all variables are independent of φ, reducing the problem to that for a single cable.

Radial Cable Systems of Circular Planform.   These systems exhibit remarkable behavior in supporting a general load. The most important type of an asymmetric live load is one involving a component proportionate to $\cos \varphi$. As is seen from the first two of equations (36), such a load, superimposed on a present symmetric load, produces a cable force increment of the form

$$t = t_0 + t_1 \cos \varphi.$$

By virtue of the fourth equation in (36),

$$I\,(t \cos \varphi) = 0, \tag{5.37}$$

expressing equilibrium in the x direction, the above coefficient t_1 turns into zero, and the force increment t does not depend on the polar angle.

Thus, in this approximation, a circular system, supporting a combination of an axisymmetric and a cosine load, has identical tension forces in all of its cables. As a result, the support ring remains momentless in its plane, which is an important advantage, taking into account the large magnitudes of cable forces

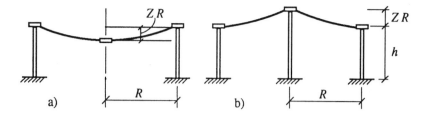

Figure 5.9. Shallow systems of a circular planform: a) concave; b) tent-shaped.

in shallow systems (Fig. 9a). This conclusion is, of course, only qualitative, since it does not reflect the nonlinear behavior of the system. However, a more accurate incremental analysis shows that the described benign behavior deteriorates only slowly in the course of loading. The reason is that the condition of horizontal equilibrium of the central node of the system, in conjunction with kinematic mobility of the cables, always entails a favorable distribution of the cable forces.

Under a similar loading, the cable force distribution in a tent-shaped radial system (Fig. 9b) is different, as is seen from the matrix equation for this system:

$$
\begin{bmatrix}
F & -C\Theta & \cos\varphi \\
C\Theta & CT & 0 \\
-I\cos\varphi & 0 & 3EI_c R/h^3
\end{bmatrix}
\begin{bmatrix}
t \\
\vartheta \\
x
\end{bmatrix}
=
\begin{bmatrix}
-ft \\
Cv_0 \\
0
\end{bmatrix}.
\tag{5.38}
$$

The equation is obtained from (36) by noticing that $z = y = 0$ and introducing the bending stiffness of the central column, EI_c. It is this additional term that, by altering equation (37), adversely affects the distribution of cable forces. As a result, in this cable system, a vertical load produces a horizontal interaction between the central and peripheral supports of the system. Thereby, large bending moments are induced in both the central and periferal columns, as well as in the contour ring. This can be avoided by either allowing the central column to tilt freely or by assuring horizontal mobility of the contour ring, which might be appropriate if the cables are sufficiently taut.

Two-Layer Radial Systems

Systems with Cable Layers Connected Only at the Center. In two-layer radial systems of the type shown in Fig. 10, the inner rings are either connected by a vertical hub or, ultimately, joined together, forming one ring. By taking into account the type of the ring connection, the basic equations obtained for a one-layer system can be adapted for the analysis of two-layer systems. It is assumed for simplicity that the system to be analyzed, including

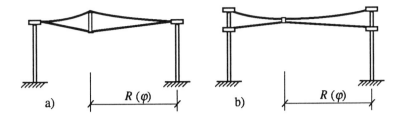

Figure 5.10. Two-layer radial cable systems with cables connected only at the center: a) with separate inner rings; and b) with one inner ring.

external loads, has at least one vertical plane of symmetry, and the bottom layer is load-free. This leads to $y_1 = y_2 = x_2 = \Theta_2 = 0$, where the subscripts 1 and 2 relate to the two cable layers. The resulting system of incremental equations based on (36) is:

t_1	t_2	ϑ_1	z	x_1	RHS
F_1	0	$-C\Theta_1$	z_1	$\cos\varphi$	$-f_1T_1$
0	F_2	0	$-Z_2$	0	$-f_2T_2$
$C\Theta_1$	0	CT_1	0	0	Cv_0 (5.39)
$-IZ_1$	IZ_2	0	IT	0	q_0
$-I\cos\varphi$	0	0	0	0	0

where $T = T_1 + T_2$ and the geometric parameters Θ_1, Z_1, and Z_2 represent, as before, the state of the system serving as a reference state for the current incremental step. Note that, depending on the geometric and stiffness parameters of the system, an incremental vertical load may cause the cable tension T_1 in the upper layer either to increase or decrease.

It is not difficult to see that the favorable behavior under a cosine live load, revealed earlier for a one-layer system, is also observed in the two-layer system under consideration. However, this is not so for the system in Fig. 10b, where the two cable layers share the inner ring and its displacements. Here $x_1 = x_2 = x$ and the condition of the ring equilibrium in the horizontal direction becomes

$$I(T\cos\varphi) = 0.$$

Thus, although the sum of the two horizontal thrust patterns transmitted by the two cable layers onto the contour rings is self-balanced, each pattern taken separately may be not self-balanced. As a result, the peripheral columns are subjected to large horizontal forces, and the contour rings to the corresponding concentrated reactions. This unfavorable effect is especially strong under a

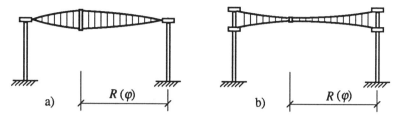

Figure 5.11. Two-layer radial cable systems with continuously connected layers: a) with two inner rings; b) with one inner ring.

cosine external load, the very load supported so favorably by radial systems with horizontally unrestrained inner rings. Therefore, for the system in Fig. 10b, it might be advisable to have two separate inner rings (perhaps, closely spaced) with a flexible connection not restraining their mutual horizontal displacement. This will make the horizontal forces in the cables of each layer self-balanced and, in the case of a circular system, uniformly distributed. Note that in systems of this type, tension in the bottom cables reduces in the course of loading, ultimately leading to their possible disengagement.

Systems with Continuously Connected Cable Layers. In such systems (Fig. 11), external loads produce an increase in tension forces in one layer and a reduction in the other. For this reason, the system must be prestressed to such a level as to preclude cable disengagement in the unloading layer. The initial forces, T_1^i and T_2^i, figure in the incremental matrix equation which is constructed on the basis of equation (36), assuming the presence of a vertical plane of symmetry.

A few differences in this matrix, as compared with the one for the previously considered systems, are due to the opposing curvatures and continuous connections of the two cable layers. This is reflected in the partial sign reversals in the first equation and in that the parameters Θ_1 and Θ_2 are incremented as $\Theta_1 - \vartheta$ and $\Theta_2 + \vartheta$, where $\vartheta = \vartheta_1 = \vartheta_2$ is one of the unknowns:

t_1	t_2	ϑ_1	z	x_2	x_1	RHS
F_1	0	$C\Theta_1$	Z_1	0	$\cos\varphi$	$-f_1(T_1 - T_1^i)$
0	F_2	$-C\Theta_2$	$-Z_2$	$\cos\varphi$	0	$-f_2(T_2 - T_2^i)$
$-C\Theta_1$	$C\Theta_2$	CT	0	0	0	Cv_0
$-IZ_1$	IZ_2	0	IT	0	0	q_0
0	$-I\cos\varphi$	0	0	0	0	0
$-I\cos\varphi$	0	0	0	0	0	0

$$(5.40)$$

As could be expected, systems of the types shown in Fig. 11a and b with a circular planform exhibit the same difference in behavior under a cosine load as their respective counterparts in Fig. 10a and b.

Design Considerations. The above incremental equations can be used for a direct design of radial cable systems. Selecting a suitable configuration for the system state under the full load allows the maximum cable forces and the required cross sections to be determined at the first step of the analysis. For a one-layer system, it only remains to evaluate the resulting initial cable lengths. For a two-layer system under full load, the forces in the prestressing cables can be assigned near to or at zero. Then the load capacity of the system is exhausted simultaneously with the cable disengagement—a combination of two failure modes. With the sections of load-carrying cables determined, the problem reduces to the evaluation of the load-free, prestressed state of the system and the required cross sections of the prestressing cables. In an incremental analysis, the prestressing cable sections are initially given large areas, so that the removal of the entire live load produces an acceptably small displacement. In the following steps, their sections are gradually reduced by incrementing the flexibility parameter F_1, until an acceptable state of prestress is reached.

Setting the stage for the analysis of a contour ring in the following section, note that a radial cable system under a given vertical load can always be designed so as to attain any desired self-balanced pattern of cable thrusts, $H(\varphi)$. This is an especially easy task for a one-layer system, taking into account equations (26) and (27). For a radial system of an arbitrary planform given by $R = R(\varphi)$, the cable slope, $\beta(\varphi)$, at the central node is

$$\beta(\varphi) = \theta - Z = V_0/H - S/R.$$

Example 5.4. Evaluate the geometric parameters of a one-layer radial system of an elliptic planform with the central node at a focus, such that under a uniform vertical load, w, and a given sag, S, the thrust $H = H(\varphi) = \text{const.}$ (As will be shown shortly this makes the elliptic contour ring momentless.)

The polar equation of an ellipse with semiaxes a and b is

$$R = p/(1 + e \cos \varphi), \quad p = b^2/a, \quad e = \sqrt{a^2 - b^2}/a. \tag{a}$$

Vertical equilibrium of the node requires [cf. (35)]

$$I\,[H\beta(\varphi)] = I\,(wR^2/6 - HS/R) = w\,\pi ab/3 - 2H\pi S/p^2 = 0. \tag{b}$$

Upon evaluating H, the cable slope, $\beta(\varphi)$, at the central node is found:

$$H = wp^2 ab/6S, \quad \beta(\varphi) = Z\,(R^3/b^3 - 1). \tag{c}$$

Since $\beta = 0$ at $R = b$, the thrust H in this system is the same as in a radial cable system of a circular form, with sag S and radius b.

The required geometry of the system is implemented by statical means (by adjusting the cable tensions under a uniform load).

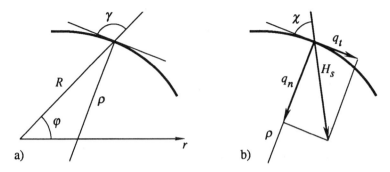

Figure 5.12. Equilibrium of an infinitesimal element of a curvilinear bar subjected to:
a) a force field of an arbitrary given direction; b) a radial force field.

5.4 Contour Ring

A cable assembly supporting a transverse surface load produces a self-balanced
system of thrust forces in the plane perpendicular to the load. For a shallow
assembly, the thrust forces are large and are best supported by a closed contour
ring having the form of the corresponding pressure curve. This poses a problem
of finding the load pattern making a given ring momentless and, conversely,
determining the pressure curve for a given force field. In the context of cable
systems, this old problem is complicated by the fact that the load pattern is the
cable thrust which, in turn, is a function of the ring shape. The resulting
coupled problem is addressed here in application to a radial cable system of
arbitrary shape.

Statics and Geometry of a Momentless Ring

Equilibrium Without Bending Moments. In the absence of bending and
shear, equilibrium conditions for an infinitesimal element of a curvilinear bar are
identical (after sign reversal) to those for a cable:

$$dN = P_t\, ds, \quad \sigma N = P_n, \tag{5.41}$$

where N is the axial compression force and σ is the curvature of the ring axis.
By eliminating N, a differential relation between the normal and tangential load
components for a momentless ring is obtained

$$d\,(P_n/\sigma)/ds = P_t. \tag{5.42}$$

The thrust H_s per unit arclength of the ring axis is considered as a force field
forming a given angle $\gamma(s)$ with the positive direction of the tangent to the
axis (Fig. 12a). Then

$$P_n = H_s \sin \gamma, \quad P_t = -H_s \cos \gamma, \tag{5.43}$$

which, upon substitution into (42), leads to a first-order ordinary differential equation in the unknown H_s. Its solution is

$$H_s \sin \gamma = C\sigma \exp\left(-\int \sigma \operatorname{ctn} \gamma \, ds\right). \tag{5.44}$$

For free-body equilibrium of the ring, it is necessary that the resultant vector and moment of the load H_s over the entire ring are zero. It can be shown that the force resultant of the load given by (44) turns into zero identically for any closed ring, whereas the moment resultant turns into zero only for some force fields. Such are, for example, central (radial) and hydrostatic (normal pressure) fields. In the latter case, $\gamma = \pi/2$, and $P_t = 0$, leading to an obvious outcome $H_s = \sigma N$.

Radial (Central) Force Field. This is the most interesting case for applications. If the field center is chosen as the origin of the polar coordinate system, the angle γ becomes the angle formed by the radius-vector, $R = R(\varphi)$, of the ring axis and the positive direction of the tangent (Fig. 12b). For an arbitrary plane curve given by its radius-vector, the following differential-geometric relations hold:

$$\sigma = (R^2 + 2R'^2 - RR'')/s'^3, \quad s' = (R^2 + R'^2)^{1/2},$$
$$R = s' \sin \gamma, \qquad R' = s' \cos \gamma, \tag{5.45}$$

where prime denotes differentiation over the polar angle φ. With the aid of these relations, the integral in (44) is evaluated in a closed form:

$$\int \sigma \operatorname{ctn} \gamma \, ds = \ln (R \sin \gamma) + \ln C.$$

This leads to a simple general formula for a radial force field rendering momentless a closed ring of an arbitrary shape:

$$H_s = C\sigma/R \sin^2 \gamma. \tag{5.46}$$

In the obtained solution, one of the geometric parameters, the curvature σ, is invariant, whereas the other two, R and γ, depend on the position of the ring relative to the center of the force field. Taking into account (41) and (43), the compression force in the ring is found

$$N = C/R \sin \gamma. \tag{5.47}$$

From here, the mechanical meaning of the constant C is established:

$$C = NR \sin \gamma = M.$$

Thus, C is the moment of the axial force in any section of the ring about the center of the radial field. For a ring to be momentless, the value of this moment must not vary along the ring axis.

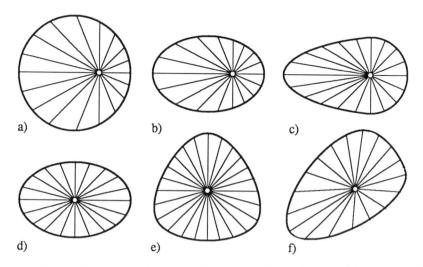

Figure 5.13. Momentless rings in radial force fields: a) – c) with one axis of symmetry; d) with two axes; e) with three axes; and f) with no axes of symmetry.

For a radial force field, it is natural to relate the field forces to a unit polar angle instead of the unit length of the ring axis:

$$H = H(\varphi) = H_s ds/d\varphi = C\sigma/\sin^3 \gamma \qquad (5.48)$$

or, taking advantage of (45),

$$H = C(R^2 + 2R'^2 - RR'')/R^3. \qquad (5.49)$$

These results apply to a contour ring of an arbitrary shape, with an arbitrary location of the force field center. The presence or absence of planes of symmetry and their number is irrelevant (Fig. 13). Furthermore, the ring axis need not necessarily be a smooth curve, and may be a polygon with either straight or curved sides. The load pattern at a polygon vertex can be analyzed by assuming two adjacent sides being connected by a smooth transition circular arc. If the arc radius and length reduce while preserving the subtended angle, the load over the arc becomes in the limit a concentrated force. It is evaluated using equations (48) and (45)$_3$:

$$H_v = \lim_{\substack{r \to 0 \\ \Delta\varphi \to 0}} \int_{\varphi}^{\varphi + \Delta\varphi} (C\sigma/\sin^3 \gamma)\, d\varphi = C(\text{ctn } \gamma_1 - \text{ctn } \gamma_2)/R, \qquad (5.50)$$

where γ_1 and γ_2 are the respective values of γ before and after the vertex. As follows from (48), in the particular case of a polygon with rectilinear sides, the distributed load along the sides must be absent and only the concentrated vertex forces remain.

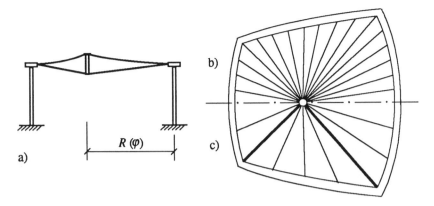

Figure 5.14. A two-layer radial cable system with a momentless support ring: a) the system geometry; b) upper layer layout; c) bottom layer layout.

As demonstrated in the previous section, a radial cable system with a given load can be designed so as to realize any desired pattern of thrust $H(\gamma)$. Using this result, in conjunction with the solution for a polygon ring, proves the feasibility of a radial cable system with a momentless contour ring of arbitrary shape. As an example, Fig. 14 shows a two-layer radial cable system with a contour ring in the form of a curvilinear trapezoid. The concentrated vertex forces, necessary for making the ring momentless, are generated in a special lower-layer cable layout (Fig. 14c), with four heavy cables anchored at the vertices.

Momentless Ring for a Given Radial Force Field. The obtained results also facilitate the solution of a reverse problem: for a given radial force field $H(\varphi)$, determine the shape of a momentless ring, $R(\varphi)$. Note, first of all, that there is no one-to-one correspondence between the force field and the ring shape: for a given ring shape, $R(\varphi)$, there is a unique $H(\gamma)$ given by (49), whereas, for a given force field, there exists a family of momentless ring shapes.

The simplest radial cable system is one where, under a uniform surface load w, each cable has a horizontal tangent at the central node. For such a system,

$$H = wR^3/6S.$$

Substituting this into (49) and lumping all constants into a constant C yields

$$RR'' - 2R'^2 - R^2 + CR^6 = 0.$$

Somewhat surprisingly, this nonlinear equation admits a closed-form solution:

$$R(\varphi) = 1/\sqrt{(\cos^2\varphi)/a^2 \pm (\sin^2\varphi)/b^2}. \tag{5.51}$$

Here a and b are arbitrary constants and the double sign corresponds to two distinct geometric configurations: an ellipse and a hyperbola with semiaxes a

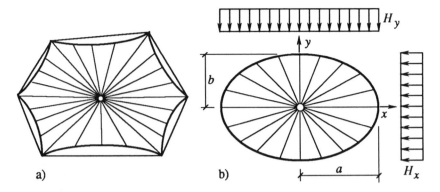

Figure 5.15. Momentless support rings of a radial system: a) an assembly of anchored hyperbolic segments (tension);   b) an ellipse (compression).

and b (Fig. 15). Obviously, a contour ring comprised of hyperbolic sections is in tension, and is feasible only when the vertices are supported in some way. Thus an ellipse represents the only possible shape of a compressed momentless ring for a radial cable system of the described type. The Cartesian projections of the cable forces found from (50) and (51) are

$$H_x = H(\partial\varphi/\partial y)\cos\varphi = wa^2/6S, \quad H_y = wb^2/6S,$$

and satisfy the known condition for the absence of bending moments in an elliptic ring. Furthermore, comparison with an orthogonal net of a similar geometry shows that the above-mentioned statical advantages of a radial layout, observed and evaluated for cable systems of a circular planform, are not limited to this form.

In the cable system just considered, the cable forces, hence, the required cross-section areas, vary from cable to cable. Therefore, it is desirable to explore the feasibility of a radial cable system where the cable forces do not vary with the polar angle, an obvious and highly advantageous property of a circular system. Assuming $H = \text{const.}$ in (49) gives

$$RR'' - 2R'^2 - R^2 + CR^3 = 0.$$

and another serendipitous closed-form solution:

$$R(\varphi) = p\,(1 + e\cos\varphi), \quad p = b^2/a, \quad e = \sqrt{a^2 - b^2}/a.$$

It describes a family of confocal ellipses with arbitrary semimajor, a, and semiminor, b, axes. (At $b > a$, it becomes a family of hyperbolas, in which case the ring degenerates into an assembly of hyperbolic segments.)

Thus, the only possible shape of a momentless ring for a radial force field of constant intensity per unit polar angle, is an ellipse with one of its foci at the field center (Fig. 16). Although the similarity between this statical fact and Kepler's law of planetary motion is striking, it appears to be purely coincidental.

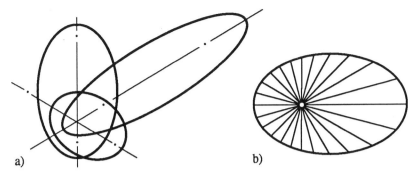

Figure 5.16. Momentless rings for a radial force field of constant intensity per unit polar angle: a) a family of ellipses; b) the system layout.

Interaction of the Radial Cable System with the Support Ring

Problem Formulation.   Elastic Foundation.   This nonlinear problem can be simplified taking the following two-step approach. First, the cable system with an undeformable support ring is analyzed and the cable forces are evaluated. Then, assuming the ring incompressible, its bending flexibility is accounted for at the second step of the analysis. At this stage, the system displacements usually are sufficiently small to justify incremental linearization. Accordingly, the response of the cable system is described by its tangent stiffness matrix, and the ring response by an appropriate linear model. Thus, in solving this nonlinear contact problem, nonlinearity is confined to the analysis of the assembly of cables, while the interaction problem per se becomes linear. For a radial cable system, this problem is rather straightforward, and for a circular system it admits a closed-form solution. This solution reveals some interesting effects in the bending and stability behavior of a circular support ring. Some of these features can be, at least qualitatively, extrapolated to support rings of cable systems of other shapes and types.

Consider a circular radial cable system under an axisymmetric transverse load. The cable forces H_s, evaluated in the assumption of a rigid ring, are supposed known. In the process of the ring deformation (and because of it), the load transferred by the cables onto the ring undergoes some changes which must be accounted for. The final value of the normal load component, P_n, acting on the ring, is evaluated as a power series in the radial displacement of the ring, v:

$$P_n = H_s + (dH_s/dv)|_0 \, v + \cdots \approx H_s - k_v \, v. \qquad (5.52)$$

Following the adopted linearized approach, only the first-order term is retained in the expansion. Since the load increment in (52) is proportionate to the corresponding displacement, it can be interpreted as a reaction of a linear elastic foundation with a modulus

$$k_v = - \, dH_s/dv \,|_{v=0} \, .$$

Note in passing that the columns supporting the contour ring can also be accounted for as horizontal elastic supports, perhaps, continuously distributed ("spread") along the perimeter.

Aside from a change in the magnitude of the load, the ring deformation also causes a change in the load direction relative to the tangent to the ring axis. To evaluate this change, note first of all that it is caused only by radial displacements of the ring. These can be represented as a finite or infinite Fourier series of the form $\Sigma\, v_n \cos n\varphi$. The series starts with $n = 2$, since the term with $n = 0$ would produce an axial compression of the ring, which is contrary to the assumption made, and the term with $n = 1$ corresponds to a free-body translation.

When a circular contour ring of a radial cable system undergoes small cyclic displacements with $n \geq 2$, the central node of the system remains equilibrated and does not move. Therefore, in the process of the ring deformation, the cable tension remains directed towards the center. As a result, rotation of the tangent to the ring axis due to bending produces a tangential load component

$$P_t = H_s v'. \tag{5.53}$$

In a similar way, it is possible to evaluate the changes that the ring deformation produces in the transverse load component, acting in the direction of the binormal to the ring axis. The original binormal load is the vertical component, V_s, of the cable forces. The final value of the load acting along the binormal to the deformed ring axis is

$$P_b \approx V_s - k_w w - H_s \tau,$$

where w is the out-of-plane displacement and τ is the ring torsion. The coefficient k_w (the second foundation modulus) is determined much like k_v:

$$k_w = - \, dV_s / dw \,|_{w=0}.$$

Evaluation of Foundation Moduli. Coefficients k_v and k_w for the particular radial cable systems are obtained with the aid of the tangent stiffness matrices formulated in the previous sections. For systems of the type shown in Fig. 11, the first two of equations (40) express the boundary conditions at the far end of the cable: $u_1(R) = u_2(R) = 0$. For the evaluation of k_v, these displacements should instead be given unit magnitudes and the resulting terms transferred to the right-hand side in these equations. Solving them together with the third of equations (40) (with a zero right-hand side) produces the two tension force increments per unit polar angle of the support ring; when combined, they yield the value of the first of the foundation moduli:

$$k_v = (t_1 + t_2)/R. \tag{5.54}$$

Similarly, k_w is evaluated by introducing a unit mutual vertical displacement of the cable ends, $zR = 1$, and the outcome is

$$k_w = (t_1 \beta_{1R} + t_2 \beta_{2R})/R, \qquad (5.55)$$

where

$$\beta_{1R} = -(Z_1 + 2\Theta_1), \quad \beta_{2R} = Z_2 + 2\Theta_2,$$

are the cable slopes at the contour.

For a circular two-layer system with cables connected only at the center, the calculation proceeds along the same lines, and is simplified by the fact that the bottom-layer cables are rectilinear. From equations (39), a solution for the cable force increments is obtained which, being substituted into (54) and (55), yields

$$k_v = (k_1 + EA_2/R)/R,$$

$$k_w = (Z_1 k_1 \beta_{1R} + Z_2^2 EA_2/R)/R.$$

Here

$$k_1 = T_1/(C \Theta_1^2 + \varepsilon_1)R$$

is a familiar parameter [cf. (33)] characterizing the tangent stiffness of a shallow cable and

$$\beta_{1R} = Z_1 - 2\Theta_1.$$

Bending and Stability of the Support Ring. The differential equation of the in-plane bending of an incompressible circular ring is

$$EI (u^{VI} + 2u^{IV} + u'') = \rho^3 [(\rho P_n)' - \rho P_t], \qquad (5.56)$$

where EI is the in-plane bending stiffness of the ring, u is the longitudinal displacement, ρ is the radius of curvature of the deformed axis, and the prime denotes differentiation with respect to the polar angle. For an incompressible ring, the relation between the tangential and normal displacements and the expression for the final radius of curvature are as follows:

$$v = u', \quad \rho = R - u''' - u'.$$

Introducing these relations along with the loads (52) and (53) into equation (56) and retaining only the first-order terms yields

$$EI u^{VI} + (2EI + H_s R^3) u^{IV} + (EI + 2H_s R^3 + k_v R^4) u'' = H_s' R^4. \quad (5.57)$$

At $H_s = $ const. (axisymmetric load) this equation becomes homogeneous and describes the in-plane buckling of the contour ring of a radial cable system.

The critical magnitude of the thrust is

$$H_s^{cr} = \frac{(n^2 - 1)^2 + c}{n^2 - 2} \frac{EI}{R^3} \qquad (c = k_v R^4/EI).$$

It attains a minimum at a wave number of the order of $c^{1/4}$ corresponding to

$$H_s^{cr} = 2[1 + (1 + c)^{1/2}]EI/R^3. \qquad (5.58)$$

The peculiarity of this result is that the foundation modulus k_v and, with it, the parameter c are increasing together with the load on the cable system, which is physically quite obvious. Thus, in a benign yet odd chain of relations, an increase in the external load leads to an increase not only in the system thrust, but also in the foundation modulus which, in turn, raises the critical value of the thrust. This effect is so strong that the resulting critical thrust magnitude is, as a rule, beyond the elastic limit of the ring material—a self-destructing feature of the presented solution. The described effect has been confirmed experimentally.

A similar favorable interaction between the contour ring and the cable system can be expected in a general compression-bending mode. As has been noted above, under nonuniform but smooth external loads, the cable forces vary little along the polar angle because of the force-leveling effect of the kinematic mobility of the central node. As a result, the variable parameters in equation (57) are almost constant. Replacing them with the respective average values,

$$\overline{H}_s = I\,(H_s)/2\pi, \quad \overline{k}_v = I\,(k_v)/2\pi,$$

produces an equation with constant coefficients. At the same time, assuming the presence of a vertical plane of symmetry, the function $H_s(\varphi)$ in the right-hand side of (57) can be presented as

$$H_s = H_n \cos n\varphi \quad (n \geq 2;\ \text{summation over } n). \qquad (5.59)$$

Then the solution has the form of a similar Fourier series; the amplitude values of the radial deflection, v_n, and the bending moment, M_n, are

$$v_n = H_n R^4/EI\,C(n), \quad M_n = (n-1)H_n R^2/C(n), \qquad (5.60)$$

where

$$C(n) = n^4 - (2 + \overline{H}_s R^3/EI)\,n^2 + 1 + 2(\overline{H}_s R^3 + \overline{k}_v R^4)/EI.$$

As was the case of in-plane buckling, the effect of the cable system on the bending of the contour ring is strong and favorable.

A similar analysis of the support ring for out-of-plane buckling or bending employs the second elastic foundation introduced above. However, the analysis shows that the effect of the interaction with the cable system in this direction, although quite tangible, is far less significant.

6
Nets

A net is an assembly of two noncoinciding arrays of linear elements (cables, wires, or fibers) located on one surface. Each element usually intersects only the elements of the other array, although singular points, like the pole in a radial-hoop system, may exist. Unless the intersections are fastened, mutual in-surface slipping (but not delamination) of the cables is assumed possible. Accordingly, the described model applies not only to cable or fiber nets proper, but also to two-ply fabrics with a conventional weave. In this case, the possibility or absence of fiber slipping are attributable, respectively, to negligible or very high friction. Regardless of whether or not the element intersections are fastened, a net is an underconstrained structural system. It allows singular configurations, of which some lack kinematic mobility (quasi-variant nets).

Representing net elements by two arrays of material lines with infinitely fine spacing leads to a continuous model of a net. Such a model allows systematic use of the mathematical concept of a net in differential geometry, along with the corresponding well-developed analytical apparatus. The effort is rewarded with an avalanch of closed-form solutions and formulas presented in this chapter, which is an updated and expanded version of Kuznetsov (1969).

The first two sections contain a general statical-geometric theory of singular nets. The key concept of the theory is the statical vector of a net: this vector being gradient is an invariant criterion of a singular net. After establishing the existence and the number of general singular nets, their most important particular classes are studied: orthogonal, conjugate, and the nets of principal curvatures. The presentation of the above material is exceptional in that it requires the reader's familiarity with tensorial differential geometry; otherwise, many of the analytical calculations, proofs, and conclusions are to be taken on faith. An alternative, conventional presentation of the theory, besides taking much more space, would not allow the results to be presented in a consise invariant form. The results most relevant for applications are elaborated on in more conventional form in the next chapter, where they are applied to axisymmetric nets.

The third section explores general and net-specific interrelations among equilibrium loads, configurations, and internal forces in nets. Particularly interesting, and relevant in the optimal design of nets, are the results obtained for geodesic and semigeodesic nets under normal surface loads. Small deformations of extensible nets are studied in the last section, leading to a basic system of governing equations of statics of nets.

6.1   Statical-Geometric Theory of Singular Nets

A continuous model of a fiber net in mechanics is practically identical to a well-developed differential-geometric model of a net. All that needs to be done to put the mathematical model to efficient use, is to incorporate into it the necessary statical concepts (external loads, internal forces, and their equilibrium). However, the geometric theory of nets relies heavily on tensor analysis. The following subsection is intended as a reference introducing the relevant terminology and definitions from differential geometry; it can be omitted by a reader familiar with the subject.

Information from Tensorial Differential Geometry

Parameters, Vectors, and Tensors Associated with a Net. Notation employed throughout this chapter is basically that found in Shulikovski (1963) which is an encyclopedia of nets. Except for subscripts u and v denoting the two line arrays of a net, all of the letter and number indices designate vector and tensor components. Partial derivatives are designated either by a subscript or by a symbol ∂_k with $k = 1, 2$. Covariant differentiation is denoted by a comma in the subscript.

In a slight departure from convention, raising and lowering of tensor indices is done with a perturbation tensor ε_{ij} introduced as

$$\varepsilon_{12} = - \varepsilon_{21} = \sqrt{g_{11}g_{22} - g_{12}^2}; \quad \varepsilon_{11} = \varepsilon_{22} = 0, \tag{6.1}$$

where g_{ij} is the metric tensor of the surface. As a result, a switch in the levels of two repeating (dummy) indices causes sign reversal of the contracted tensor.

A net on a surface is determined by two independent vector fields, u_i and v_i. If these are unit length vectors, the angle between them is given by

$$\cos (u,v) = \cos \omega = g_{ij}\, u^i v^j = - g_j^i u_i v^j = \tilde{u}^j v_j = - \tilde{u}_j v^j, \tag{6.2}$$

$$\sin (u,v) = \sin \omega = g_{ij}\, \tilde{u}^i \tilde{v}^j = - g_j^i \tilde{u}_i v^j = u_j v^j = - u^j v_j, \tag{6.3}$$

where

$$\tilde{u}_i = g_i^j u_j, \quad \tilde{u}^i = g_j^i \tilde{u}^j. \tag{6.4}$$

Vector $\tilde{u}_i$ is orthogonal to u_i and has a unit length, so that

$$v^i = u^i \cos \omega + \tilde{u}^i \sin \omega, \quad u^i = v^i \cos \omega - \tilde{v}^i \sin \omega. \tag{6.5}$$

The direction vectors u_i and v_i of a net satisfy the equation

$$(u_i v_j + u_j v_i)\, x^i x^j \equiv b_{ij} x^i x^j = 0. \tag{6.6}$$

The symmetric tensor b_{ij} defines two line arrays on the surface. Multiplying this tensor by such a scalar function that makes its discriminant equal to the discriminant of the metric tensor, leads to the metrically normalized tensor of the net, a_{ij}. Tensors ε_{ij}, a_{ij}, and the Kronecker tensor δ_j^i can be expressed in terms of the unit direction vectors of the net and the net angle as follows:

$$\varepsilon_{ij} = (u_i v_j - u_j v_i)/\sin \omega \equiv 2u_{[i} v_{j]}/\sin \omega,$$

$$a_{ij} = (u_i v_j + u_j v_i)/\sin \omega \equiv 2u_{(i} v_{j)}/\sin \omega,$$

$$\delta_j^i = \varepsilon^{ik} \varepsilon_{jk} = -(u^i v_j - u_j v^i)/\sin \omega,$$

$$a_j^i = \varepsilon^{ik} a_{jk} = (u^i v_j + u_j v^i)/\sin \omega.$$

(6.7)

Note the parenthesized and bracketed subscripts denoting, respectively, symmetric and skew-symmetric objects; this notation is occasionally used below. If the chosen coordinate system coincides with the net lines, the components of the introduced vectors and tensors become

$$u^i \ (1/A, \ 0); \quad v^i \ (0, \ 1/B); \quad u_i \ (0, \ B \sin \omega), \quad v_i \ (-A \sin \omega, \ 0);$$

(6.8)

$$a_1^1 = -a_2^2 = -1; \quad a_2^1 = a_1^2 = 0; \quad \varepsilon_{12} = -\varepsilon_{21} = 1/\varepsilon^{12} = -1/\varepsilon^{21} = AB \sin \omega;$$

where A and B are the Lamé parameters (coefficients of the first quadratic form) of the surface:

$$ds^2 = A^2 du^2 + 2AB \cos \omega \, du \, dv + B^2 dv^2.$$

(6.9)

A covariant derivative of a unit vector u^i is representable as

$$u^i{}_{,j} = \tilde{u}^i t_{uj},$$

(6.10)

where t_{uj} is called the transversal vector of field u^i. When a vector field is rotated through an angle ω, the tranversal vector becomes

$$t_{vi} = t_{ui} + \omega_i.$$

(6.11)

Contraction of the transversal vectors with the direction vectors of the net produces the following invariants:

$$t_{ui} u^i = \kappa_u = \frac{1}{A} \left(\frac{B_1 \cos \omega - A_2}{B \sin \omega} - \omega_1 \right), \quad t_{vi} u^i = \lambda_u = \frac{B_1 \cos \omega - A_2}{AB \sin \omega},$$

$$t_{vi} v^i = \kappa_v = \frac{1}{B} \left(\frac{B_1 - A_2 \cos \omega}{A \sin \omega} + \omega_2 \right), \quad t_{ui} v^i = \lambda_v = \frac{B_1 - A_2 \cos \omega}{AB \sin \omega}.$$

(6.12)

Parameters κ_u, κ_v are the geodesic curvatures and λ_u, λ_v the Chebyshev curvatures of the net. The condition $\kappa_u = \kappa_v = 0$ describes a geodesic net, while $\lambda_u = \lambda_v = 0$ characterizes an equidistant (Chebyshev) net. The latter is remarkable in that all of its cells are rhombic. The parameters λ_u and λ_v figure in the Chebyshev and rhombic vectors of a net given, respectively, by

$$a_i = (\lambda_u u_i + \lambda_v v_i)/\sin^2 \omega,$$
$$R_i = (\lambda_v \tilde{u}_i - \lambda_u \tilde{v}_i)/\sin^2 \omega. \tag{6.13}$$

Extrinsic Geometry of a Net. All of the net properties, as well as vectors and scalar parameters considered so far, are associated only with the intrinsic geometry of the net surface, that is, with the metric tensor and the corresponding first quadratic form. A complete description of a surface also requires its second tensor, h_{ij}, or the associated coefficients of the second quadratic form. Contracting the second tensor with the direction vectors of a net produces the following invariants:

$$h_{ij}u^i u^j = \sigma_u, \quad h_{ij}v^i v^j = \sigma_v, \quad h_{ij}u^i \tilde{u}^j = \tau_u, \quad h_{ij}v^i v^j = \tau_v. \tag{6.14}$$

Here σ_u, σ_v are the normal curvatures and τ_u, τ_v the geodesic torsions of the net lines. In turn, the second tensor can be expressed in terms of these invariants and the unit direction vectors:

$$h_{ij} = [\sigma_u v_i v_j - 2\chi u_{(i}v_{j)} + \sigma_v u_i u_j]/\sin^2 \omega, \tag{6.15}$$

where

$$\chi = \sigma_u \cos \omega + \tau_u \sin \omega = \sigma_v \cos \omega - \tau_v \sin \omega = h_{ij}u^i v^j. \tag{6.16}$$

Turning into zero of the invariant χ is a sign of a conjugate net, that is, a net whose lines follow two mutually conjugate directions on the surface.

Lines with zero normal curvatures are called asymptotic and form a net of the same name that exists only on surfaces of negative total curvature. On developable surfaces, this net degenerates into the array of linear generators, and on surfaces of positive curvature it does not exist.

The above invariants appear in the Gauss formulas expressing the derivatives of the net unit direction vectors and the unit normal vector **n** in three-dimensional space (in this case the vectors are designated by bold type):

$$\mathbf{u}_1 = A (\sigma_u \mathbf{n} + \kappa_u \tilde{\mathbf{u}}), \quad \mathbf{u}_2 = B (\chi \mathbf{n} + \lambda_v \tilde{\mathbf{u}}),$$
$$\mathbf{v}_1 = A (\chi \mathbf{n} + \lambda_u \tilde{\mathbf{v}}), \quad \mathbf{v}_2 = B (\sigma_v \mathbf{n} + \kappa_v \tilde{\mathbf{v}}), \tag{6.17}$$
$$\mathbf{n}_1 = -A (\sigma_u \mathbf{u} + \tau_u \tilde{\mathbf{u}}), \quad \mathbf{n}_2 = -B (\sigma_v \mathbf{v} + \tau_u \tilde{\mathbf{v}}).$$

The most prominent and best-known net on a surface is the net of principal curvatures, whose tensor, c_{ij}, can be expressed in terms of the two fundamental tensors of the surface. Contraction of these three tensors engenders the three basic invariants of a surface—the total (Gaussian) curvature, K; the mean curvature, H; and Euler's difference, E:

$$2K = h_{ij}h^{ij}, \quad 2H = g_{ij}h^{ij}, \quad 2E = -c_{ij}c^{ij}. \tag{6.18}$$

These invariants are not independent; they obey the relation

$$E = H^2 - K, \tag{6.19}$$

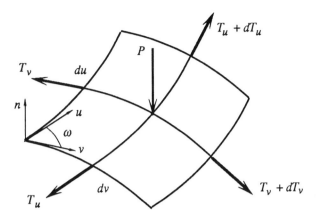

Figure 6.1. Infinitesimal element of a net.

and can be expressed in terms of the two principal curvatures, σ_α and σ_β:

$$K = \sigma_\alpha \sigma_\beta, \quad 2H = (\sigma_\alpha + \sigma_\beta), \quad 4E = (\sigma_\alpha - \sigma_\beta)^2. \qquad (6.20)$$

As is customary in differential geometry, the forthcoming analysis is local and the functions employed in it are assumed sufficiently smooth and differentiable the necessary number of times.

Statical Vector of Net

Equilibrium Equations and Their Reduction. A singular net is one that admits a nontrivial solution to the homogeneous equilibrium equations. In terms of structural mechanics, it means the existence of virtual self-stress, i.e., statical indeterminacy. The net is considered as unbounded; however, if there is an edge, the required forces are assumed to be acting there, either as a support reaction or a contour equilibrium load. Virtual self-stress is stable and the net prestressable when all of the cable forces are tensile; this simple observation makes a stability analysis unnecessary.

To develop equilibrium equations, consider an infinitesimal element of a net bounded by the coordinate lines u and v coinciding with the two fiber arrays (Fig. 1). In vectorial form, the condition of equilibrium reads

$$\frac{\partial}{\partial u}(\mathbf{N}_u \, ds_v) \, du + \frac{\partial}{\partial v}(\mathbf{N}_v \, ds_u) \, dv = \mathbf{P} \, ds_u ds_v \sin \omega. \qquad (6.21)$$

Here $\mathbf{N}_u = N_u \mathbf{u}$ and $\mathbf{N}_v = N_v \mathbf{v}$ are the vector forces in the net related, as in the statics of membranes, to the respective unit increments of linear elements $ds_v = B \, dv = 1$ and $ds_u = A \, du = 1$; $\mathbf{P}$ is the vector of external load per unit area of the surface. Note that in the chosen reference system, the elementary strips $du = 1$ and $dv = 1$ contain a certain constant number of fibers, whereas the number of fibers within the unit width strips $ds_u = 1$ and $ds_v = 1$ varies along the length. To go over from the membrane forces N_u and N_v to the forces T_u and T_v acting in a certain constant numbers of fibers, the former must be multiplied, respectively, by the complementary coefficients of the metric form (the Lamé parameters):

$$T_u = BN_u, \quad T_v = AN_v. \tag{6.22}$$

Evaluating the derivatives in (21) and taking into account equations (17) gives

$$(N_{u,1} + N_u B_1/B)\, \mathbf{u}/A + N_u(\sigma_u \mathbf{n} + \kappa_u \widetilde{\mathbf{u}})$$
$$+ (N_{v,2} + N_v A_2/A)\, \mathbf{v}/B + N_v(\sigma_v \mathbf{n} + \kappa_v \widetilde{\mathbf{v}}) = \mathbf{P} \sin \omega.$$

This equation is obtained with reference to a coordinate system coinciding with the net lines. An invariant tensor version of the equation, valid in any coordinate system, is

$$(N_{u,i} - N_u R_i) u^i \mathbf{u} + N_u(\sigma_u \mathbf{n} + \kappa_u \widetilde{\mathbf{u}})$$
$$+ (N_{v,i} - N_v R_i) v^i \mathbf{v} + N_v(\sigma_v \mathbf{n} + \kappa_v \widetilde{\mathbf{v}}) = \mathbf{P} \sin \omega. \tag{6.23}$$

That this version is correct follows from the fact that upon substituting the expressions for u_i, v_i, and R_i given by (8) and (13), the coordinate form of the equation is recovered. As shown in tensor calculus, this guarantees validity of a tensorial equation in any coordinate system.

A characteristic sign of a singular net is found from the system of three homogeneous $(\mathbf{P} = 0)$ equilibrium equations obtained by scalar multiplication of equation (23), respectively, by $\widetilde{\mathbf{v}}$, $\widetilde{\mathbf{u}}$, and $\mathbf{n}$:

$$(N_{u,i} - N_u R_i)\, u^i \sin \omega - N_u \kappa_u \cos \omega \; - \; N_v \kappa_v = 0,$$
$$(N_{v,i} - N_v R_i)\, v^i \sin \omega + N_v \kappa_v \cos \omega \; + N_u \kappa_u = 0, \tag{6.24}$$
$$N_u \sigma_u + N_v \sigma_v = 0.$$

After introducing a solving function N by

$$N_u = N \sigma_v, \quad N_v = -N \sigma_u, \tag{6.25}$$

the third equation in (24) is satisfied identically while the first two become

$$(N_i u^i)/N = (R_i - \sigma_{v,i}/\sigma_v)\, u^i + (\kappa_u \cos \omega - \kappa_v \sigma_u/\sigma_v)/\sin \omega,$$
$$(N_i v^i)/N = (R_i - \sigma_{u,i}/\sigma_u)\, v^i + (\kappa_v \cos \omega - \kappa_u \sigma_v/\sigma_u)/\sin \omega. \tag{6.26}$$

Integrability Condition. Multiply the first equation (26) by $v_k/\sin\omega$ and subtract from it the second one multiplied by $u_k/\sin\omega$. Then taking into account the expression for the Kronecker tensor in (7) gives

$$N_k/N \equiv \partial_k \ln N = S_k, \tag{6.27}$$

where

$$S_k = \frac{1}{\sin^2 \omega} \left[\left(\frac{\sigma_v}{\sigma_u} \kappa_u - \kappa_v \cos \omega \right) u_k - \left(\frac{\sigma_u}{\sigma_v} \kappa_v - \kappa_u \cos \omega \right) v_k \right]$$

$$- \frac{1}{\sin \omega} \left(\frac{\sigma_{u,i}}{\sigma_u} v^i u_k - \frac{\sigma_{v,i}}{\sigma_v} u^i v_k \right) + R_k. \qquad (6.28)$$

The obtained vector, being a linear combination of the net direction vectors u_k and v_k, belongs to the same surface and is uniquely determined by the geometric parameters of the net.

Definition. Vector S_k engendered by a given net in accordance with (28) is called the statical vector of the net.

Invariant Criterion of a Singular Net. For a net to be singular, it is necessary and sufficient that its statical vector is gradient.

Indeed, since the vector in the left-hand side of (27) is the gradient of a scalar function $\ln N$, this equation and, with it, the homogeneous equilibrium equations (24) admit a nontrivial solution if and only if the statical vector S_k is also a gradient, i.e.,

$$\varepsilon^{ij} S_{i,j} = \varepsilon^{ij} \partial_i S_j = S_{i,}{}^i = 0. \qquad (6.29)$$

This condition, in turn, is necessary and sufficient for the existence of a scalar function S such that $S_k = \partial_k S$.

Definition. The potential S of a gradient statical vector is called the statical potential of the net.

By virtue of (27), the solving function N introduced by (25) is a function of the statical potential:

$$N = C e^S. \qquad (6.30)$$

Thus, knowing the statical potential makes the pattern of initial forces in the net obtainable immediately from (22) and (25):

$$T_u = CN \sigma_v, \qquad N_v = - CN \sigma_u. \qquad (6.31)$$

The constant C is the measure of intensity of these forces.

The condition $S_k = $ grad. in (29) is the integrability condition for the overdetermined system of homogeneous equilibrium equations (24). In what follows, this condition is repeatedly used in the synthesis of particular types of singular nets. This is done by specifying the properties of a net to within one function and using condition (29) as an equation for evaluating this function.

The uniqueness of the statical vector of a net implies that both the statical potential S (if it exists) and the corresponding virtual self-stress pattern (31) are also unique for the net. Hence, a singular net is statically indeterminate to the first degree. As is seen from (25), when the normal curvatures σ_u and σ_v are of the same sign, the initial forces have opposite signs and vice versa. Since stability of self-stress requires tensile forces in both of the net arrays, a singular net is quasi-variant only when the curvatures are of opposite sign. This is possible only on surfaces of negative total curvature and, even then, only for some of the singular nets.

In a coordinate system coinciding with the net lines, condition (29) for a statical vector (28) to be gradient acquires the form

$$\frac{\partial}{\partial u}\left[\frac{B}{\sin\omega}\left(\frac{\sigma_v}{\sigma_u}\kappa_u - \kappa_v\cos\omega\right)\right] + \frac{\partial}{\partial u}\left[\frac{A}{\sin\omega}\left(\frac{\sigma_u}{\sigma_v}\kappa_v - \kappa_u\cos\omega\right)\right] = \frac{\partial^2}{\partial u\partial v}\ln\left(\frac{A\sigma_u}{B\sigma_v}\right).$$

Net properties associated with the intrinsic geometry of the surface are preserved in isometric surface bending. Such are, for instance, the properties of being geodesic or Chebyshev. Since the statical vector of a net depends on the normal curvatures of the net lines, the property of being singular, generally, is not preserved in bending.

Existence and the Number of Singular Nets

Problem Statement and the Solving Equation. The problem of the existence and the number of singular nets is addressed by posing the following question: Is it possible to complement a given one-parametric array of lines on a surface with another array so as to obtain a singular net?

Let u_i be the unit direction vector of the first array, and the second array forms with it the as-yet unknown angle ω. According to (5) and (14), the normal curvature of the second array is

$$\sigma_v = h_{ij}\,(u^i\cos\omega + \tilde{u}^i\sin\omega)\,(u^j\cos\omega + \tilde{u}^j\sin\omega)$$

$$= \sigma_u\cos^2\omega + \sigma_{\tilde{u}}\sin^2\omega + \tau_u\sin 2\omega.$$

Differentiation yields

$$\sigma_{v,i} = [2\tau_u\cos 2\omega - (\sigma_u - \sigma_{\tilde{u}})\sin 2\omega]\,\omega_i + \cdots,$$

where the ellipses stand for the terms not containing derivatives of ω, as these terms are not needed for the analysis. The geodesic curvature κ_v and the vector R_k can be expressed in a similar way using equations (11)–(13):

$$\kappa_v = (t_{ui} + \omega_i)\,v^i = v^i\,\omega_i + \cdots,$$

$$R_k = (t_{ui}\,v^i\,\tilde{u}_k - t_{vi}\,u^i\tilde{v}_k)/\sin\omega = -u^i\tilde{v}_k\,\omega_i/\sin\omega + \cdots.$$

The above four expressions are introduced into formula (28) for the statical vector, whereupon the latter is required to be gradient by implementing condition (29). The result is a second-order partial differential equation whose leading term, i.e., the term containing the second derivatives of ω, is

$$(1/\sigma_v)\, \varepsilon^{kj} \left\{ (\sigma_u\, v^i v_k - \sigma_v\, v^i u_k \cos \omega)/\sin^2\omega \right.$$

$$\left. + ([2\tau_u \cos 2\omega - (\sigma_u - \sigma_{\widetilde u})\sin 2\omega]\, u^i v_k - \sigma_v\, u^i \widetilde v_k)/\sin \omega \right\} \omega_{ij} + \cdots = 0.$$

By virtue of the symmetry of ω_{ij} with respect to the two subscripts,

$$\varepsilon^{kj} u^i v_k\, \omega_{ij} = -u^i v^i\, \omega_{ij} = -(u^{(i}v^{j)} + u^{[i}v^{j]})\, \omega_{ij} = -u^{(i}v^{j)}\, \omega_{ij}.$$

Note also that

$$\varepsilon^{kj} u^i \widetilde v_k = -u^i \widetilde v^j = u^i (u^j - v^j \cos \omega)/\sin \omega.$$

With the aid of these relations, the leading term can be transformed as follows:

$$[\sigma_u\, v^i v^j - 2(\sigma_u \cos \omega + \tau_u \sin \omega)u^{(i}v^{j)} + \sigma_v\, u^i u^j]\, \omega_{ij}/\sigma_v \sin^2\omega + \cdots = 0. \quad (6.32)$$

Provided that neither σ_v nor ω is zero, and taking into account (15), this equation is equivalent to

$$h^{ij}\, \omega_{ij} + \cdots = 0. \qquad (6.33)$$

The discriminant of the form in the right-hand side is

$$\delta = \det [h^{ij}] = K/g. \qquad (6.34)$$

Since the discriminant g of the metric tensor is positive-definite, the sign of δ and, thereby, the type of equation (33), is determined by the sign of the total curvature K. Accordingly, the equation can be elliptic, parabolic, or hyperbolic. It has the characteristic directions coincident with the asymptotic lines and acquires the canonical form when the coordinate net either is asymptotic (then $h_{12} = 0$) or is the net of the principal curvatures (then $h_{11} = h_{22} = 0$).

In view of the existence theorems for second-order quasi-linear partial differential equations, the performed analysis provides an answer to the stated question on singular nets.

Existence and the Number of Singular Nets. Any nonasymptotic one-parametric array of lines on a surface can be complemented [modulo boundary or initial conditions on function ω in equation (6.33)] with another array such that the formed net is singular.

Asymptotic Nets. The special role of the normal curvatures in the net statics sets apart this class of nets, while at the same time making their singularity immediately apparent. At $\sigma_u = \sigma_v = 0$ (an asymptotic net, including its trivial particular case, an arbitrary plane net), the last of equations (24) vanishes. The remaining two equations form a hyperbolic system with the characteristic directions coinciding with the net lines. The resulting Cauchy problem has a solution determined by assigning net forces N_u and N_v along a suitable initial curve or a pair of intersecting characteristics. This means that the number of virtual self-stress patterns is infinite and ensures that the net is prestressable. Hence, any asymptotic or plane net is singular, quasi-variant (type III) and, unlike all other singular nets, is statically indeterminate to an infinite degree. The acquisition of this degree of statical indeterminacy is accompanied by the corresponding reduction in the number of equilibrium loads: an asymptotic net, in contrast to all other nets, singular or nonsingular, has no equilibrium loads normal to the surface.

Note that a net with one asymptotic array may have self-stress only if the other array is force-free. This is seen from the equilibrium condition in the normal direction given by the last of equations (24). Trerefore, in what follows, both of the normal curvatures of a net are assumed to be different from zero.

6.2   Particular Classes of Singular Nets

Two important classes of nets will be investigated for singularity: orthogonal and conjugate. Orthogonal nets are relevant to the statics of membranes, as the nets of principal force trajectories. Conjugate nets are interesting in that on a surface with $K < 0$, the curvatures of the two arrays are of opposite sign. For a singular net this entails tensile self-stress, hence, prestressability. Finally, the most prominent net of any surface—the net of principal curvatures—is the one, and the only one, that is simultaneously orthogonal and conjugate.

Orthogonal Singular Nets

Statical Vector of an Orthogonal Net.   As is seen from (12), (14), and (18), an orthogonal net is characterized by

$$\kappa_u = \lambda_u, \quad \kappa_v = \lambda_v, \quad \sigma_u + \sigma_v = \sigma_\alpha + \sigma_\beta = 2H.$$

Accordingly, the statical vector of this net is obtained from (28) as

$$S_k = 2H\left(\frac{\kappa_u}{\sigma_u} u_k + \frac{\kappa_v}{\sigma_v} v_k\right) - \frac{\sigma_{u,i}}{\sigma_u} v^i u_k + \frac{\sigma_{v,i}}{\sigma_v} u^i v_k. \tag{6.35}$$

Orthogonal Singular Nets of a Surface.   Singular orthogonal nets of a given surface are sought as the result of rotation of the net of principal curvatures (α, β) through an as-yet unknown angle γ. Thus, each direction vector of the resulting net forms an angle γ with the direction vector of the respective line of principal curvature (Fig. 2a). Then the normal curvatures σ_u and σ_v can be evaluated with the aid of Euler's formula relating the curvature of an arbitrary line on a surface to the principal curvatures:

$$\sigma_u = \sigma_\alpha \cos^2\gamma + \sigma_\beta \sin^2\gamma, \quad \gamma = \widehat{\alpha, u}. \tag{6.36}$$

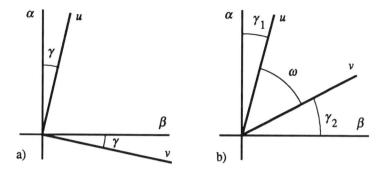

Figure 6.2. Analytical construction of: a) orthogonal net; b) conjugate net.

From here,

$$\sigma_{u,i} = (\sigma_\beta - \sigma_\alpha) \sin 2\gamma\, \gamma_i + \cdots, \quad \sigma_{v,i} = (\sigma_\alpha - \sigma_\beta) \sin 2\gamma\, \gamma_i + \cdots,$$

where the ellipses, as before, denote the terms not containing derivatives of γ. The geodesic curvatures are evaluated using (11) and (12):

$$\kappa_u = u^i \gamma_i + \cdots, \quad \kappa_v = v^i \gamma_i + \cdots.$$

The above expressions are introduced into formula (35) of the statical vector and the latter is required to be gradient as in (29). The result is a second-order quasilinear differential equation in γ with the leading term

$$\varepsilon^{kj} \big[2H\, (\sigma_v\, u^i u_k + \sigma_u\, v^i v_k)/\sigma_u \sigma_v$$

$$+ (\sigma_\alpha - \sigma_\beta) \sin 2\gamma\, (\sigma_v\, v^i u_k + \sigma_u\, u^i v_k)/\sigma_u \sigma_v \big] \gamma_{ij} + \cdots = 0.$$

Since the object γ_{ij} is symmetric,

$$\varepsilon^{kj} (\sigma_v\, v^i u_k + \sigma_u\, u^i v_k)\, \gamma_{ij} + \cdots = -2H u^{(i} v^{j)}\, \gamma_{ij}.$$

Furthermore, for the net in consideration,

$$(\sigma_\alpha - \sigma_\beta) \sin 2\gamma = -2\tau_u = -2\chi,$$

and the equation becomes

$$-2\,(H/\sigma_u \sigma_v)\,(\sigma_u\, v^i v^j - 2\chi u^{(i} v^{j)} + \sigma_v\, u^i u^j)\, \gamma_{ij} + \cdots = 0.$$

Upon multiplying by $\sigma_u\, \sigma_v / 2$ and taking into account (15), it can be written

$$H\, h^{ij}\, \gamma_{ij} + \cdots = 0. \tag{6.37}$$

Its discriminant is

$$\delta = H^2 K / g. \tag{6.38}$$

Thus, with the exception of the case of $H = 0$, everything said earlier about equation (33) holds. The result can be summed up as follows:

Existence and the Number of Orthogonal Singular Nets. Orthogonal singular nets exist on any surface and are determined modulo boundary or initial conditions for a second-order partial differential equation.

Exceptional Cases. As to the case of zero mean curvature (minimal surface), the condition $H = 0$ entails $\sigma_u = -\sigma_v = \sigma$ for any orthogonal net. The statical vector (35) becomes

$$S_k = (u^i v_k - u_k v^i)\, \sigma_i/\sigma = -\delta_k^i\, \partial_i \ln \sigma = -\partial_k \ln \sigma, \qquad (6.39)$$

yielding

$$N = Ce^S = C/\sigma, \qquad N_u = N_v = C. \qquad (6.40)$$

Since the forces are of the same sign, self-stress is stable. Thus, any orthogonal net of a minimal surface is singular and prestressable. This outcome could be expected in light of the fact that a uniform initial stress field of a minimal surface is its definitive sign established in the classical Plateau experiments with soap films. However, the result is more involved and informative for orthogonal nets of minimal surfaces. As follows from (40) and (22), the prestressing force pattern in such a net is given by

$$T_u = B, \qquad T_v = A, \qquad (6.41)$$

so that it depends on the surface metric in the coordinate system coinciding with the net lines.

Minimal surfaces are of negative total curvature, they have asymptotic nets and these are known to be orthogonal. Although solution (41) is still valid, in this case it represents just one of the infinite number of statically possible prestress patterns for such a net.

A sphere represents another exceptional case where the above analysis is inapplicable: here the net of principal curvatures is undefined. By virtue of $\sigma_u = \sigma_v = $ const. and in view of (13), for an orthogonal net on a sphere,

$$S_k = 2\, (\kappa_u u_k + \kappa_v v_k) = 2a_k. \qquad (6.42)$$

The Chebyshev vector a_k being gradient is known to be a characteristic sign of an isothermic net, that is, one for which the first quadratic form reduces to

$$ds^2 = e^{-2a}\, (du^2 + dv^2) \qquad (6.43)$$

with $a = a\, (u, v)$. Thus, according to (42), any isothermic net on a sphere is singular and, conversely, any orthogonal singular net on a sphere is isothermic. The statical potential of such a net is $S = 2a$ and, since $A = B = e^{-a}$, the virtual self-stress is described by

$$N_u = -N_v = Ce^{2a}, \qquad T_u = -T_v = Ce^a. \qquad (6.44)$$

On surfaces other than a sphere, orthogonal nets may have equal, but not constant, normal curvatures of the two arrays. For example, according to (36), for a bisector net of any net of principal curvatures,

$$\sigma_u = \sigma_v = \sigma = H. \tag{6.45}$$

With this, the statical vector (35) becomes

$$S_k = 2(\kappa_u u_k + \kappa_v v_k) + (u^i v_k - u_k v^i)\partial_i \ln \sigma = 2a_k - \partial_k \ln H. \tag{6.46}$$

Since the last term is gradient, it makes sense to look for bisector nets with a gradient Chebyshev vector a_k. Remarkably, the Chebyshev vector is preserved when the net (i.e., both of its direction vectors) rotates through a constant angle. Hence, the vector a_k of the bisector net is the same as that of the net of principal curvatures (α, β). To take advantage of this, consider an isothermic surface, that is, one whose net of principal curvatures is isothermic. This prominent class includes all surfaces of revolution, quadrics, surfaces of constant mean curvature, Liouville surfaces, and many others. The described property of the Chebyshev vector means that on any isothermic surface the bisector net of the net of principal curvatures is isothermic; it has a gradient a_k and, as seen from (46), is a singular orthogonal net. According to (25) and (45), its statical potential and virtual self-stress are

$$S = 2a - \ln H, \quad N_u = -N_v = C e^{2a} \sigma/H = C e^{2a}. \tag{6.47}$$

In particular, for surfaces of constant mean curvature H, including minimal surfaces $(H = 0)$,

$$S = 2a = \ln \sqrt{E}, \tag{6.48}$$

where E is the invariant defined in (19).

Note that in isothermal bisector nets, equilibrium in the normal direction is satisfied in a trivial way: $N_u = -N_v$ while $\sigma_u = \sigma_v$. Because of this, such nets are singular but never prestressable.

Conjugate Singular Nets

Statical Vector of a Conjugate Net. Specification of the statical vector (28) for a conjugate net requires evaluating the coordinate derivatives of the normal curvatures of the net lines. Covariant differentiation of the first expression in (14) gives

$$\sigma_{u,i} = h_{jk,i} u^j u^k + 2h_{jk} u^j u^k{}_{,i}$$

and, taking into account (10),

$$\sigma_{u,i} v^i = h_{jk,i} u^j u^k v^i + 2h_{jk} u^j \tilde{u}^k t_{ui} v^i. \tag{6.49}$$

The first term in the right-hand side can be transformed using the analytical definition of a conjugate net, that is, the zero value of the invariant χ in (16). Differentiating this expression yields

$$(h_{jk} u^j v^k)_{,i} u^i = 0$$

or, taking advantage of (5) and (10),

$$(h_{jk} u^j v^k)_{,i} u^i = h_{jk,i} u^j v^k u^i + (\sigma_v \kappa_u - \sigma_u \lambda_u)/\sin \omega = 0.$$

Because of the symmetry of the tensor $h_{ij,k}$ with respect to all three indices

$$h_{jk,i} u^j u^k v^i = (\sigma_u \lambda_u - \sigma_v \kappa_u)/\sin \omega.$$

The evaluation of the last term in (49) is straightforward [cf. (12)]:

$$2 h_{jk} u^j \tilde{u}^k t_{ui} v^i = 2 h_{jk} u^j (v^k - u^k \cos \omega) \lambda_v/\sin \omega = - 2\sigma_u \lambda_v \operatorname{ctn} \omega.$$

Thus,

$$\sigma_{ui} v^i = (\sigma_u \lambda_u - \sigma_v \kappa_u - 2\sigma_u \lambda_v \cos \omega)/\sin \omega$$

and, by analogy,

$$\sigma_{vi} u^i = - (\sigma_v \lambda_v - \sigma_u \kappa_v - 2\sigma_v \lambda_u \cos \omega)/\sin \omega.$$

Introducing these results along with expression (13) for R_k into (28) gives

$$S_k = 2 \left(\frac{\sigma_v}{\sigma_u} \kappa_u u_k + \frac{\sigma_u}{\sigma_v} \kappa_v v_k \right)/\sin^2 \omega$$

$$+ [(\lambda_v - \kappa_v) u_k + (\lambda_u - \kappa_u) v_k] \cos \omega/\sin^2 \omega.$$

The last term can be transformed with the aid of (11):

$$\lambda_u - \kappa_u = t_{vi} u^i - t_{ui} u^i = \omega_i u^i, \quad \lambda_v - \kappa_v = - \omega_i v^i,$$

leading to

$$- \operatorname{ctn} \omega \, \omega_i (v^i u_k - u^i v_k)/\sin \omega = - \operatorname{ctn} \omega \, \omega_i \delta_k^i = - \omega_k \operatorname{ctn} \omega.$$

The final result is the desired formula for the statical vector of a conjugate net:

$$S_k = \frac{2}{\sin^2 \omega} \left(\frac{\sigma_v}{\sigma_u} \kappa_u u_k + \frac{\sigma_u}{\sigma_v} \kappa_v v_k \right) - \omega_k \operatorname{ctn} \omega. \qquad (6.50)$$

Conjugate Singular Nets of a Surface. To establish the existence and the number of conjugate singular nets of a surface, such a net will be constructed by skewing the net of principal curvature through angles γ_1 and γ_2 (Fig. 2b).

The geodesic curvatures figuring in the statical vector (50) can be expressed as

$$\kappa_u = \gamma_{1i} u^i + \cdots, \quad \kappa_v = \gamma_{2i} v^i + \cdots. \qquad (6.51)$$

Evaluating the derivatives of γ_1 and γ_2 in terms of the net angle ω requires additional information. One relation is immediately available (Fig. 2b):

$$\omega = \pi/2 - \gamma_1 - \gamma_2 .$$

Obtaining another one requires specifying the net as conjugate. For this net,

$$\sigma_u + \sigma_v = 2H \sin^2 \omega, \qquad \sigma_u \sigma_v = K \sin^2 \omega ,$$

so that

$$\sigma_{u\,(v)} = (k \sin \omega \pm \Omega) \sqrt{E} \sin \omega, \quad k = H/\sqrt{E}, \quad \Omega = \sqrt{1 - k^2 \cos^2 \omega} . \quad (6.52)$$

On the other hand, the normal curvatures of lines forming angles γ_1 and γ_2 with the lines of principal curvatures can be evaluated using (36):

$$\sigma_u = \sqrt{E} \, (k - \cos 2\gamma_1), \quad \sigma_v = \sqrt{E} \, (k - \cos 2\gamma_2).$$

Upon elimination of σ_u and σ_v and some trigonometric transformations, the second equation relating γ_1 and γ_2 to the net angle is obtained:

$$\sin (\gamma_1 - \gamma_2) = k \cos \omega .$$

The necessary derivatives are

$$\gamma_{1i} = [k_i \cos \omega - \omega_i \, (\Omega + k \sin \omega)]/2\Omega,$$
$$\gamma_{2i} = - [k_i \cos \omega + \omega_i \, (\Omega - k \sin \omega)]/2\Omega. \qquad (6.53)$$

Everything is now in place for an evaluation of the statical vector (50). After introducing geodesic curvatures (51), it can be rearranged as follows:

$$S_k = \left[\left(\sigma_v u^i u_k + \sigma_u v^i v_k \right) \left(\frac{\gamma_{1i}}{\sigma_u} - \frac{\gamma_{2i}}{\sigma_v} \right) \right] / \sin^2 \omega$$
$$+ \left[\left(\sigma_v u^i u_k - \sigma_u v^i v_k \right) \left(\frac{\gamma_{1i}}{\sigma_u} + \frac{\gamma_{2i}}{\sigma_v} \right) \right] / \sin^2 \omega - \omega_k \operatorname{ctn} \omega .$$

As is seen from (52) and (53),

$$\frac{\gamma_{1i}}{\sigma_u} = \frac{k_i \cos \omega/(\Omega + k \sin \omega) - \omega_i}{2\Omega \sqrt{E} \sin \omega} , \quad \frac{\gamma_{2i}}{\sigma_v} = \frac{k_i \cos \omega/(\Omega - k \sin \omega) + \omega_i}{2\Omega \sqrt{E} \sin \omega} .$$

The difference of these quantities contains ω_i:

$$\frac{\gamma_{1i}}{\sigma_u} - \frac{\gamma_{2i}}{\sigma_v} = - \frac{\omega_i}{\Omega \sqrt{E} \sin \omega} + \cdots ,$$

whereas their sum does not. Taking into account (15), the above S_k becomes

$$S_k = - h_k^i \, \omega_i/\Omega \sqrt{E} \sin \omega - \omega_k \operatorname{ctn} \omega + \cdots .$$

Substitution into (29) produces a second-order quasi-linear equation

$$S_{k,}{}^k = \varepsilon^{kj} \partial_j S_k = - h^{ij} \, \omega_{ij}/\Omega \sqrt{E} \sin \omega + \cdots = 0,$$

and once again the sign of the discriminant coincides with the sign of total curvature K of the surface.

Existence and the Number of Conjugate Singular Nets. Conjugate singular nets exist on any surface and are determined modulo boundary or initial conditions for a second-order partial differential equation.

Particular Classes of Conjugate Singular Nets. As is seen from (52), on surfaces with $K < 0$, conjugate lines have normal curvatures of opposite sign. Hence, on these surfaces, conjugate singular nets are prestressable, whereas on surfaces with $K > 0$ they are not.

One of the most prominent nets of this class is the *Voss net.* This net is conjugate and geodesic $(\kappa_u = \kappa_v = 0)$ and preserves this property in surface bending. According to (50), its statical vector is

$$S_k = - \omega_k \operatorname{ctn} \omega = - \partial_k \ln \sin \omega,$$

and is a function of only the intrinsic geometry of the net. The absence of extrinsic geometric parameters, represented by the normal curvatures, suggests that the statical singularity of this net is also preserved in surface bending. If the net is taken as coordinate, the normal curvatures acquire the form

$$\sigma_u = U(u) \sin \omega / A, \qquad \sigma_v = V(v) \sin \omega / B,$$

and, according to (25) and (22), the fiber forces are

$$T_u = BCe^S \sigma_v = CV(v), \quad T_v = - CU(u). \tag{6.54}$$

Thus, the fiber forces do not vary along the length which, as shown below, is characteristic of all geodesic nets.

A rich class of conjugate nets are *translation nets*. They are Chebyshev nets formed by mutually translating two arbitrary curves. When the curves are flat and lie in mutually perpendicular planes, the corresponding surface equation in Cartesian coordinates is

$$z = X(x) + Y(y),$$

where X and Y are arbitrary functions of their arguments. The variables needed for the evaluation of the statical vector are

$$A^2 = 1 + X'^2, \quad B^2 = 1 + Y'^2, \quad \cos \omega = X'Y'/AB,$$

$$\kappa_x = - \frac{1}{A} \frac{\partial \omega}{\partial x}, \quad \kappa_y = \frac{1}{B} \frac{\partial \omega}{\partial y}, \quad \sigma_x = \frac{X''}{A^3 B \sin \omega}, \quad \sigma_y = \frac{Y''}{AB^3 \sin \omega},$$

$$\tag{6.55}$$

leading to

$$S_1 = \frac{2X'X''}{1 + X'^2 + Y'^2} - \frac{\partial}{\partial x} \ln \sin \omega$$

$$= (1/2) \partial_x \ln [(1 + X'^2)(1 + Y'^2)(1 + X'^2 + Y'^2)],$$

and a similar expression for the component S_2. Obviously, this is a singular net with the statical potential

$$S = (1/2) \ln [(1 + X'^2 + Y'^2)(1 + Y'^2)(1 + X'^2)].$$

The net forces are

$$T_x = CY'' \sqrt{1 + X'^2}, \quad T_y = -CX'' \sqrt{1 + Y'^2}, \tag{6.56}$$

and have projections onto the plane x, y

$$H_x = T_x \cos \beta_x = T_x / \sqrt{1 + X'^2} = CY'', \quad H_y = -CX''.$$

Probably, the simplest surfaces of this class are paraboloids. For the translation net of a hyperbolic paraboloid, $2z = by^2 - ax^2$, the above solution becomes

$$T_x = Cb\sqrt{1 + a^2 x^2}, \quad T_y = Ca\sqrt{1 + b^2 x^2}, \quad H_x = CB, \quad H_y = Ca.$$

A less trivial example is a translation net generated by two circular arcs:

$$z = R - \sqrt{\xi} - (r - \sqrt{\eta}),$$

where $\xi = R^2 - x^2$ and $\eta = r^2 - y^2$, with R and r being the radii of the two arcs. The net forces and their horizontal projections are found from (56):

$$T_x = C \frac{Rr^2}{\eta \sqrt{\xi \eta}}, \quad T_y = C \frac{R^2 r}{\xi \sqrt{\xi \eta}}, \quad H_x = C \frac{R^2}{\eta^{3/2}}, \quad H_y = C \frac{r^2}{\xi^{3/2}}.$$

It is now possible to evaluate the vertical contact pressure between the elements of the two arrays. Its intensity per unit length of the horizontal projection is

$$V_x = H_y (\partial^2 z / \partial y^2) = - CR^2 r^2 / (\xi \eta)^{3/2}.$$

Curiously, under this mutual vertical contact pressure, every fiber has a circular equilibrium profile. For example, for the fiber $x = x_0$, the load is

$$V_y(y) = C R^2 r^2 / (\xi_0 \eta)^{3/2} = C_1 / (r^2 - y^2)^{3/2}.$$

Singular Nets of Principal Curvatures

Statical Vector of a Net of Principal Curvatures. This most prominent net of a surface is the one (and the only one) that is simultaneously orthogonal and conjugate. The expression for its statical vector is easily obtained from (50) by setting $\omega = \pi/2$:

$$S_k = 2(\alpha_k \kappa_\alpha \sigma_\beta / \sigma_\alpha + \beta_k \kappa_\beta \sigma_\alpha / \sigma_\beta). \tag{6.57}$$

The geodesic and normal curvatures of the lines α, β are closely related to the surface invariants H, K, and E. Note first that, according to (20),

$$\sigma_\alpha = H + \sqrt{E}, \quad \sigma_\beta = H - \sqrt{E}.$$

Taking into account (12) and (13),

$$S_k = [a_k\,(\sigma_\alpha^2 + \sigma_\beta^2) + a_i\,(\beta^i\,\alpha_k + \alpha^i\,\beta_k)\,(\sigma_\beta^2 - \sigma_\alpha^2)]/K$$

$$= 2\,[a_k\,(H^2 + E) - 2a_i\,a_k^i\,H\sqrt{E}]/K.$$

Very useful for further transformation is the fourth tensor of a surface, one engendered by the first two tensors as

$$c_{ij} = h_i^k\,g_{jk} + \varepsilon_{ij}H.$$

(The third tensor of a surface is the metric tensor of its spherical image.) The fourth tensor differs by a scalar multiplier $\sqrt{E}$ from the metrically normalized tensor of the net of principal curvatures, so that

$$c_k^i = \sqrt{E}\,a_k^i, \quad c_j^i\,c_k^j = E\delta_k^i. \tag{6.58}$$

The Chebyshev vector of the net of principal curvatures can also be expressed in terms of the fourth tensor:

$$2a_k = \partial_k \ln \sqrt{E} + c_k^i\,H_i/E.$$

With the aid of these relations, the above expression for the statical vector is transformed as follows:

$$S_k = [(H^2 + E)E_k/2 - 2EHH_k + (H^2H_i + EH_i - HE_i)\,c_k^i]/EK$$

$$= \partial_k \ln (\sqrt{E/K}) + (H/E)\,c_k^i\,\partial_i \ln (K/H). \tag{6.59}$$

This formula, giving the statical vector of the net of principal curvatures in terms of the surface invariants, also happens to describe the Chebyshev vector of the spherical image of the net:

$$S_k = 2c_k. \tag{6.60}$$

Recalling that a gradient Chebyshev vector characterizes an isothermal net leads to the following result.

Invariant Criterion of a Singular Net of Principal Curvatures. For a net of principal curvatures to be singular, it is necessary and sufficient that its spherical image is an isothermic net.

Surfaces with a Singular Net of Principal Curvatures. The obtained criterion allows, in principle, finding all surfaces whose net of principal curvatures is singular. It is known that an arbitrary isothermic net on a sphere is obtainable by rotating any isothermic net (say, the net of meridians and parallels) through an angle given by an arbitrary harmonic function γ. The Chebyshev vector of the thus obtained isothermic net,

$$c_i = \overset{*}{c_i} + g_i^j\, \gamma_j,$$

determines all four of the so-called tangential coordinates of the sought surface. Thus, each harmonic function ultimately defines some surface with a singular net of principal curvatures along with its statical potential, $S = 2c$.

A readily available example is a class of surfaces with two plane arrays of lines of principal curvatures. The spherical image of such a net consists of two mutually orthogonal arrays of circles forming on a sphere an isothermal net (called a Bonnet net). One of the better known surfaces of this class is a cyclide of Dupin, where all lines of curvature are circles.

Another example is a general quadric surface. Its invariant description is

$$\partial_k \ln \sqrt{K} = (H/E)\, c_k^i\, (K_i/K - 2H_i/H).$$

After contracting both sides of the equation with c_j^k [cf. (58)], it becomes

$$c_j^k\, K_k/2K = H\, (K_j/K - 2H_j/H).$$

Combining this with the original equation, the second addend in the statical vector (59) is evaluated as

$$K_k/4K + (H/E)\, c_k^i\, K_i/2 = K_k/4K + (H^2/E)\, (K_k/K - 2H_k/H).$$

Introducing this into (59) shows this statical vector to be gradient:

$$S_k = K_k/4K - E_k/2E.$$

Thus, the net of principal curvatures of a quadric is singular, with

$$S = \ln \sqrt{\sqrt{|K|}\,/E}.$$

This result holds for both central quadrics and paraboloids. For a central quadric referred to its lines of principal curvatures,

$$\sigma_\alpha = abc/\alpha \sqrt{\alpha\beta}, \quad \sigma_\beta = abc/\beta \sqrt{\alpha\beta},$$

$$K = (abc/\alpha\beta)^2, \quad E = [abc\,(\alpha - \beta)]^2/(\alpha\beta)^3,$$

so that

$$S = \ln\,[\alpha\beta/(\alpha - \beta)].$$

The coefficients of the first quadratic form are

$$A = \frac{1}{2}\sqrt{\frac{\alpha\,(\alpha - \beta)}{(a^2 - \alpha)(b^2 - \alpha)(c^2 - \alpha)}}, \quad B = \frac{1}{2}\sqrt{\frac{\beta\,(\beta - \alpha)}{(a^2 - \beta)(b^2 - \beta)(c^2 - \beta)}},$$

and, in accordance with (22) and (25), the fiber forces are

$$T_\alpha = CB\sqrt{\alpha\beta}/\beta(\alpha - \beta), \quad T_\beta = CA\sqrt{\alpha\beta}/\alpha(\alpha - \beta). \tag{6.61}$$

For every surface, depending on the parameters a, b, and c, the domain of α and β is such that A and B are positive and all other quantities are real.

For an elliptic or hyperbolic paraboloid,

$$A = \frac{1}{2}\sqrt{\frac{\alpha\,(\alpha - \beta)}{(a^2 - \alpha)(b^2 - \beta)}}, \qquad B = \frac{1}{2}\sqrt{\frac{\beta\,(\beta - \alpha)}{(a^2 - \beta)(b^2 - \beta)}},$$

while the normal curvatures and the statical potential differ from the above only in inconsequential constants. As a result, formulas (61) for the net forces require only the replacement of parameters A and B.

6.3   Equilibrium Loads and Configurations

The purpose of this section is to establish statical-geometric interrelations among external loads, equilibrium configurations, and internal forces in a net. Both the generic and particular, net-specific, aspects of the problem are addressed. As with any underconstrained system, a given load uniquely determines the corresponding equilibrium configuration of a net, whereas the number of equilibrium loads for a given configuration is infinite. A simple mental experiment clarifies this observation. If all nodes of a net in a given configuration are fixed in space, tension forces T_u and T_v can be assigned arbitrarily and then the resulting nodal reactions represent an equilibrium load. Thus, three equilibrium load components are determined by two arbitrary functions of two variables—the assigned forces in the net.

Singular Versus Nonsingular Nets

Normal Equilibrium Loads of a Net.   The most interesting case for applications is normal surface loading. To establish the number of normal equilibrium loads, imagine that the net is spread over a smooth frictionless rigid surface. Then, arbitrarily pulling at the net edge produces tension forces in the net and a distributed normal interaction between the net and the surface at the interface. This interaction represents a normal equilibrium load for the given configuration of the net. As is readily seen, such a load is determined to within two arbitrary functions of one variable (the chosen tension forces T_u and T_v along the net edge).

Studying net equilibrium under a normal load requires going back to equations (24) written in membrane forces. After restoring the load term in the third equation, it acquires the form

$$N_u\,\sigma_u + N_v\,\sigma_v = P\sin\omega.$$

A solving function N satisfying this equation is introduced by

$$N_u = (F + N)\sigma_v, \quad N_v = (F - N)\sigma_u, \quad F = P \sin \omega / 2\sigma_v \sigma_u. \quad (6.62)$$

In the absence of tangential loads, the first two equations remain intact and, upon substitution of (62), can be presented as one vector equation:

$$N_k - N S_k = F Q_k + F_i a_k^i, \quad (6.63)$$

where Q_k is some vector determined, like vector S_k, by the net geometry. Since N is a scalar, N_k is gradient so that

$$N_{k,}^{\ k} = - N_k S^k + N S_{k,}^{\ k} - F_k Q^k + F Q_{k,}^{\ k} + (F_i a_k^i)_{,}^{\ k} = 0.$$

Replacing N_k with the aid of (63) and noticing that $S_k S^k = 0$ [cf. (3)] gives

$$N = [F_k Q^k - (F Q_{k,}^{\ k} - S^k Q_k) + F_i a_k^i S^k - (F_i a_k^i)_{,}^{\ k}]/S_{k,}^{\ k}. \quad (6.64)$$

On the other hand, scalar multiplication of both sides in (63) by the vector S_k provides another equation relating the functions N and F:

$$(N_k - F Q_k - F_i a_k^i) S_k = 0.$$

From here, upon the elimination of N using expression (64), a third-order differential equation in the unknown F is obtained. This equation, together with (62), determines all equilibrium normal loads for a given net.

As is seen from (64), generally, each equilibrium load gives rise to a unique solving function N and the corresponding tension force pattern in the net. In other words, a solution to the inhomogeneous statical problem for a net, as a rule, is unique. At the same time, expression (64) singles out singular nets by exposing their statical indeterminacy. Indeed, according to the invariant criterion (29), the denominator in (64) equals zero. Therefore, when a solution to the inhomogeneous problem exists, it is not unique.

This observation is consistent with the characteristic property of singular nets—the existence of a nontrivial solution to the homogeneous statical problem (virtual self-stress). For a solution to the inhomogeneous problem to exist, the numerator in (64) must also turn into zero. This leads to a second-order partial differential equation in F determining normal equilibrium loads for singular nets. The equation is hyperbolic, with the characteristics coinciding with the net lines. If a zero load is assigned along the initial curve, then $P = 0$ within the entire influence domain. Note that the reduction in the equation order compared to a nonsingular net, from third to second, indicates a reduction in the number of equilibrium loads. This is a usual consequence of an underconstrained system becoming singular and, thereby, acquiring statical indeterminacy.

To verify that the net forces under a given equilibrium load are tensile, it is convenient to check the sign of the product

$$N_u N_v = (F^2 - N^2) \sigma_u \sigma_v > 0.$$

Equilibrium Loads for Particular Nets.   An overview of the above equations shows that equilibrium normal loads for a given net are determined by the vectors S_k and Q_k, which turn out to be very similar in structure. As a result, some simple particular solutions of the above equations can be obtained by inspection and verified by substitution. For example, for an orthogonal net, such a particular solution is

$$F = C_1 H/\sigma_u \sigma_v, \quad N = C_1 (\sigma_u - \sigma_v)/2\sigma_u \sigma_v, \quad P = 2C_1 H,$$

$$T_u = C_1 B, \quad T_v = C_1 A,$$

meaning that a normal load proportionate to the mean curvature of the net surface is an equilibrium load for any orthogonal net (singular or not).

For a conjugate net,

$$Q_k = a_k^i \omega_i \operatorname{ctn} \omega.$$

In the case of a singular conjugate net, a readily available particular solution is

$$F = C_1/\sin \omega, \quad N = Ce^S, \quad P = 2C_1 K, \quad T_u = (F + N)B\sigma_v, \quad T_v = (F - N)A\sigma_u.$$

It must be emphasized that this solution is valid only for singular nets; otherwise it is meaningless, as it gives a zero value to N in (64) which is impossible for a net supporting an external load. Thus, a normal load proportionate to the Gaussian curvature K is an equilibrium load for a singular conjugate net.

For a geodesic conjugate net (the Voss net) the solution becomes

$$T_u = (C_1 + C)V(v), \quad T_v = (C_1 - C)U(u).$$

For a translation net [cf. (55)],

$$T_x = (C_1/W + C) Y'' \sqrt{1 + X'^2},$$

$$W = 1 + X'^2 + Y'^2.$$

$$T_y = (C_1/W - C) X'' \sqrt{1 + Y'^2},$$

A net of principal curvatures, being simultaneously orthogonal and conjugate, leads to the ultimate simplification: $Q_k = 0$. The entire chain of pertinent equations, starting with (63), simplifies accordingly. For example, for a singular net of principal curvatures, setting the numerator in (64) to zero gives an equation for F, which in the coordinate parameters of the net becomes

$$2F_{12} - S_1 F_2 - S_2 F_1 = 0.$$

Even trivial particular solutions of this equation might be meaningful. For example, for the net of principal curvatures of a quadric, described earlier, the normal equilibrium load and the resulting net forces, corresponding to a solution $F = C_1$, are

$$P = C_1/\alpha^2 \beta^2, \quad T_\alpha = B\sigma_\beta (C_1 + Ce^S), \quad T_\beta = A\sigma_\alpha (C_1 - Ce^S).$$

Geodesic and Semigeodesic Nets

It is convenient to investigate these nets starting with equilibrium equations. Scalar multiplication of equation (21), respectively by u, v, and n while taking into account (17) and (22), produces the following system of equations

$$T_{u,1} + T_{v,2} \cos \omega - BT_v \, \kappa_v \sin \omega = 0,$$

$$T_{u,1} \cos \omega + T_{v,2} - AT_u \, \kappa_u \sin \omega = 0, \qquad (6.65)$$

$$AT_u \sigma_u + BT_v \sigma_v = PAB \sin \omega.$$

Analyzing this system leads to several interesting conclusions on the statics of nets involving one or two geodesic arrays.

Before proceeding with the analysis, note that a net member (fiber or cable) with a tension force not varying along the length is called *isotensoid*. An already familar example of a net with all members isotensoid is a conjugate geodesic net (the Voss net) whose virtual self-stress is given by (54). An *isotensoid array* is one where the tension force is constant both along each member and among the members. Obviously, an isotensoid member or an array may be force-free.

Consider a net under a normal load with one force-free array: $T_u = 0$. Since $\omega \neq 0$, the first two equations (65) yield

$$T_{v,2} = 0, \quad T_v = T_v \,(u), \quad T_v \kappa_v = 0.$$

Turning into zero of the second force, T_v, is impossible in a net supporting a load. Hence, the geodesic curvature κ_v must be zero, leading to the following conclusion.

Property of Nets with One Force-Free Array. If a net carries a normal load and one fiber array is force-free, the other array is geodesic and isotensoid.

Corollary. A one-parametric array of fibers carrying a normal surface load is geodesic and isotensoid.

A typical example of such an array is a radial cable system. The resulting surface force pattern is called in the literature a *tension field* [see, for example, Steigmann and Pipkin (1989a); and Steigmann (1990)].

Isotensoid Nets. Now suppose that in a net under a normal load, tension forces in both arrays are constant:

$$T_u = T_u \,(v), \qquad T_v = T_v \,(u).$$

It follows immediately from the first two equations (65) that both of the geodesic curvatures are zero. The converse is also true: at $\kappa_u = \kappa_v = 0$, the equations become a linear homogeneous algebraic system in the force derivatives, with a nonvanishing determinant $\sin^2 \omega$. Hence, both of the forces remain constant along the length. Both the direct and converse statements are also valid for initial forces, i.e., for a virtual self-stress pattern. Thus, in the absence of tangential load components, a geodesic net (and only this net) is isotensoid. Furthermore, if the nodal connections develop only normal (but not tangential) interaction between the arrays, the net is geodesic, and conversely: in a geodesic net, the nodal connections do not develop tangential interaction (even when this is physically possible due to the fastened intersections). From here it follows that the possibility of mutual fiber slipping, as in a loosely woven and frictionless fabric, makes the net geodesic regardless of its original geometry. On the other hand, since fibers of a geodesic net do not interact in the tangential direction, there is no need in any tangential fasteners at the intersections if the net is to be maintained geodesic and isotensoid. The presented observations can be summed up in a concise form.

Characteristic Properties of a Geodesic Net. If, under a normal surface load and an edge load, a net meets *one* of the three conditions: (1) the net is geodesic; (2) member intersections do not transfer tangential forces; (3) the net is isotensoid; then *all three* conditions are met.

Corollary. A net with all members yielding (a plastic net) is geodesic.

This follows from the simple observation that comprehensive yielding renders the net isotensoid. Interestingly, the net fibers need not be identical; it is only necessary that under some normal load or prestress all of them yield. In the case of a plane plastic net, all fibers must be rectilinear, no matter what has been the initial configuration of the net. Take, for example, a plane net with one or both of the arrays curvilinear (say, a radial-hoop net or a net of confocal conics) stretched by an arbitrary edge load. Such a net must undergo considerable elasto-plastic deformation before all of the fibers become straight, and only then the yielding becomes comprehensive.

As follows from (63) and the last of equations (65), normal loads of the form

$$P = [A\sigma_u V(v) + B\sigma_v U(u)]/AB \sin \omega, \qquad (6.66)$$

and only such loads, are the normal equilibrium loads for a geodesic net.

A Net with One Isotensoid Array. A net formed by a geodesic array and an array orthogonal to it is called semigeodesic. Specializing equilibrium equations (65) in the absence of tangential loads for an orthogonal net ($\omega = \pi/2$) reveals the following.

A Property of a Semigeodesic Net. If in an orthogonal net forces in one array are constant, then the other array (assuming nonzero forces in it) is geodesic—the net is semigeodesic. Conversely, in an orthogonal semigeodesic net, the nongeodesic array is isotensoid.

This property may be useful in a realization of orthogonal semigeodesic nets, like the net of meridians and parallels of a surface of revolution. The non-geodesic array (parallels) must be able to slip at some fixed, say, equidistant, points along each meridian, thereby guaranteeing constant (but not necessarily equal) forces in the hoop fibers or cables.

Design Implications. Aside from providing general information on equilibrium loads and configurations, the above results are useful in a conceptual design of cable nets. The characterization of nets with one force-free array applies to the limit state of a net caused by disengagement of the prestressing array. At the moment of disengagement, the load-carrying cables must be directed along the geodesics of the deformed surface while supporting the maximum load. On the other hand, under a given load, the cable forces are inversely proportional to the normal curvatures of the cables. Hence, it makes sense to maximize the normal curvature by arranging the load-supporting cables along the lines of principal curvature. Both of the requirements can be met by assuring that the deformed surface has lines of principal curvature that are geodesic. Only one of the two principal curvature arrays can be geodesic; such arrays are known to exist on certain surfaces and, remarkably, consist of plane lines. Examples include surfaces of revolution (meridians are geodesic lines of curvature); cyclides (plane circular geodesic lines of curvature); so-called carved surfaces; and some others. In a direct design procedure, such surfaces can be chosen as the deformed, limit state, configurations.

For an entire net under a normal load to be uniformly stressed (isotensoid), the net must be geodesic. Only in this net may the onset of yielding occur simultaneously in all cables. At the same time, it must be remembered that, due to the phenomenon of geometric hardening, the ultimate load for a net is determined by the ultimate strain, not the yield stress. The resulting reserve of load capacity may be considerable and the characteristic properties of a geodesic net will be maintained throughout the entire range of plastic yielding, up to the state with the ultimate strain.

It is known that neither orthogonal nor Chebyshev nets can be geodesic, unless the surface is developable. Thus, in general, these two most important nets cannot be fully (or just uniformly) stressed under any normal load. .

6.4 Statics of Elastic Nets

Starting with early works on Chebyshev nets by Rivlin (1955, 1959), statics of elastic nets developed fast. Papers by Pipkin (1984) and, especially, Steigmann and Pipkin (1991) are representative of the current state of the art. For a comprehensive statical analysis of a net, equations of equilibrium must be complemented with strain–displacement and constitutive relations. Consistent with the main emphasis of this book, most attention is devoted here to the kinematic aspect of the problem, while leaving the constitutive relations in a generic form. The resulting closed system of the equations of statics of elastic nets is obtained in a natural coordinate system, then presented in an invariant form and, finally, specified for an important particular class of shallow nets.

Small Deformations of a Net

Displacements and Strains. Equations relating displacements and strains in a net are conveniently obtained in the natural coordinate system coinciding with the net lines. The radius-vector $\mathbf{r}^*$ of the deformed surface is the sum of the original radius-vector and the displacement vector:

$$\mathbf{r}^* = \mathbf{r} + \Delta, \quad \Delta = U\mathbf{u} + V\mathbf{v} = W\mathbf{n}.$$

Here the displacement vector Δ is represented by its coordinate components—tangential displacements, U and V, and normal displacement, W. The fiber strains are expressed in terms of the coefficients of the first fundamental forms of the original and deformed surfaces

$$\varepsilon_u = (ds_u^* - ds_u)/ds_u = (A^* - A)/A, \quad \varepsilon_v = (B^* - B)/B,$$

with

$$A^2 = r_1^2, \quad B^2 = r_2^2, \quad A^{*2} = r_1^{*2}, \quad B^{*2} = r_2^{*2}.$$

Assuming that strains are small $(\varepsilon_{u,v} \ll 1)$, they can be approximated as

$$\varepsilon_u = \frac{A^{*2} - A^2}{A\,(A^* + A)} = \frac{2A\mathbf{u}\Delta_1 + \Delta_1^2}{A^2(2 + \varepsilon_u)} \approx \frac{\mathbf{u}\Delta_1}{A} + \frac{\Delta_1^2}{2A^2},$$

$$\varepsilon_v \approx \frac{\mathbf{v}\Delta_2}{B} + \frac{\Delta_2^2}{2B^2}. \tag{6.67}$$

The derivatives of the displacement vector are evaluated with the aid of (17):

$$\begin{aligned}
\Delta_1 &= U_1\mathbf{u} + UA\,(\sigma_u\mathbf{n} + \kappa_u\widetilde{\mathbf{u}}) + V_1\mathbf{v} + VA\,(\chi\mathbf{n} + \lambda_u\widetilde{\mathbf{v}}) \\
&\quad + W_1\mathbf{n} - WA\,(\sigma_u\mathbf{u} + \tau_u\widetilde{\mathbf{u}}), \\
\Delta_2 &= V_2\mathbf{v} + VB\,(\sigma_v\mathbf{n} + \kappa_v\widetilde{\mathbf{v}}) + U_2\mathbf{u} + UB\,(\chi\mathbf{n} + \lambda_v\widetilde{\mathbf{u}}) \\
&\quad + W_2\mathbf{n} - WB\,(\sigma_v\mathbf{v} + \tau_v\widetilde{\mathbf{v}}).
\end{aligned} \tag{6.68}$$

Substituting these into the above formulas for strains leads to expressions containing terms of different magnitudes, some being so small that their retainment would be inconsistent with the accuracy adopted in (67). For a typical, membrane-like, deformation of a net, the largest quantities are the derivatives of deflection W, that is, the rotations of the surface normal. These will be the only terms whose squares are retained along with the linear terms in the final expressions for strains:

$$\begin{aligned}
A\varepsilon_u &= U_1 + V_1 \cos\omega - VA\lambda_u \sin\omega - AW\sigma_u + W_1^2/2A, \\
B\varepsilon_v &= V_2 + U_2 \cos\omega + UB\lambda_v \sin\omega - BW\sigma_v + W_2^2/2B.
\end{aligned} \tag{6.69}$$

Equilibrium Equations for a Deformed Configuration. Going back to the equilibrium equations, it is necessary to refine them taking into account that they describe equilibrium in the final, deformed (but yet unknown), configuration of the net. In doing so, the level of accuracy adopted earlier will be consistently maintained.

The unit direction vectors of the deformed net are

$$\mathbf{u}^* = \frac{\mathbf{r}_1}{A^*} = \frac{(\mathbf{r}_1^* + \Delta_1)}{A(1 + \varepsilon_u)} \approx \mathbf{u} + \frac{\Delta_1}{A}, \qquad \mathbf{v}^* \approx \mathbf{v} + \frac{\Delta_2}{B}.$$

When introducing the derivatives (68), only the rotations are retained

$$\mathbf{u}^* = \mathbf{u} + (W_1/A)\,\mathbf{n}, \qquad \mathbf{v}^* = \mathbf{v} + (W_2/B)\,\mathbf{n}.$$

The necessary derivatives of these vectors are evaluated to the same accuracy:

$$\mathbf{u}_1^* \approx A\,(\sigma_u \mathbf{n} + \kappa_u \widetilde{\mathbf{u}}) + (W_1/A)_1\,\mathbf{n}, \qquad \mathbf{v}_2^* \approx B\,(\sigma_v \mathbf{n} + \kappa_v \widetilde{\mathbf{v}}) + (W_2/B)_2\,\mathbf{n}.$$

Equilibrium equation (21) is now adapted to the deformed state. After switching to the net forces, it acquires the form

$$T_{u,1}\mathbf{u}^* + T_u\,\mathbf{u}_1^* + T_{v,2}\mathbf{v}^* + T_v\,\mathbf{v}_2^* = PAB\,\sin\omega.$$

Multiplying this equation by $\widetilde{\mathbf{v}}$, $\widetilde{\mathbf{u}}$, and $\mathbf{n}$ produces the following three scalar equations of equilibrium (tangential load components are assumed absent):

$$
\begin{aligned}
&T_{u,1}\sin\omega - AT_u\,\kappa_u\cos\omega - BT_v\kappa_v = 0, \\
&T_{v,2}\sin\omega + AT_u\,\kappa_u - BT_v\kappa_v\cos\omega = 0, \\
&T_{u,1}W_1/A + T_u\,[A\sigma_u + (W_1/A)_1] \\
&\quad + T_{v,2}W_2/B + T_v\,[B\sigma_v + (W_2/B)_2] = PAB\,\sin\omega.
\end{aligned}
\tag{6.70}
$$

These equations, together with the kinematic equations (69) and the constitutive equations relating the fiber forces to strains, form a closed system of equations of statics of nets. The boundary conditions of the problem describe the forces or displacements along the net edge. For example, if the edge is rigidly supported against all displacements, the boundary conditions are $U = V = W = 0$; their number corresponds to the order of the system of equations—a necessary condition for a correctly posed boundary value problem. Alternatively, the boundary conditions can be assigned in terms of the forces acting at the edge, or as equations relating the edge forces and displacements (elastic supports).

The obtained system of equations is nonlinear. If it is linearized, the highest-order terms in the last of equations (70) do not survive linearization. The resulting drop in the equation order is a sign of this continuous structural system being underconstrained. Recall that the parallel feature in a discrete underconstrained system is overdeterminacy of the algebraic system of equilibrium equations, making it meaningful only in the context of equilibrium loads. As with discrete systems, the presence of an equilibrium load (perhaps, prestress) makes the linearization of equations (70) possible.

The correctness of the above system of equations has been confirmed by its alternative derivation employing the construction of the energy functional. Yet another criterion of correctness of any system of equations describing a physical phenomenon is its invariance in coordinate transformations. Invariance of equations (69)–(70) follows from the possibility of their presentation in a tensor form, which requires going over to the membrane forces N_u and N_v in accordance with (22):

$$(N_{u,i} - N_u R_i) u^i \sin \omega - N_u \kappa_u \cos \omega - N_v \kappa_v = 0,$$

$$(N_{v,i} - N_v R_i) v^i \sin \omega + N_v \kappa_v \cos \omega - N_u \kappa_u = 0,$$

$$\left[(N_{u,i} - N_u R_i) W_j u^j + N_u (W_j u^j)_{,i} \right] u^i$$

$$+ \left[(N_{v,i} - N_v R_i) W_j v^j + N_v (W_j v^j)_{,i} \right] v^i = P \sin \omega, \qquad (6.71)$$

$$(U_i + V_i \cos \omega) u^i - V \lambda_u \sin \omega - W \sigma_u = \varepsilon_u, \quad \varepsilon_u = \varepsilon_u (T_u - T_u^0),$$

$$(U_i \cos \omega + V_i) v^i + U \lambda_v \sin \omega - W \sigma_v = \varepsilon_v, \quad \varepsilon_v = \varepsilon_v (T_v - T_u^0).$$

Here the strains ε_u and ε_v are assumed known functions of the net forces as given by the pertinent constitutive relations. Note that the strains, as well as the displacements, are measured from the state of the system chosen as a reference state. It may be a prestressed state, as indicated in the last two of equations (71).

The tensor version of the basic equations is valid in any coordinate system, which is important in applications where the natural coordinates are not the most convenient ones. Such is the case, considered later, of a skew axisymmetric net, where the best coordinate system is the geographical net of the surface.

Statics of Shallow Nets

Many, if not most, engineering applications of nets deal with shallow nets, for which the above equations can be considerably simplified. The simplifications are based on identifying the intrinsic geometry of a shallow net with the plane geometry. Quantifying the introduced error requires some measure of the net shallowness. The most suitable one is the angle β between the tangent plane, at some prominent point of the net (say, the center), and a tangent plane at any other point. If, within a designated accuracy, $\cos \beta \approx 1$ everywhere, the net is considered shallow. Then the first fundamental form of the net surface can be represented by the metric form of the plane, with the projections of the net lines on the plane serving as coordinates. As a result, all metric parameters of the net (lengths, angles, geodesic and Chebyshev curvatures, and so on) coincide with the respective parameters of the net projection. In addition, the loads normal to the net and to the base plane become indistinguishable, and so are the net forces and their projections on the plane.

A Shallow Net of Translation. For this net, considered in Section 2,

$$\cos \beta_x = \sqrt{1 + X'^2}, \quad \cos \beta_y = \sqrt{1 + Y'^2}. \tag{6.72}$$

Consistent with the adopted accuracy, it follows from (55) that

$$X'^2 \ll 1, \quad Y'^2 \ll 1, \quad A \approx B \approx 1. \tag{6.73}$$

The net projection is an orthogonal Cartesian net, so that

$$\omega = \pi/2, \quad \kappa_x = \lambda_x = \kappa_y = \lambda_y = 0, \tag{6.74}$$

i.e., in this approximation, the net is simultaneously orthogonal, geodesic, and Chebyshev. This is the result of identifying the intrinsic geometry of the net with the plane geometry, as follows from approximations (73). The exact expressions for the normal curvatures simplify as well:

$$\sigma_x = X'', \quad \sigma_y = Y''. \tag{6.75}$$

Similar simplifications are observed in the net statics. Taking into account (72), the state of prestress (56) becomes

$$T_x = H_x = CY'', \quad T_y = H_y = - CX''. \tag{6.76}$$

Furthermore, the possibility of considering this net approximately as geodesic, makes formula (66) applicable to it. Hence, in the light of (73) and (75), the class of equilibrium normal loads for this net is confined to

$$P = X'' Y_0 + Y'' X_0, \tag{6.77}$$

where $X_0(x)$ and $Y_0(y)$ are arbitrary functions of their arguments. Note that, since a translation net is conjugate and approximately orthogonal, it can also be considered, with the same accuracy, as the net of principal curvatures.

To obtain a few representative numbers characterizing this accuracy, consider a hyperbolic paraboloid

$$2z = a \, (y^2 - x^2).$$

According to (55), the exact parameters are

$$A^2 = 1 + a^2 x^2, \quad B^2 = 1 + a^2 y^2, \quad \cos \omega = - a^2 xy/AB,$$
$$A \kappa_x \sin \omega = \partial \, (\cos \omega)/\partial x = - a^2 y/AB^3, \quad A\sigma_x = a/A \sqrt{A^2 + B^2 - 1}. \tag{6.78}$$

Let the net border be a square $x = \pm 1$, $y = \pm 1$, and each of the generating parabolas has an aspect (sag-to-chord) ratio of $1/12$, corresponding to $a = 1/3$. It is easy to see that the above approximations are exact at the apex of the paraboloid and produce the maximum error at the corners. The exact and the respective approximate values at the corner are

$$A^2 = B^2 = 10/9 \approx 1, \quad \cos \omega = 0.1 \approx 0, \quad \omega = 84° \approx 90°,$$
$$A \kappa_x \sin \omega = 0.09 \approx 0, \quad A\sigma_x = 3/\sqrt{10 \cdot 11} \approx 1/3.$$

Basic Equations for a Shallow Net. The system of basic equations (69)–(70) simplifies considerably for a shallow net of translation. According to (73) and (74), the system becomes

$$T_{x,1} = 0, \qquad T_{y,2} = 0,$$

$$T_x (\sigma_x + W_{11}) + T_y (\sigma_y + W_{22}) = P,$$

$$U_1 - W\sigma_x + W_1^2/2 = \varepsilon_x, \quad \varepsilon_x = \varepsilon_x (T_x - T_x^0),$$

$$V_2 - W\sigma_y + W_2^2/2 = \varepsilon_y, \quad \varepsilon_y = \varepsilon_y (T_y - T_y^0).$$

$$(6.79)$$

The first two equations, expressing equilibrium in the tangential directions, yield

$$T_x = T_x (y), \quad T_y = T_y (x). \tag{6.80}$$

This result is an obvious consequence of a shallow net of translation being approximately geodesic; in the absence of tangential loads, the net members are isotensoid. Thus, in this case, the system reduces to the last three equations: one condition of equilibrium in the normal direction, and two kinematic equations combined with constitutive relations.

7
Axisymmetric Nets

The general results of the preceding chapter are elaborated on and applied here to axisymmetric nets. This class of nets is segregated for two main reasons. First, axisymmetric nets are widely used in engineering applications, including situations where a suitable segment of an axisymmetric net approximates a general-type surface with an acceptable precision. Second, their analysis reduces to ordinary differential equations, admitting in many cases closed-form first integrals and making unnecessary the use of tensors. The simpler analysis facilitates better understanding and provides additional insights into the structural behavior of all nets, not just axisymmetric ones.

This chapter follows the same general outline as the preceding one. The first section is an analysis of a singular axisymmetric net. Generally, this net is skew, i.e., it does not possess reflection symmetry with respect to a meridian plane. An interesting class of such nets—skew, yet torque-free in prestress—is identified and investigated. Particular attention is paid to geodesic and Chebyshev nets as easily realizable and most important for applications. Furthermore, the characteristic properties of these nets preserve in kinematic deformations and, for a geodesic net under a normal load, in elastic deformations as well. As to other nets, such as orthogonal, reflection-symmetric, and asymptotic, their characteristic properties, generally, do not preserve even in kinematic deformations. Still, investigation of these nets has interesting implications, especially in the case of orthogonal nets, because of their relevance to the theory of membranes, in particular, wrinkling ones.

The next section is concerned with equilibrium loads and configurations of axisymmetric nets. General statical-geometric relations are established for such nets under normal surface loads, which paves the way for an analysis of net-stiffened elastic membranes in Chapter 9. Closed-form first integrals are obtained for several particular classes of nets, thereby reducing their shape finding to forward integration.

An analysis of net displacements in the last section is based on three invariant geometric parameters, two of which preserve in kinematic deformations of a net and the third in both kinematic and elastic deformations. Kinematic displacements are evaluated for a reflection-symmetric Chebyshev net, and a more comprehensive elastic analysis is carried out for a skew geodesic net. The analysis reveals some unexpected (and beneficial) features of the structural behavior of skew torque-free nets under a uniform surface pressure, in particular, a strong force-leveling effect between the two geodesic arrays.

7.1   Singular Axisymmetric Nets

General Statical-Kinematic Relations

Net of Meridians and Parallels. On a surface of revolution, $r = r(z)$, referred to a cylindrical coordinate system (r, φ, z), an axisymmetric net is one whose geometric parameters are independent of the polar angle φ. The geographical net (the net of meridians and parallels), of course, is not the only axisymmetric net; however, it is the most important one, since it happens to be the net of principal curvatures of the surface. If it is selected as a coordinate net, the Lamé parameters of the surface can be chosen as

$$A = \sqrt{1 + r'^2}, \quad B = r, \tag{7.1}$$

where the prime denotes differentiation with respect to z. The geodesic and normal curvatures of the net lines are

$$\kappa_1 = 0, \quad \kappa_2 = r'/rA, \quad \sigma_1 = -r''/A^3, \quad \sigma_2 = 1/rA, \tag{7.2}$$

with the latter two being the principal curvatures of the surface.

Substituting (2) into formula (6.57) of the statical vector gives

$$S_1 = S' = 2r'\,r''/(1 + r'^2) = \partial_z \ln (1 + r'^2) = \partial_z \ln \csc^2\theta, \quad S_2 = 0, \tag{7.3}$$

where θ is the angle between the normal to the surface and the axis of revolution. Thus, the geographical net on any surface of revolution is singular, with the statical potential

$$S = \ln (1 + r'^2). \tag{7.4}$$

The net forces are found with the aid of (6.30) and (6.31):

$$T_S = C\sqrt{1 + r'^2}, \quad T_\varphi = Cr''. \tag{7.5}$$

Of course, this particular, rather simple result can be obtained without involving the statical vector. The taken aproach just capitalizes on the availability of the statical vector and provides an additional illustration of its use.

As should be expected, the expression for the meridional force T_S coincides with that for tension in a cable supporting a system of parallel forces proportionate to $T_\varphi(z)$. By a suitable selection of the prestressing hoop forces T_φ, it is possible to obtain a quasi-variant net in the form of any surface of revolution of negative total curvature or its segment of an arbitrary shape (e.g., a curved trapezoid bounded by two meridians and two parallels).

An Analytical Criterion for a Singular Axisymmetric Net. A skew (i.e., most general) axisymmetric net (Fig. 1a) is investigated here by analyzing its statical vector S_k. This is the case when special, in this instance, geographical, coordinates are preferable to the natural coordinates coinciding with

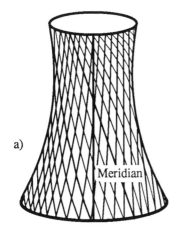 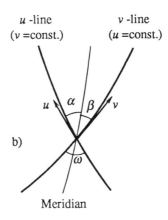

Figure 7.1. A skew axisymmetric net: a) isometric view; b) infinitesimal element and its geometric description.

the net lines. In geographical coordinates, all geometric attributes of an axisymmetric net are independent of the polar angle. This includes the statical vector of the net and its statical potential if the latter exists. Accordingly, a singular axisymmetric net is characterized by a vanishing second component of the statical vector, which leads to an ordinary differential equation obtained by specifying the first component:

$$S_2 = 0, \quad S_1 = \partial S / \partial z = S', \quad S = \int S' \, dz. \tag{7.6}$$

An axisymmetric net is identified by the angles α and β formed by the net lines with the meridian (Fig. 1b). The direction unit vectors of the net lines are

$$u_1 = -A \sin \alpha, \quad u^1 = \cos \alpha / A, \quad u_2 = B \cos \alpha, \quad u^2 = \sin \alpha / B, \tag{7.7}$$

v_i and v^i being obtainable by replacing α with β.

The intrinsic (i.e., geodesic and Chebyshev) curvatures of the net lines can also be evaluated in terms of α and β. Since the meridians and parallels are orthogonal, their transversal vectors, determined by equation (6.10), are equal; they are given by

$$t_{si} = t_{\varphi i} = (0, \ B'/A).$$

According to (6.11), the transversal vectors of the net lines are

$$t_{ui} = t_{si} + \alpha_i = (\alpha', \ B'/A), \quad t_{vi} = (\beta', \ B'/A),$$

and the intrinsic curvatures are expressed by formulas (6.12) with the aid of (7):

$$\kappa_u = [\alpha' \cos \alpha + (B'/B) \sin \alpha]/A, \quad \kappa_v = [\beta' \cos \beta + (B'/B) \sin \beta]/A, \tag{7.8}$$

$$\lambda_u = [\beta' \cos \alpha + (B'/B) \sin \alpha]/A, \quad \lambda_v = [\alpha' \cos \beta + (B'/B) \sin \beta]/A. \tag{7.9}$$

Introducing the above vectors and parameters into (6.28) at $k = 2$ and setting the result to zero, produces an ordinary differential equation relating α and β. Upon some rearrangement, it acquires the form

$$\frac{(B \sin \beta)'}{\cos \beta} (\rho - \cos \omega) + \frac{(B \sin \alpha)'}{\cos \alpha} (1/\rho - \cos \omega) + \frac{\rho'}{\rho} B \cos \omega = 0, \quad (7.10)$$

where

$$\rho = \sigma_u \cos \beta / \sigma_v \cos \alpha, \quad \omega = \beta - \alpha, \quad (7.11)$$

and ω, as before, is the net angle. Employing Euler's formula (6.36) brings equation (10) to a form allowing a solution

$$(\sigma_1/\sigma_2) \operatorname{ctn} \alpha \operatorname{ctn} \beta + C_t \, \sigma_2 (\operatorname{ctn} \alpha + \operatorname{ctn} \beta) = 1 \quad (7.12)$$

which can be checked by back substitution. The meaning of the constant of integration C_t will transpire shortly.

By relating the direction angles α and β to the surface invariants (the two principal curvatures), the obtained solution provides an analytical criterion for a singular axisymmetric net and leads to the following conclusion.

Existence and the Number of Singular Axisymmetric Nets. On a surface of revolution, any one-parametric array of lines intersecting a meridian under a given angle $\alpha = \alpha(z)$, can be complemented by another axisymmetric array to form a singular net. The second array has a direction angle $\beta = \beta(z)$ defined by (12), upon specifying the constant C_t.

As to the statical vector (6.28), its first component for an axisymmetric net is evaluated with the aid of (7)–(9):

$$S' = \{[\cos \omega \sin \beta - (\sigma_v/\sigma_u) \sin \alpha] (B \sin \alpha)'$$

$$+ [\cos \omega \sin \alpha - (\sigma_u/\sigma_v) \sin \beta] (B \sin \beta)'\}/B \sin^2 \omega \quad (7.13)$$

$$- \frac{1}{\sin \omega} \left(\frac{\sigma_v'}{\sigma_v} \sin \beta \cos \alpha - \frac{\sigma_u'}{\sigma_u} \sin \alpha \cos \beta \right) + \frac{\omega'}{\sin \omega} \cos \alpha \cos \beta - \frac{B'}{B}.$$

From here, the statical potential of the net and the corresponding self-stress pattern could be evaluated in the usual way. However, for an axisymmetric net the statical vector is mostly of theoretical interest since the initial forces are conveniently obtained from equilibrium equations, as shown in Section 2.

Torque-Free Singular Axisymmetric Nets

An Analytical Criterion.   For a singular axisymmetric net to be prestressable, both of the initial forces must be tensile; hence, the corresponding meridional membrane force and its axial resultant must be different from zero. However, it is possible, and would be advantageous, if the membrane shearing

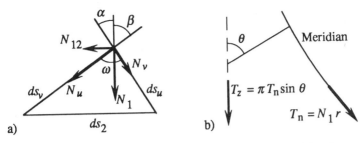

Figure 7.2. Equilibrium of a net element: a) membrane forces and fiber forces acting on a net element; b) meridional force and its axial resultant.

forces, along with the resulting axial torque moment, are absent. This would mean that the azimuthal components of the two net forces cancel each other and the force resultant has meridional direction, so that the net as a whole torque-free.

To evaluate the membrane shearing force, N_{12}, consider equilibrium in the hoop direction of a net element consisting of two intersecting fibers (Fig. 2a):

$$N_{12}\,ds_2 = N_u\,ds_v \sin \alpha + N_v\,ds_u \sin \beta.$$

As seen from the figure,

$$ds_u/\cos \beta = ds_v/\cos \alpha = ds_2/\sin \omega,$$

so that

$$N_{12} = (N_u \sin 2\alpha + N_v \sin 2\beta)/2 \sin \omega. \tag{7.14}$$

Setting this force to zero and combining the result with the equilibrium equation in the normal direction (6.24), gives for a torque-free net

$$\sigma_u/\sigma_v = \sin 2\alpha/\sin 2\beta. \tag{7.15}$$

Using Euler's formula (6.36) for the normal curvatures yields the sought analytical criterion, along with an answer to the existence question:

$$\sigma_1/\sigma_2 = \tan \alpha \tan \beta. \tag{7.16}$$

Existence and the Number of Singular Torque-Free Axisymmetric Nets. Such nets on a given surface of revolution are uniquely determined by the direction angle, $\alpha = \alpha(z)$, of a one-parametric array of lines; the second array has a direction angle $\beta = \beta(z)$ defined by (16).

Comparing (16) to (12) shows that for a torque-free net

$$C_t\, \sigma_2\, (\text{ctn } \alpha + \text{ctn } \beta) = 0.$$

The case $\text{ctn } \alpha + \text{ctn } \beta = 0$ describes a reflection-symmetric net; it has $\sigma_u = \sigma_v$ and, according to (15), cannot be torque-free unless both of the curvatures are zero (an asymptotic net). Hence, there must be $C_t = 0$, indicating that this constant, evolved in solution (12), is associated with torque.

Forces of Self-Stress. For a torque-free net, expression (13) of the statical vector can be rearranged and simplified, taking into account (1) and (15):

$$S' = -\frac{r'}{r} \tan \alpha \tan \beta - \frac{\sigma_v'}{\sigma_v} + \frac{\beta' \cos \beta}{\sin \beta} - \frac{\alpha' \sin \alpha}{\cos \alpha} - \frac{r'}{r} .$$

Only the first term in the right-hand side does not look like a derivative but, in fact, it is. Indeed, the principal curvatures of a surface of revolution can be expressed as

$$\sigma_1 = d (\sin \theta)/dr, \qquad \sigma_2 = \sin \theta/r, \tag{7.17}$$

where the angle θ, between the normal and the axis of revolution, is also the meridian slope (ctn $\theta = r'$). Substitution into (16) leads to

$$(r'/r) \tan \alpha \tan \beta = (\sin \theta)'/\sin \theta. \tag{7.18}$$

As a result, the statical potential of a torque-free net is obtained in a closed form:

$$S = \ln (\sin \beta/r \sigma_v \sin \theta \cos \alpha) \quad \text{or} \quad S = \ln (\sin \alpha/r \sigma_u \sin \theta \cos \beta),$$

and the membrane forces are found in the usual way, with the aid of equations (6.25) and (6.30):

$$N_u = C \sin \beta/r \sin \theta \cos \alpha, \qquad N_v = - C \sin \alpha/r \sin \theta \cos \beta.$$

These forces are related to unit increments the of the respective linear elements ds_v and ds_u. As shown in the previous chapter, more meaningful for a net are forces T_u and T_v per unit coordinate increments $dv = 1$ and $du = 1$. Although the width of these coordinate strips varies, each contains a fixed number of fibers or cables. In addition, for an axisymmetric net, parameters u and v are independent of the polar angle and vary only with z. To go over to T_u and T_v, the membrane forces must be multiplied by the complementary Lamé parameters, B_n and A_n, of the first quadratic form of the net. Introducing the coordinate differentials as

$$du = (A/B) \tan \beta \, dz - d\varphi, \qquad dv = (A/B) \tan \alpha \, dz - d\varphi,$$

gives

$$dz = \frac{B(du - dv)}{A(\tan \beta - \tan \alpha)}, \qquad d\varphi = \frac{du \tan \alpha - dv \tan \beta}{\tan \beta - \tan \alpha} .$$

The transformation formulas for the metric tensor

$$A_n^2 = A^2 (\partial z/\partial u)^2 + B^2 (\partial \varphi/\partial u)^2, \qquad B_n = A^2 (\partial z/\partial v)^2 + B^2 (\partial \varphi/\partial v)^2,$$

yield the following expressions for the Lamé parameters of the net:

$$A_n = r \cos \beta/\sin \omega, \qquad B_n = r \cos \alpha/\sin \omega . \tag{7.19}$$

The net forces T_u and T_v can now be evaluated:

$$T_u = B_n N_u = C \sin \beta/\sin \omega \sin \theta, \qquad T_v = - C \sin \alpha/\sin \omega \sin \theta. \tag{7.20}$$

From here, the axial resultant of the net forces is found by projection (Fig. 2b):

$$T_z = 2\pi (T_u \cos\alpha + T_v \cos\beta) \sin\theta = 2\pi T_n \sin\theta = 2\pi C, \qquad (7.21)$$

where T_n is the meridional force in the net per unit polar angle. This confirms that the axial force T_z is constant (as it should be in the absence of surface loads) and reveals the physical meaning of the arbitrary constant C.

Note that the evaluation of the net forces is but a by-product of the foregoing analysis; its main objective and significance are in identifying the singular nets. The obtained analytical criteria (12) and (16) establish the existence and the number of singular axisymmetric nets, in particular, torque-free ones. Further investigation is carried out for the most interesting special classes of these nets.

Geodesic Nets

Analytical Criterion, Statical Potential, and Net Forces. Geodesic nets on a surface of revolution obey the Clairaut formula:

$$r \sin\alpha = a, \qquad r \sin\beta = b, \qquad (7.22)$$

where each of the constants a and b defines a one-parametric axisymmetric array of lines. These relations can be verified by introducing the metric parameters (1) into expressions (8) for κ_u and κ_v, setting them to zero,

$$\kappa_u = \alpha' \cos\alpha + r' \sin\alpha/r = 0, \qquad \kappa_v = \beta' \cos\beta + r' \sin\beta/r = 0, \quad (7.23)$$

and integrating.

By virtue of (22), equation (10) simplifies drastically: its first two terms vanish (recall that $B = r$; see also the original expression of the statical vector (6.28) containing geodesic curvatures). In the last term, neither B nor $\cos\omega$ can be zero (geodesic nets can be orthogonal only on developable surfaces). Hence, the solution (i.e., an analytical criterion of a singular geodesic net) is

$$\rho = \sigma_u \cos\beta/\sigma_v \cos\alpha = C_1. \qquad (7.24)$$

Of all axisymmetric geodesic nets, those and only those satisfying relation (24) are singular.

The first component of the statical vector (6.28) for this net acquires the form

$$S' = \frac{1}{\sin\omega} \left(\frac{\sigma_u'}{\sigma_u} \sin\alpha \cos\beta - \frac{\sigma_v'}{\sigma_v} \sin\beta \cos\alpha \right) + \frac{\omega'}{\sin\omega} \cos\alpha \cos\beta - \frac{B'}{B}.$$

Taking into account (1) and (22), the solution of this equation—the statical potential of a singular axisymmetric geodesic net—is found:

$$S = \ln (\sin\omega/r\sigma_v \cos\alpha) \quad \text{or} \quad S = \ln (\sin\omega/r\sigma_u \cos\beta).$$

According to (6.25), (6.30), and (24), the initial forces in the net are

$$N_u = C \sin\omega/r \cos\alpha, \qquad N_v = - CC_1 \sin\omega/r \cos\beta, \qquad (7.25)$$

and, in view of (19), the fiber forces in both arrays are constant:

$$T_u = B_n N_u = C, \quad T_v = A_n N_v = -C_1 C. \tag{7.26}$$

Thus, the virtual self-stress pattern in a singular axisymmetric geodesic net represents two isotensoid arrays. The reason is physically obvious: the two arrays of the net interact only by pressing against each other in the normal direction. The intensity of this contact pressure is

$$P = \frac{N_u \sigma_u}{\sin \omega} = -\frac{N_v \sigma_v}{\sin \omega} = \frac{C \sigma_u}{r \cos \alpha} = \frac{C_1 C \sigma_v}{r \cos \beta}.$$

By comparing (24) and (26), a statical-geometric interrelation for singular axisymmetric geodesic nets is established:

$$\sigma_u \cos \beta / \sigma_v \cos \alpha = -T_v/T_u = C_1. \tag{7.27}$$

Torque-Free Geodesic Net. The membrane shear produced by the initial forces of a singular geodesic net is evaluated by introducing forces (25) into (14):

$$N_{12} = C(\sin \alpha - C_1 \sin \beta)/r = C(a - C_1 b)/r^2.$$

The resulting axial torque moment is

$$M_z = 2\pi r B N_{12} = 2\pi C (a - C_1 b);$$

naturally, it does not vary over the net and equals (in magnitude) the moments acting in the opposite directions at the two edge parallels. As is readily seen, a torque-free singular net is feasible. Indeed, assuming $\beta > 0$ and both initial forces tensile, it follows that a prestressable net is torque-free at

$$\alpha < 0, \quad a < 0, \quad C_1 = a/b = -T_v/T_u.$$

An explicit geometric description of a torque-free singular net is obtained by integrating equation (18). Its left-hand side, transformed with the aid of relations (23), becomes a derivative, and the equation admits a closed-form integral

$$r \sin (\alpha + \beta)/\sin \theta \equiv t(r)/\sin \theta = \text{const}. \tag{7.28}$$

where $t(r)$ is, according to (22), a known function of r. At the surface equator,

$$\theta_e = \pi/2, \quad t(r_e) = r_e \sin (\alpha_e + \beta_e) = t_e,$$

and the above solution can be rewritten as

$$\sin \theta = t(r)/t_e. \tag{7.29}$$

It leads to a differential relation,

$$dz = \tan \theta \, dr = t/\sqrt{t_e^2 - t^2} \, dr, \tag{7.30}$$

between the axial coordinate z and radius r of the net surface, thereby reducing its determination to a quadrature.

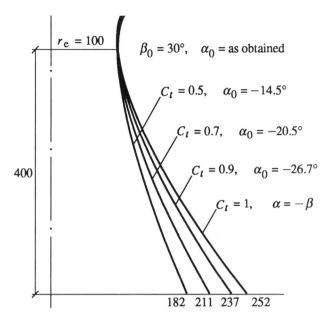

Figure 7.3. Profiles of a prestressable torque-free geodesic net.

Case Study: Torque-Free Geodesic Net. In this parametric study, the shape of a skew torque-free geodesic net is obtained explicitly by numerical integration of equation (30). The surface equator, being the plane of symmetry and the origin of the axial coordinate z, is chosen as a reference plane. A computationally convenient set of initial data includes $r_e = r(0)$, $\beta_e = \beta(0)$, and the ratio C_1. The radius r_e of the equator determines only the size of the net, thus being merely a scale factor. The net configuration is governed by the remaining two parameters which provide a variety of feasible shapes. The initial value β_e was found to affect strongly the radius of revolution: the larger β_e, the more rapidly the radius increases. The angles α and β vary considerably along the meridian, but their ratio is almost invariant and approaches the value of C_1 for sufficiently small angles.

The role of the force ratio C_1 is illustrated in Fig. 3 showing the net profiles for fixed values of $r_e = 100$, $\beta_e = 30°$, and several values of C_1. As the latter increases, the resulting profile widens and angles α and β become closer to each other in absolute value. This causes the normal curvatures of the members to decrease as the net lines gradually approach the asymptotic (zero normal curvature) directions on the surface. Accordingly, the net resistance to normal surface loads (the lateral stiffness) and the normal contact pressure between the two fiber arrays also diminish. At $C_1 = -1$, corresponding to $T_u = T_v$, the net lines become asymptotic [cf. (27)] and the surface profile degenerates into a limiting hyperbola; the fibers are the linear generators of a one-sheet hyperboloid of revolution and exert zero mutual pressure. In this configuration, the lateral stiffness reduces drastically: the first-order stiffness is of a statical-kinematic nature, and is proportional to the prestress level. Obviously, from the standpoint of the lateral stiffness, the optimal configuration of a net is one with large member curvatures.

Note that a solution for a reflection-symmetric geodesic net $(\alpha = -\beta)$ does not follow from (28). The reason is that this net is asymptotic and necessitates a different analytical approach, implemented below.

Chebyshev Nets

Analytical Criterion, Statical Potential, and Net Forces. This is one of the most common nets: all of its cells are rhombuses. Its geometric theory is a by-product of Chebyshev's work (for a textile industrialist) on the optimization of cut-out patterns in clothes design. Chebyshev proved that due to, and at the expense of, the varying net angle, the net is applicable to any smooth surface and is determined to within two arbitrary functions of one variable. This means, for example, that two of the net lines may be directed along two designated lines on the surface.

The net is characterized by constant Lamé parameters (19) or by zero Chebyshev curvatures (9). The second condition leads to

$$r'/r = -\alpha' \operatorname{ctn} \beta, \quad r'/r = -\beta' \operatorname{ctn} \alpha. \qquad (7.31)$$

Solving these equations simultaneously yields

$$r \cos \beta / \sin \omega = a, \quad r \cos \alpha / \sin \omega = b, \qquad (7.32)$$

where a and b are arbitrary constants. Formulas (32) play the same key role for a Chebyshev net as do the Clairaut formulas (22) for a geodesic net. Comparing (32) with (19) confirms constancy of the Lamé parameters of the net: $A_n = B_n = 1$. Accordingly, for a Chebyshev net, and only for it, the membrane forces and the corresponding net forces are identical: $N_{u(v)} = T_{u(v)}$.

To establish some global properties of an axisymmetric Chebyshev net, let $b > a$ and $\beta > 0$. Then, according to (32), $\beta > |\alpha|$, so that

$$a \cos \alpha = b \cos \beta \qquad (7.33)$$

and the net radius is

$$r = b \sin \beta - a \sin \alpha. \qquad (7.34)$$

From here it can easily be found that, consistent with the above inequalities, the minimum and maximum radii of the net are, respectively,

$$r_1 = r \,|_{\alpha = \beta = \pi/2} = b - a, \quad r_2 = r \,|_{\alpha = -\beta = -\pi/2} = b + a. \qquad (7.35)$$

Another important radius found from (33) and (34) is

$$r_0 = r \,|_{\alpha = 0} = \sqrt{(b^2 - a^2)} = \sqrt{r_1 r_2}, \qquad (7.36)$$

and from equations (32),

$$\sin \alpha = (r_0^2 - r^2)/2ar, \quad \sin \beta = (r_0^2 + r^2)/2br. \tag{7.37}$$

The obtained relations are sufficient for establishing the statical-geometric interrelations for singular axisymmetric Chebyshev nets. This is done by equating to zero the second component of the statical vector. After setting $k = 2$ in (6.28), and rearranging it using appropriate substitutions, the following equation is obtained:

$$r' (\rho - 1/\rho) - r (\sigma_u'/\sigma_u - \sigma_v'/\sigma_v) \cos \alpha \cos \beta = 0,$$

where ρ is given by (11). With the aid of (31), its solution (i.e., an analytical criterion of a singular Chebyshev net), is found

$$\rho = (C_1 \sin \beta + \sin \alpha) / (C_1 \sin \alpha + \sin \beta), \tag{7.38}$$

where C_1 is an arbitrary constant. Of all axisymmetric Chebyshev nets, those and only those satisfying relation (38) are singular.

Employing this relation allows the first component of the statical vector (6.28) of an axisymmetric Chebyshev net to be presented as

$$S_1 = S' = \frac{\beta' \rho}{\sin \alpha \cos \beta} - \frac{\sigma_v'}{\sigma_v} + \frac{\omega' \cos \omega}{\sin \omega}.$$

Integration yields two equivalent versions of the statical potential:

$$S = \ln \frac{C_1 \sin \alpha + \sin \beta}{\sigma_v \cos \alpha \sin \omega} \quad \text{or} \quad S = \ln \frac{C_1 \sin \beta + \sin \alpha}{\sigma_u \cos \beta \sin \omega}.$$

The net forces, equal in this case to the membrane forces, are [cf. (11) and (38)]

$$T_u = N_u = C \frac{C_1 \sin \alpha + \sin \beta}{\cos \alpha \sin \omega}, \quad T_v = N_v = - C \frac{C_1 \sin \beta + \sin \alpha}{\cos \beta \sin \omega}. \tag{7.39}$$

As before, the most interesting case is a torque-free net. It is obtained by evaluating and then setting to zero the membrane shear force that produces torque. Introducing (39) and (20) into (14) gives

$$N_{12} = C C_1 (\sin^2 \alpha - \sin^2 \beta)/\sin^2 \omega = 0.$$

The only meaningful way this expression can turn into zero is with $C_1 = 0$; hence, the initial forces in a torque-free net are

$$T_u = C \sin \beta/\cos \alpha \sin \omega, \quad T_v = - C \sin \alpha/\cos \beta \sin \omega. \tag{7.40}$$

Torque-Free Chebyshev Nets. Like a geodesic net, a singular torque-free Chebyshev net allows a closed-form first integral leading to an explicit geometric description of the net. Substituting expressions (31), one at a time, into (18) produces either of the following:

$$\sin \theta/\cos \alpha = \text{const.}, \quad \sin \theta/\cos \beta = \text{const.} \tag{7.41}$$

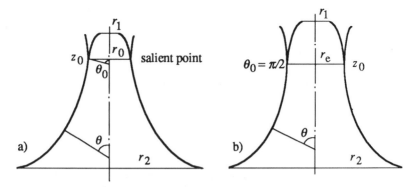

Figure 7.4. Profiles of a singular torque-free Chebyshev net: a) without an equator; b) with an equator.

Taking into account (32), this result can be presented as

$$r \sin \theta / \sin \omega = s(r) \sin \theta = \text{const.}, \tag{7.42}$$

where $s(r)$, in view of (32), can be considered as a known function of r. At the surface equator, $s(r_e) = s_e$, so that

$$\sin \theta = s_e / s(r), \tag{7.43}$$

and the surface profile is determined by forward integration from the equation

$$dz = \tan \theta \, dr = s_e / \sqrt{s^2 - s_e^2} \, dr. \tag{7.44}$$

The integration can be done upon assigning an appropriate set of four arbitrary parameters. At least one of the parameters must be a linear dimension (e.g., r_0, a, or b) thus determining the actual size of the net or serving as a size scale factor. The remaining parameters can be other dimensions, meridian slope θ, angles α and β, force ratio T_u/T_v, etc. These may be assigned at one or more locations along the z-axis although a computationally convenient assignment is the location $z = z_0$, where, by virtue of (36), (38), and (40),

$$\alpha_0 = 0, \quad \beta_0 = \cos^{-1}(a/b), \quad \sigma_u^0 = 0, \quad T_v^0 = 0, \quad T_u^0 = C.$$

As follows from (42), at this location,

$$r'' = (\tan \theta)' = 0,$$

i.e., (r_0, z_0) is an inflection point on the surface meridian, and the surface segment above the reference parallel $z = z_0$ has positive total curvature (Fig. 4). Here the net cannot be prestressed since one of the forces would have to be compressive as is also seen from (40) at $\alpha > 0$. In order to extend the net beyond z_0, the mirror image of the lower segment of the surface can be used. Generally, this will give rise to a salient point on the surface meridian (as is shown in Fig. 4a) in which case a structural ring is needed to support the radial resultant of the net forces, $T_r^0 = T_u^0 \cos \theta$. However, if $\theta_0 = \pi/2$, the inflection parallel becomes an equator, the meridian is smooth (Fig. 4b) and then, according to (41), $\sin \theta = \cos \alpha$ over the entire net.

Note that $z = z_0$ is the plane of symmetry of the net surface but not necessarily of the net itself. It is both geometrically and statically possible that the net changes the direction of twist when passing through $z = z_0$, i.e., α and β either retain their signs or change them simultaneously.

At the planes $z = z(r_1)$ and $z = z(r_2)$ of the extreme radii (35), the surface meridian has a cusp with a zero slope, so that both T_u and T_v become infinite; a net containing such parallels cannot be prestressed.

Other Singular Axisymmetric Nets

Orthogonal Nets. On any surface, orthogonal nets are determined to within one arbitrary function of two variables. For example, if this function defines a one-parametric array of lines, these can always be complemented with orthogonal trajectories to form a net. In contrast, a geodesic or a Chebyshev net is defined by two functions of one variable. Hence, there exist "many more" orthogonal nets on a surface than geodesic or Chebyshev nets, and this is also true for particular subclasses of the respective nets, such as singular or axisymmetric.

The calculation employed in establishing the existence of orthogonal singular nets in the preceding chapter can be used for their identification on a surface of revolution. Specifically, a singular orthogonal axisymmetric net is sought as the result of rotating the net of meridians and parallels through an unknown angle $\beta = \beta(z)$. The geodesic curvatures of the net lines are evaluated according to (6.12) and (6.8) while taking advantage of the axial symmetry:

$$\kappa_u = B' \sin \beta/AB + \beta' \cos \beta/A, \quad \kappa_v = B' \cos \beta/AB - \beta' \sin \beta/A.$$

The normal curvatures of the net lines found from the Euler formula are

$$\sigma_u = \sigma_1 \cos^2 \beta + \sigma_2 \sin^2 \beta, \quad \sigma_v = \sigma_1 \sin^2 \beta + \sigma_2 \cos^2 \beta, \quad \sigma_u + \sigma_v = 2H,$$

where H is the mean curvature of the surface. As a result,

$$\sigma_u - \sigma_v = (\sigma_1 - \sigma_2) \cos 2\beta = 2(H - \sigma_2) \cos 2\beta.$$

Substituting these expressions along with the unit direction vectors (6.8) into (6.35) at $k = 2$ gives after some transformations

$$S_2 = [4H\sigma_2\gamma' + (H'\sigma_2 - 2H\sigma_2') \sin 4\gamma] B/A\sigma_u\sigma_v.$$

For S_2 to vanish, the bracketed expression must be zero,

$$4\beta'/\sin 4\beta + H'/H - 2\sigma_2'/\sigma_2 = 0.$$

The solution of this ordinary differential equation is

$$\tan 2\beta = C_1 \sigma_2^2/H. \tag{7.45}$$

Thus, on any surface of revolution, there exists a one-parametric family of singular orthogonal axisymmetric nets determined by the constant C_1 in (45). As before, minimal surfaces ($H = 0$), in this case represented by the only minimal axisymmetric surface—a catenoid of revolution, constitute an exception: here every orthogonal net is singular.

Using the preceding formulas, the first component of the statical vector can be transformed in a similar way, yielding the statical potential

$$S = \ln (\sec 2\beta \csc^2 \theta).$$

As shown in the next chapter, the corresponding initial forces, given by (6.25),

$$N_u = C\, \sigma_v \sec 2\beta \csc^2\theta, \qquad N_u = -\,C\, \sigma_u \sec 2\beta \csc^2\theta,$$

coincide with the principal forces in the membrane of revolution subjected to an axial force and a torque moment. Hence, in contrast to their geodesic and Chebyshev counterparts, all orthogonal singular nets (not just axisymmetric ones) are obtainable by means of the membrane shell analysis. Specifically, they are nets of the principal force trajectories in membranes subjected to edge loads and, as such, can be identified by solving a homogeneous problem of membrane statics.

Going back to axisymmetric singular orthogonal nets, a somewhat unexpected situation is encountered with their torque-free subclass. With the obvious (and, in this context, trivial) exception of geographical nets, torque-free orthogonal nets exist on only one surface of revolution. Indeed, the membrane shearing force (14) for an orthogonal singular net is

$$N_{12} = (N_u - N_v) \sin 2\beta = CH \tan 2\beta \csc^2 \theta \qquad (7.46)$$

and, for a skew net ($\beta \neq 0$), it vanishes only together with the mean curvature H, i.e., only on a catenoid. Thus, a catenoid is the only surface of revolution admitting torque-free singular orthogonal nets. Moreover, as with any minimal surface, all orthogonal nets of a catenoid are singular: they have $\sigma_u = -\sigma_v$ and, therefore, $N_u = N_v$, hence, all of them are prestressable and torque-free.

Once again, a universal exception is the asymptotic net. Recall that this net on minimal surfaces (and only there) happens to be orthogonal and bisects the net of principal curvatures. Because of the zero normal curvatures of the asymptotic net lines, equation (46) is inapplicable and the initial forces may produce some torque. Thus, the exceptional "wealth" of singular orthogonal torque-free nets on a catenoid is accompanied by the extreme scarcity of such nets with a possibility of torque: only the asymptotic net of a catenoid allows virtual self-stress patterns with a nonzero resultant torque moment.

Reflection-Symmetric and Asymptotic Nets. In addition to axial symmetry, these nets are symmetric relative to a meridian plane, i.e., they possess both rotation and reflection symmetry. As a result,

$$\alpha = -\beta, \quad \omega = 2\beta, \quad k_u = -k_v, \quad \lambda_u = -\lambda_v, \quad \sigma_u = \sigma_v. \qquad (7.47)$$

The equality of the normal curvatures immediately entails $T_u = -T_v$. The corresponding meridional membrane force N_1 and, with it, the axial resultant T_z are zero, whereas the membrane shear given by (14) is not:

$$N_1 = 0, \quad T_z = 0, \quad N_{12} = 2N_u \sin 2\beta/2 \sin \omega = N_u.$$

The resulting torque moment

$$M_z = 2\pi r^2 N_{12} = 2\pi r^2 N_u$$

does not vary along the z-axis and represents an arbitrary constant in the expressions for the membrane forces of the virtual self-stress pattern

$$N_u = -N_v = M_z/2\pi r^2.$$

According to (6.22) and (19), the corresponding net forces are

$$T_u = -T_v = M_z/4\pi r \sin \beta \tag{7.48}$$

and are always of the opposite sign. Thus, reflection-symmetric nets are singular but unprestressable and are "pure torsion" nets.

Asymptotic nets exist only on surfaces of negative total curvature and are characterized by $\sigma_u = \sigma_u = 0$. Because of that, the homogeneous equilibrium equation in the normal direction is satisfied identically, and the remaining two differential equations in two unknowns always admit a nontrivial solution. Accordingly, asymptotic nets are singular and infinitely statically indeterminate; they have an infinite number of linearly independent virtual self-stress patterns but no equilibrium loads normal to the surface.

On a surface of revolution, the asymptotic net is reflection-symmetric, so that relations (47) apply. However, other than that, this unique net is very different from a general reflection-symmetric net. For the latter, the algebraic equation of normal equilibrium is meaningful and leads to the above simple "pure torsion" solution. For an asymptotic net, the condition of normal equilibrium does not provide any information, and the two differential equations of equilibrium in tangential directions must be solved. Taking advantage of the symmetry, this problem can be reduced to an evaluation of two independent self-stress patterns: symmetric, $T_u = T_v = T_+$, and antisymmetric, $T_u = -T_v = T_-$, with the respective axial force and torque moment resultants. In the first case, equilibrium of the net in the axial direction requires

$$T_+ = T_z/4\pi \cos \beta \sin \theta. \tag{7.49}$$

In the same way, the already familiar antisymmetric solution (48) is found. Combining the two solutions gives

$$T_{u(v)} = \frac{1}{4\pi r} \left(\frac{T_z}{\cos \beta \sin \theta} \pm \frac{M_z}{r \sin \beta} \right). \tag{7.50}$$

This combined axisymmetric solution contains only two arbitrary parameters, T_z and M_z, thereby describing just two out of the infinite number of virtual self-stress patterns of an asymptotic net. All other patterns are not axisymmetric and their evaluation would require solving simultaneous partial derivative equations.

The obtained solution leads to the following proposition:

Statical-Geometric Property of Reflection-Symmetric Nets. If a net with rotation and reflection symmetry has a statically possible state (virtual self-stress) under an axial force and a nonzero torque moment, it is the asymptotic net of the surface.

The proof is simple. The stipulated type of the net and load symmetry entails

$$\sigma_u = \sigma_v, \quad T_{u(v)} = T_+ \pm T_-.$$

But then the equilibrium condition in the normal direction described by $(6.24)_3$ requires $\sigma_u = \sigma_v = 0$, meaning that the net is asymptotic.

Asymptotic nets of surfaces of revolution are well studied. Therefore, given some intrinsic geometric properties of the net, the corresponding surface might be identifiable immediately. For example, if an axisymmetric asymptotic net is geodesic, its two arrays do not interact statically either in the normal or tangential directions. As a result, each of the net forces must be constant and, according to (49), this net is characterized by

$$\cos \beta \sin \theta = \cos \beta_e = \text{const.} \tag{7.51}$$

where subscript e refers to the surface equator. It is easy to verify that this net is the net of linear generators of a one-sheet hyperboloid of revolution given by

$$r^2/r_e^2 - z^2/c^2 = 1, \quad \cos \beta_e = c/\sqrt{r_e^2 + c^2}.$$

When an asymptotic net is orthogonal, the surface is minimal. The only minimal surface of revolution is a catenoid; its axisymmetric orthogonal nets, including the asymptotic net, have been discussed in detail above.

Probably, the most interesting case is an axisymmetric asymptotic Chebyshev net. According to (18) and (31), for this net,

$$-\beta' \tan \beta = (\sin \theta)'/\sin \theta,$$

leading to a simple closed-form solution:

$$\cos \beta/\sin \theta = \text{const.} \tag{7.52}$$

Chebyshev nets are asymptotic on surfaces of constant negative total curvature. Among surfaces of revolution, such are pseudospherical surfaces. Thus, a segment of a Chebyshev net in the form of a hose (like a basketball net), stretched between two parallel and coaxial rings, acquires the form of a hyperbolic, parabolic, or elliptic pseudosphere (Fig. 5). The particular outcome depends, respectively, on whether the angle θ is greater, equal to, or smaller than half the net angle, β. (As is seen from (52), the relation between θ and β is valid throughout the entire net.) This curious result, obtained by simple

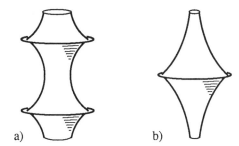

a)   b)   c)

Figure 7.5. Equilibrium shapes of a reflection-symmetric Chebyshev net are three pseudospherical surfaces: a) hyperbolic; b) parabolic; and c) elliptic.

statical means, has, apparently, been overlooked in differential geometry, although a similar problem on the shape of a liquid film—the classic Plateau problem—was solved long ago (its axisymmetric solution is a catenoid of revolution).

7.2   Equilibrium Loads and Configurations

General Statical-Geometric Relations

External Loads and Net Forces.   The following analysis is concerned with statical-geometric interrelations for a skew axisymmetric net under a normal surface load. The analysis proceeds from the overdeterminate system of three equilibrium equations in two unknown net forces. The third equation, which is now inhomogeneous, can be rewritten in terms of the membrane forces and principal curvatures:

$$N_1 \sigma_1 + N_2 \sigma_2 = P. \tag{7.53}$$

Taking into account (14) and (19), the three membrane forces can be related to the net forces (Fig. 2) as follows:

$$rN_1 = T_u \cos \alpha + T_v \cos \beta \equiv T_n,$$
$$rN_{12} = T_u \sin \alpha + T_v \sin \beta, \tag{7.54}$$
$$rN_2 = T_u \sin \alpha \tan \alpha + T_v \sin \beta \tan \beta.$$

Note the different dimensionality of the membrane and net forces: (force/length) for the former and force per unit polar angle for the latter. Recall that a strip $d\varphi = 1$ contains a fixed number of fibers of each array.

The membrane forces produce the axial force and torque moment resultants

$$T_z = 2\pi r N_1 \sin \theta = 2\pi r^2 \sigma_2 N_1, \quad M_z = 2\pi r^2 N_{12}. \tag{7.55}$$

In terms of these resultants, the forces in the two arrys of the net are expressed this way:

$$2\pi r T_u \sin \omega = r T_z \sin \beta / \sin \theta - M_z \cos \beta,$$
$$2\pi r T_v \sin \omega = M_z \cos \alpha - r T_z \sin \alpha / \sin \theta,$$

(7.56)

where $\omega = \beta - \alpha$ is the net angle. The torque moment is constant throughout the net and equals its value at the edges of the net. The axial force T_z varies along the z-axis due to the presence of a surface pressure and is evaluated from the condition of axial equilibrium

$$2\pi r P = d T_z / dr.$$

(7.57)

Integration yields

$$T_z = T_{z0} + 2\pi \int_{r_0}^{r_1} P(r) \, dr = T_{z0} + 2\pi \int_{z_0}^{z_1} P(z) \operatorname{ctn} \theta \, dz,$$

where T_{z0} is the axial force at the parallel $z = z_0$. If pressure P is known as a function of r, the first version of this equation leads to an explicit expression for $T_z(r)$. For a uniform pressure,

$$T_z = T_{z0} + \pi P (r^2 - r_0^2).$$

When P is given as a function of z, the axial force cannot be obtained explicitly. In this case, choosing between z and r as an independent variable becomes a matter of computational convenience and depends on the load and support conditions of the net.

Since the three membrane forces in equations (54) are produced by only two net forces, the membrane forces are not independent. This allows the hoop force to be expressed in terms of the meridional and shearing forces:

$$N_2 = N_{12} (\tan \alpha + \tan \beta) - N_1 \tan \alpha \tan \beta.$$

Introducing this into equation (53) and going over to the net force resultants (55) while taking into account (17) gives

$$T_z (\sigma_1 / \sigma_2 - \tan \alpha \tan \beta) + M_z \sigma_2 (\tan \alpha + \tan \beta) = 2\pi r^2 P.$$

(7.58)

With the aid of (17) and (57), this equation reduces to

$$\frac{d}{dr} \frac{T_z}{\sin \theta} + \frac{T_z \tan \alpha \tan \beta}{r \sin \theta} - \frac{M_z (\tan \alpha + \tan \beta)}{r^2} = 0.$$

(7.59)

The obtained first-order differential equation interrelates the statical and geometric parameters of equilibrium configurations of an axisymmetric net. Recall that the net is subjected to an axial force, torque moment, and normal surface pressure, although the pressure figures in this equation only implicitly, by means of (57). A general solution of equation (59) is

$$T_z / \sin \theta = \left[M_z \int (g/f) \, dr + C \right] f = 2\pi T_n,$$

(7.60)

where

$$f = f(r) = \exp\left[-\int (\tan\alpha\,\tan\beta/r)\,dr\right], \quad g = g(r) = (\tan\alpha + \tan\beta)/r^2,$$

and T_n is the meridional force in the net introduced in (21). Explicit solutions for particular types of nets can be obtained from here as soon as the angles α and β are specified as functions of the radius r.

Equilibrium Configurations and Their Evolution. The equilibrium shape of a net under a given load can be determined by forward integration of the equation obtained from (60)

$$\operatorname{ctn}\theta \equiv dr/dz = \sqrt{(2\pi T_n/T_z)^2 - 1}. \tag{7.61}$$

In the process of integration, relation (57) is used to go over from a given surface pressure P to the axial force T_z. However, if the pressure is known as a function of z (like, e.g., hydrostatic pressure), equation (57) is replaced by

$$2\pi r P = (dT_z/dz)\tan\theta. \tag{7.62}$$

It follows from (58) that in the absence of torque moment

$$\sigma_1/\sigma_2 = 2\pi r^2 P/T_z + \tan\alpha\,\tan\beta. \tag{7.63}$$

The first integral (60) reduces for a torque-free net to

$$T_z/f\,\sin\theta = 2\pi T_n/f = C = T_{z0}/f_0\,\sin\theta_0. \tag{7.64}$$

Combining this with (63) produces a statical-geometric relation characterizing axisymmetric torque-free nets:

$$T_{n0}\,(\sigma_1 - \sigma_2\tan\alpha\,\tan\beta)\,f/rf_0 = P. \tag{7.65}$$

The above relations allow some qualitative observations to be made on the shape evolution of a torque-free net with a constant edge tension T_{z0} when the net is subjected to a gradually increasing pressure loading. According to equation $(56)_2$, for a net in tension,

$$\tan\alpha\,\tan\beta < 0.$$

As is seen from (63), under an external pressure ($P < 0$), as well as in the absence of pressure (a state of prestress), there must be $\sigma_1 < 0$, i.e., the meridian is concave. When an internal pressure is applied, σ_1 gradually reduces in absolute value, and a point on the meridian where it first reaches zero becomes an inflection point with

$$2\pi r^2 P = -T_z\tan\alpha\,\tan\beta. \tag{7.66}$$

As the pressure increases, the inflection parallel shifts along the net, leaving behind a segment of the net with a convex meridian. A cylindrical shape is possible only if the two edge rings of the net are of the same radius and condition (66) is satisfied.

For a reflection-symmetric net, $\alpha = -\beta$, and solution (64) is valid for a net *with or without torque* since the last term in equation (59) vanishes anyway. Thus, for reflection-symmetric nets, the part of the problem associated with torque completely uncouples. The net forces induced by a torque moment are given by (48). Since these forces are equal and opposite, whereas the normal cuvatures of the two member arrays are equal, there is no norml interaction between the arrays. Therefore, in contrast to a skew net, the torque-induced forces do not affect the equilibrium configuration geometry which is, in fact, the reason for the problem uncoupling.

The equilibrium configuration, as before, is obtainable by forward integration. However, torque produces compression forces in one of the arrays, so caution must be exercised in order to prevent fiber disengagement when the compression exceeds the tension forces induced by the surface loads and prestress.

An interesting general property of skew axisymmetric torque-free nets is the fact that an axisymmetric load is not necessarily an equilibrium load. Specifically, this is the case when one of the two edge rings is not constrained against axial rotation. For such a net, any axisymmetric surface pressure contains a perturbation load component which will cause the unrestrained ring to rotate before arriving at a deformed torque-free configuration.

Further investigation requires knowledge of the intrinsic geometric properties of the net obtainable only upon specifying the net type. In what follows, the three most important particular types of nets are studied—geodesic, Chebyshev, and orthogonal nets. As to an asymptotic net, it does not have equilibrium loads normal to the surface. When subjected to such a load, the net, like any other underconstrained system under a perturbation load, must change its configuration before coming to equilibrium.

Geometric Invariants of Axisymmetric Nets. There exist three invariant parameters associated with a segment of an axisymmetric net contained between the edge parallels, z_0 and z_1. The first two invariants are the natural (unstretched) cable lengths:

$$L_u = \int_{z_0}^{z_1} \frac{dz}{\sin\theta \cos\alpha}, \qquad L_v = \int_{z_0}^{z_1} \frac{dz}{\sin\theta \cos\beta}. \qquad (7.67)$$

The third invariant is the angular distance, Φ (Fig. 6a), at the terminal parallel, z_1, between a u-line and a v-line originating at one and the same point on the initial parallel, z_0. The angle Φ is evaluated via the respective cable winding angles, Φ_u and Φ_v. Although both of these angles may change in the net deformations, their difference

$$\Phi = \Phi_v - \Phi_u = \int_{z_0}^{z_1} \frac{\tan\beta - \tan\alpha}{r \sin\theta} dz \qquad (7.68)$$

is preserved. The parameters L_u, L_v, and Φ are the only geometric invariants of a net; all other parameters, including the axial length $L_z = z_1 - z_0$ and the radii of the edge parallels, may change in the net deformations.

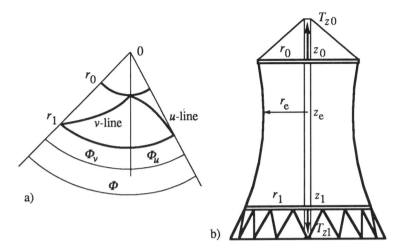

Figure 7.6. A skew axisymmetric net: a) illustration of the angular invariant, Φ; b) geometric parameters of a geodesic net analyzed in Example 1 below.

Particular Axisymmetric Nets

Geodesic Nets. The function $f(r)$ in (60) is evaluated for a geodesic net with the aid of relations (22) and (23); omitting an inessential constant, the result is

$$f(r) = r \sin \omega. \tag{7.69}$$

As a result, the first integral is obtained in a closed form:

$$T_z/\sin\theta - M_z (\cos\alpha + \cos\beta)/(a+b) = Cr\sin\omega. \tag{7.70}$$

The integration constant C is evaluated using equations (56):

$$C = 2\pi (T_u - T_v)/(a+b).$$

Furthermore, expressing the net forces in terms of the constants C and M_z confirms that, as expected, both arrays are isotensoid:

$$2\pi T_u = M_z/(a+b) + Cb, \quad 2\pi T_v = M_z/(a+b) - Ca. \tag{7.71}$$

After replacing trigonometric functions in (70) by means of (22), a first-order equation in the unknown r is obtained. From it, the shape of the net under a given normal load, axial force, and torque moment is found by forward integration. In particular, for a torque-free geodesic net,

$$\operatorname{ctn}\theta \equiv dr/dz = \sqrt{(Cr\sin\omega/T_z)^2 - 1}, \tag{7.72}$$

where the axial force is evaluated with the aid of one of the alternative equations (57) and (62).

The foregoing general observations on the evolution of the shape of a tensile axisymmetric net under a uniform pressure can be now specified for a geodesic net. As with any axisymmetric net, in the state of prestress or under an external pressure, the meridian is concave. In accordance with equations (63) and (22), under a uniform internal pressure the meridian gradually flattens. Eventually an inflection point sets in at the larger of the edge rings and then moves toward the smaller ring. At this stage the meridian is S-shaped and stays this way until the inflection point reaches the smaller ring, whereupon the entire meridian becomes convex. A conical profile is not feasible, but a cylindrical one is possible if $r_0 = r_1$ and both α and β are constant (but not necessarily equal).

For a reflection-symmetric geodesic net, $\beta = -\alpha = \omega/2$, $a = -b$, and the axial and torsional aspects of the problem uncouple. The function $f(r)$ becomes

$$f(r) = \cos \beta \qquad (7.73)$$

leading to [cf. solution (64) for a torque-free net]

$$T_z/\cos \beta \sin \theta = 2\pi T_{\mathrm{n}}/\cos \beta = 2\pi(T_u + T_v) = C_z. \qquad (7.74)$$

This relation is known since the pioneering works of Pipkin and Rivlin (1963) and Read (1963). The constant C_z is evaluated in terms of the axial force and angles θ and β at the equator or another reference parallel. The net forces are found from the independent conditions of axial and torsional equilibrium:

$$2\pi T_{u(v)} = (C_z \pm M_z/b)/2. \qquad (7.75)$$

Chebyshev Nets. For this net, the function $f(r)$ in (60) is evaluated using relations (31)–(36):

$$f(r) = r/\sin \omega \qquad (7.76)$$

and results in the first integral

$$T_z/\sin \theta - M_z(a^2 \tan \beta - b^2 \tan \alpha)/r \,(b^2 - a^2) = Cr/\sin \omega. \qquad (7.77)$$

Once again, relations (56) are used in evaluating the integration constant C,

$$C = 2\pi r \,(T_u \sin \beta + T_v \sin \alpha)/(b^2 - a^2),$$

and then the net forces are expressed in terms of constants C and M_z:

$$2\pi T_u = r \,[C \sin \beta - M_z \sin \alpha /(b^2 - a^2)]/\sin^2 \omega,$$
$$2\pi T_v = r \,[M_z \sin \beta/(b^2 - a^2) - C \sin \alpha]/\sin^2 \omega. \qquad (7.78)$$

As is readily seen, these forces, unlike forces in a geodesic net, vary along the fiber lengths. The equilibrium shape of a Chebyshev net can be determined explicitly by forward integration; for a torque-free net, the equation is

$$\mathrm{ctn}\,\theta \equiv dr/dz = \sqrt{(Cr/T_z \sin \omega)^2 - 1}, \qquad (7.79)$$

and is integrated together with relation (57) or (62).

The evolution of the shape of a Chebyshev net in loading is also different from that of a geodesic net. According to equations (63) and (32), in a tensioned Chebyshev net subjected to an increasing uniform internal pressure, meridian inflection first sets in at the smaller edge ring and then propagates toward the larger ring. Once again, a conical shape is infeasible while a cylindrical shape may occur under $r_0 = r_1$ and constant α and β. In the latter case, the net is simultaneously geodesic and Chebyshev, which is possible only on developable surfaces. When subjected to axisymmetric internal pressure, such a net will twist, unless it is reflection-symmetric.

For a reflection-symmetric Chebyshev net, $\beta = -\alpha = \omega/2$, $a = b$, and the function $f(r)$, evaluated with the aid of (32), is

$$f(r) = 1 / \cos \beta. \tag{7.80}$$

As a result, the first integral (60) becomes

$$T_z \cos \beta / \sin \theta = 2\pi T_n \cos \beta = 2\pi (T_u + T_v) \cos^2 \beta = C_z. \tag{7.81}$$

Once again, the problem uncouples, and the net forces, determined from the two independent conditions of equilibrium, are

$$2\pi T_{u(v)} = (C_z / \cos^2 \beta \pm M_z / 2b \sin^2 \beta)/2. \tag{7.82}$$

An interesting particular case is a cylindrical woven hose under a uniform internal pressure. No externally applied axial tension is assumed to be present, whereas the presence or absence of a torque moment does not affect the equilibrium configuration (until one of the fiber arrays disengages). In this net, the condition of equilibrium in the normal direction requires

$$2\pi r^2 P = T_z \tan^2 \beta. \tag{7.83}$$

On the other hand, the axial force induced by the pressure is $T_z = \pi r^2 P$, so that

$$\tan^2 \beta = 2, \quad \beta = \tan^{-1} \sqrt{2} = 54.7°. \tag{7.84}$$

This well-known value of the angle between the fiber trajectories and the meridian (generator) of the cylinder is the property of the equilibrium configuration of the net under a uniform pressure. When pressurized, a woven hose with unrestrained ends will expand or contract axially such that the angle β attains this value.

Orthogonal Nets. After setting $\tan \alpha = -\operatorname{ctn} \beta$ in the general equation (59), it simplifies considerably:

$$\frac{d}{dr} \frac{T_z}{\sin \theta} - \frac{T_z}{r \sin \theta} + \frac{2}{r^2} M_z \operatorname{ctn} 2\beta = 0 \tag{7.85}$$

and yields $f(r) = r$ in the first integral (60). To advance the investigation, further specification of the intrinsic geometric properties of the net is required. An important subclass is an orthogonal semigeodesic net, where one of the arrays (say, β) is geodesic. In this case, according to (22), $r \sin \beta = b$ and the solution of (85) is

$$bT_z / \sin \theta - M_z \cos \beta = C / \sin \beta. \qquad (7.86)$$

According to (56), the net forces are

$$2\pi b T_u = C, \quad 2\pi b T_v = C \operatorname{ctn} \beta + M_z. \qquad (7.87)$$

The constant force in the nongeodesic array has been expected in the light of observations made on semigeodesic nets in Chapter 6.

7.3 Kinematic and Elastic Deformations

Classification of nets is usually based on their geometric attributes in an undeformed, reference configuration. These descriptive properties may or may not be preserved in the net deformations. To begin with, a net with the member intersections not fastened is least constrained and possesses higher kinematic mobility as compared to any net with the intersections fastened. Under any normal surface load and an edge load, such a net is, and remains, geodesic in both kinematic (inextensible) and elastic deformations. This is in spite (and, actually, because) of the fact that the form of an elementary cell, as well as the shape of the net surface, change in both deformations. Such structural behavior defies conventional analytical means including the finite element method. A particular stumbling block is the necessity to satisfy the preservation conditions (67)–(68).

A Chebyshev net has rhombic cells and must have fastened intersections in order to preserve this characteristic property in kinematic deformations. The property is lost in elastic deformations, except for the trivial case of a uniform stretching of the net.

As to orthogonal nets, since the net angle generally does not preserve in either kinematic or elastic deformations, the significance of this class of nets is mostly theoretical. It lies, in particular, in an obvious analogy between a material orthogonal net and a net of the principal force trajectories of a membrane.

General deformations of axisymmetric nets are governed by the applicable equations of the preceding chapter. Simplifications are possible in two cases: when the deformation is small and when it is axisymmetric. A small deformation of an originally axisymmetric configuration has been analyzed in Rivlin (1959) by seeking cyclic (Fourier series) solutions to the linearized equations. Analysis of axisymmetric deformations depends on the net type, as illustrated in the following examples.

Example 7.1 (Geodesic Net). This example (Kuznetsov, 1984a) is an analysis of a skew geodesic net for a 250 m high prestressed, tensile structure, cooling tower with two rigid edge rings of radii $r_0 = 75$ m and $r_1 = 100$ m (Fig. 6b). The net is torque-free, i.e., no external torque is applied and the edge rings are not constrained against mutual rotation about the z-axis. In this case, the equilibrium configuration of the net is obtained explicitly from (72) by forward integration employing equations (22).

The analysis (Landmann, 1985) starts with determining the prestressed state. In this case, $P = 0$, $T_z = T^* = $ const., and only three initial parameters are needed for forward integration: a, b, and C/T^*. These are selected and then adjusted by trial and error until a satisfactory prestressed state (the net shape, as well as cable forces, lengths, and winding angles) is obtained. At this stage the invariants L_u, L_v, and Φ are evaluated

$$L_u = L_u^* - T_u^* /EA_u, \quad L_v = L_v^* - T_v^* /EA_v, \qquad (a)$$

where L_u^* and L_v^* are the pretensioned cable lengths obtained by integration, and EA_u and EA_v are the cable stiffnesses. The third invariant, given by (68), is determined solely by the chosen net geometry.

The deformed state of the net under an applied pressure is determined by a set of four parameters, say, a, b, T_u, T_v, all of them as-yet unknown. Their values must be such that the sought state is consistent with the three net invariants and the preservation of the axial length, L_z, due to incompressibility of the central column. The latter condition is equivalent to

$$r_1 = r_0 + \int_{z_0}^{z_1} \operatorname{ctn} \theta \, dz, \qquad (b)$$

and can be treated in this problem as a fourth invariant. This gives rise to a system of four simultaneous equations of the form (67), (68), and (b) in which the left-hand side values are already known while the four unknown parameters appear implicitly under the sign of a definite integral.

In solving this system of equations for several pressure levels, a shooting technique has been employed in conjunction with a fourth-order Runge–Kutta integration scheme. Using trial values of the unknown parameters, first approximations for the four invariants are obtained. The errors in the invariants are then used to refine the solution iteratively. With pressure increments sufficiently small, the net forces and geometry change gradually. As a result, the unknown parameters for each incremental pressure level are fairly accurately predicted by extrapolation, thus reducing the required number of iterations. To further the investigation, unlimited elasticity was assumed and the analysis was carried out beyond the realistic load levels. For the purposes of the parametric study, two different ratios of the elastic stiffnesses of the cable arrays were employed .

As expected, the analysis shows that the edge rings of a skew net mutually rotate under axisymmetric loads, such as axial tension and a uniform surface pressure. The effect is a typical nonlinear response to a perturbation load (Fig. 7a) and in the absence of prestress the graph would have a vertical tangent at the onset of loading. On the other hand, because of the polar symmetry of the net, there must be no edge ring rotation under any cyclic load, in particular, wind load, at least inasmuch as it can be represented by trigonometric series.

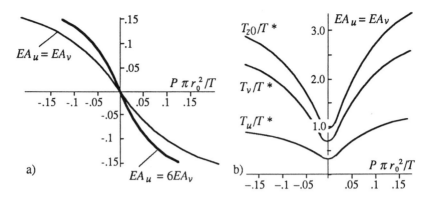

Figure 7.7. Graphs illustrating the structural behavior of a prestressed elastic geodesic net under a uniform surface pressure: a) rotation, γ, of the upper ring; b) normalized cable forces and their axial resultant.

Under a uniform external pressure, the net equator shrinks, shifts downward, and for both of the used ratios of cable stiffnesses, asymptotically approaches almost the same location at about $0.4L_z$ from the upper ring. Under an internal pressure, the equator expands, shifts toward the smaller ring, reaches it, and exits the net leaving behind a convex net.

Quite unexpected was the evolution of the cable forces in loading. In conventional, saddle-shaped, prestressed cable systems under a transverse load, the elastic deformation always causes tension to increase in one array of cables and to decrease in the other. The only way of preventing the unloading cables from disengagement is to increase the prestress, but this is counterproductive for the load-carrrying cables. Surprisingly, the axisymmetric geodesic net does not behave this way; under a uniform pressure, *either internal or external*, tension increases in both cable arrays (Fig. 7b). The explanation lies in the mutual rotation of the edge rings required by torsional equilibrium of the net. Although very small, this kinematic displacement has a strong force-leveling effect offsetting the usual, unfavorable outcome of the elastic deformation. As a result, both arrays share in supporting the applied load and the prestress requirement is appreciably reduced.

Example 7.2 (Chebyshev Net). In contrast to geodesic nets, a prestressed Chebyshev net can be analyzed by conventional means, including an appropriate variant of a nonlinear finite element method. For this reason, the following analysis is confined to kinematic (inextensible) deformations where the net cells remain rhombic, with unchanged side lengths. Consider a long (semi-infinite) segment of a Chebyshev net suspended, like a basketball net, from a ring of radius r_3 (Fig. 8a). The net is under tension T_{z0} and an axisymmetric solid is being pushed through. The net is reflection-symmetric ($\beta = -\alpha = \omega/2$, $a = b$) so that, according to equations (34),

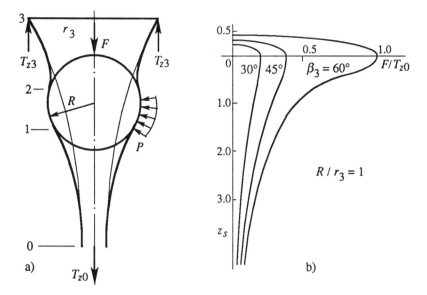

Figure 7.8. A Chebyshev net interacting with a solid: a) the problem geometry; b) normalized axial force as a function of the axial displacement of the solid.

$$r = 2b \sin \beta, \quad 2b = r_3 / \sin \beta_3. \tag{a}$$

Treating the net as infinitely long implies that initially it has the form of a parabolic pseudosphere and, by virtue of (41),

$$\sin \theta = \cos \beta, \tag{b}$$

with $\theta_0 = \pi/2$ and $\beta_0 = 0$ at $z \to \infty$. Since relation (81) is valid for an arbitrary normal surface load, it applies to the entire net, including the prestressed bottom segment (0–1) which is free of any surface load. Combining (81) with (b) gives

$$T_z = T_{z0} \sin \theta / \cos \beta \tag{c}$$

for the entire net, regardless of its deformed shape and the surface pressure pattern. Thus, in a contact problem with a solid of a known shape, the axial force is obtained immediately for the entire net from equations (a) and (c). For example, if the solid is a sphere, $r = R \sin \theta$, and the force F required for equilibrium is

$$F = T_{z2} - T_{z1} = T_{z0} (c \tan \beta_2 - 1), \quad c = \tan \theta_1 = \operatorname{ctn} \beta_1 = 2b/R. \tag{d}$$

This force is the resultant of the normal contact pressure P at the net–solid interface; it depends only on the prestress level and the axial location of the solid. The axial distance, z_s, between the support ring and the center of the sphere is evaluated as a function of the net geometry:

$$z_s = - R \cos \theta_2 + \int_{r_2}^{r_3} dr \, \tan \theta = - R \cos \theta_2 + 2 b k^2 \int_{\theta_2}^{\theta_3} \frac{\sin^2 \theta \, d\theta}{\sqrt{1 - k^2 \sin^2 \theta}} \, , \qquad (e)$$

where

$$k^2 = (\tan \beta_1 / \tan \beta_2)^2 = 1 + \operatorname{ctn}^2 \theta_2 - \operatorname{ctn}^2 \theta_1 . \qquad (f)$$

Unlike the simple formula for the force F, expression (e) for the distance z_s contains an elliptic integral.

The graph of the normalized force F as a function of the axial distance z_s is presented in Fig. 8b for several net geometries. Aside from illustrating the problem in consideration, the graph provides a solution to a reverse, perhaps more interesting, problem. It gives the magnitude of the axial tension that must be applied to the bottom part of the net in order to sustain a solid of a given weight at a desired height. The required tension force has a minimum at $z_s = 0$.

The presented solution does not account for friction at the net–solid interface. However, the friction force is not difficult to evaluate, since the contact zone geometry (segment 2–3) is known and the contact pressure is obtainable with the aid of equations (57) and (c).

8
Membranes

In structural mechanics, a membrane is modeled analytically as a material surface devoid of bending stiffness and resisting only tangential (in-surface) membrane forces—normal and shearing. The absence of bending stiffness, strictly speaking, implies no resistance to compression stress as well. However, a momentless shell supporting compression is a common model in structural analysis. It is sometimes referred to as a membrane shell, especially when it is necessary to distinguish it from a true, or ideal, membrane, which is a system with unilateral constraints in tension.

Both membrane shells and true membranes, as continuous underconstrained systems, fall within the scope of the statical-kinematic analysis of Part I. This approach provides a new perspective on the structural behavior of membranes and the related limitations of the linear membrane theory. Among the obtained results are the resolution of an apparent paradox in the statics of membranes and an explanation of the peculiar behavior of toroidal membranes with a noncircular profile, a problem with a long history (Libai and Simmonds, 1988).

Another topic of this Chapter stems from the identity of analytical models of an orthogonal fiber net and a geometric net of the principal force trajectories of a membrane. This makes possible a two-way transfer of some theoretical and analytical results between the two areas of study and paves the way to a statical-geometric theory of wrinkling membranes. The theory stems from the analysis of semigeodesic nets in conjunction with the condition of preservation of geodesics in wrinkling. This condition is derived by considering the geometric microstructure of a wrinkled surface as the limiting case of isometric bending. The condition provides the only link between the intrinsic geometries of the original and wrinkled membrane surfaces.

An analytical application of the theory is developed in detail for axisymmetric membranes, where geodesics are described by a simple closed-form equation. An analogy is established between the profile of a pressurized wrinkled membrane of revolution and Euler's *elastica*. This immediately provides a wealth of readily available results and paves the way for an analysis of partially wrinkled membranes. The latter is illustrated by a case study involving a pressurized ellipsoid of revolution with two opposing axial forces at the poles.

The concluding section addresses two, more complex problems in membrane wrinkling. The first deals with wrinkling of pressurized toroidal membranes, including one with an asymmetric profile, the second is concerned with wrinkling of pressurized membranes with an axial torque moment.

8.1 The Membrane Shell as an Underconstrained Structural System

In the context of statical-kinematic analysis, a membrane is characterized by a single differential constraint equation

$$A^2 du^2 + 2AB \cos \omega \, du \, dv + B^2 \, dv^2 = ds^2, \tag{8.1}$$

which is nothing but the first fundamental quadratic form of the surface. In kinematic terms, this equation means simply that in any kinematically possible deformation of a membrane, all elementary lengths (hence, all distances, angles, and areas) are preserved. Such a membrane is called inextensible and, obviously, does not allow compression or shearing strains as well.

Constraint idealization, whereby the membrane is assumed inextensible, reduces membrane deformations to isometric bending, smooth or piecewise smooth. The latter type of bending is characterized by angular discontinuities over isolated lines or a network of lines; it occurs, for example, in a local snap-through or overall buckling of rigid shells and in wrinkling of true membranes. Since piecewise smooth bending cannot be prevented by any contour supports, a membrane always possesses this form of kinematic mobility. Thus, a membrane is an underconstrained structural system, either geometrically variant or quasi-variant. In this light, the ability of a properly supported membrane to equilibrate in its original configuration arbitrary smooth loads demonstrates only that these are equilibrium loads, whereas discontinuous and concentrated (line and point) loads are not.

Within the class of smooth bending, a membrane is topologically adequate and, with proper geometry and support conditions, will be geometrically invariant. Note that smooth loading alone does not guarantee equilibrium in the original membrane configuration. The fact is that smooth bending is usually precluded by attaching a rigid bar to the membrane or by constraining its edge. However, a rigid bar or edge supports generally produce a line reaction with a normal component which cannot be equilibrated by the membrane without a change in geometry. Such a membrane still does not satisfy the statical criterion of invariance and, therefore, is at best quasi-invariant.

Implications for the Linear Membrane Theory

Degeneration of the Tangent Operator. The above observations are analytically reflected in the inherent nonlinearity and nonlinearizability of the equations of membrane statics, a characteristic feature of underconstrained systems. Indeed, in the simplest, von Karman-type, nonlinear formulation (small displacements, finite rotations of a normal), the equilibrium equations for a membrane are

$$(BN_1)_u - B_u N_2 + (AS)_v + A_v S = 0,$$

$$(AN_2)_v - A_v N_1 + (BS)_u + B_u S = 0, \tag{8.2}$$

$$(BN_1)_u W_u/A + (AN_2)_v W_v/B + (SW_v)_u + (SW_u)_v$$

$$+ BN_1 [A\sigma_1 + (W_u/A)_u] + AN_2 [B\sigma_2 + (W_v/B)_v] = PAB.$$

The equations are referred to the lines of principal curvatures and employ the previously used notation, except that S is now the membrane shear force; subscripts u and v denote partial derivatives.

The first two of the above equations are linear and the third one, expressing equilibrium in the normal direction, can be linearized as follows:

$$[(n_{1u}W_u + N_{1u}w_u)B + (n_1 W_u + N_1 w_u)B_u]/A$$

$$+ [(n_{2v}W_v + N_{2v}w_v)A + (n_2 W_v + N_2 w_v)A_v]/B$$

$$+ s_u W_v + S_u w_v + s_v W_u + S_v w_u + 2s W_{uv} + 2S w_{uv}$$

$$+ \underline{AB\sigma_1 n_1} + BN_1 (w_u/A)_u + \underline{AB\sigma_2 n_2} + AN_2 (w_v/B)_v = pAB, \tag{8.3}$$

where, as before, lowercase letters designate small increments of the forces and displacements denoted by the respective capital letters.

For an initially stress-free membrane, only the two underscored terms survive linearization and appear in the resulting algebraic equation of the linear membrane theory. Without the highest derivative terms in the last equation, the system of equations degenerates and fails to produce the number of arbitrary elements needed to satisfy all boundary conditions. In terms of discrete statical-kinematic analysis, this corresponds to an overdetermined system of equilibrium equations for an underconstrained structural system. This means that the analytical model of a true membrane is, generally, nonlinearizable: the linear theory it is confined to the class of problems where the surface and edge loads, including the support reactions, constitute equilibrium load combinations.

On the other hand, there is no such limitation for a membrane with pre-existing internal forces since, as seen from (3), the highest, second derivatives of the displacements are multiplied by these forces. The entire situation parallels the one discussed in Chapter 4 when considering the generic varied equilibrium equations (4.9) for a discrete underconstrained system.

The degeneration of the tangent operator in the statics of membranes prompted Marsden and Hughes (1983) to call membrane theory "linearization-unstable." From the viewpoint of statical-kinematic analysis, this is just a manifestation of the membrane being an underconstrained structural system (Kuznetsov, 1989).

Conjugate Statical and Geometric Problems.   One of the cornerstones of the linear membrane theory is the following theorem (Gol'denveizer, 1953; Vekua, 1959).

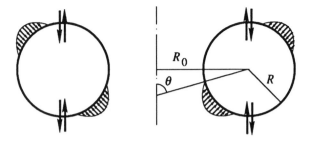

Figure 8.1. A smooth self-equilibrated load that a toroidal membrane cannot balance in its original configuration.

Theorem. If the homogeneous geometric problem for a membrane has J linearly independent nontrivial solutions U_j, V_j, W_j ($j = 1, 2, ... , J$), the conjugate statical problem of the membrane theory can have a solution only if J integral conditions

$$\int\int_G (XU_j + YV_j + ZW_j)\, AB\, d\alpha\, d\beta - \int_g LD_j\, ds = 0 \qquad (8.4)$$

are satisfied. (Here X, Y, and Z are surface loads, U_j, V_j, and W_j are displacements, L is a tangential load applied along the edge in a given direction l, and D_j is the projection of the edge displacements on the direction l.)

According to the theorem, for membranes satisfying its condition, the statical problem only can (but not always does) have a solution. Specifically, for membranes of positive Gaussian curvature, a solution (perhaps, nonunique) was shown to exist, while in a general case the question remains open. An example (or, rather, a counterexample) of a toroidal membrane (Fig. 1) was used to illustrate the general case. In differential geometry a torus is shown not to allow even infinitesimal smooth bending. By implication, the number of nontrivial solutions to the homogeneous geometric problem is zero, so that conditions (4) are assuredly met and the membrane should be able to support any smooth self-balanced surface load. Yet, as is easily seen from Fig. 1, equilibrium in the original configuration is impossible for an axisymmetric load with equal and opposite axial resultants for the inner and outer segments of the torus. (Such a load would produce a transverse shearing force along the parallel circles separating the two segments.)

However, a closer look at the conditions (4) shows these to be nothing but the implementation for an inextensible membrane of the principle of virtual work. This principle is known to be both necessary *and sufficient* for equilibrium. Then how could it be possible that the statical problem does not *always* have a solution, if conditions (4) are satisfied?

The resolution of this apparent paradox lies in a subtle point concerning the nature of nontrivial solutions to the homogeneous geometric problem premised in the theorem. The fact is that the displacements figuring in the equation of virtual work (4) are *virtual*: they are solutions of the *linearized* constraint equations. On the other hand, in differential geometry, the impossibility of displacements was established by analyzing the rigorous, *nonlinear* geometric equations. The irony of the situation is that this, *exact*, solution of the nonlinear problem is inapplicable to the theorem based on the principle of *virtual* work; in fact, using the terminology of the statical-kinematic analysis, the absence of kinematic displacements tells nothing about the virtual displacements.

As it happens, a toroidal membrane exibits the behavior of a quasi-invariant system. Specifically, it allows infinitesimal, piecewise smooth displacements at the expense of second-order strains. These displacements satisfy the linearized constraint equation, but not the original, nonlinear equation. Thus, virtual displacements do exist for a torodial membrane and one of them is axisymmetric. It is not difficult to identify *smooth* surface loads which perform work over this virtual displacement and, therefore, cannot be equilibrated by the membrane. Obviously, one of them is the axisymmetric load shown in Fig. 1. Incidentally, this demonstrates that even in the absence of line constraints (e.g., supports) smooth surface loads can give rise to nonsmooth infinitesimal bending.

As follows from the foregoing reasoning, a stronger version of Gol'denveizer's theorem would read:

If the *linearized* homogeneous geometric problem has J linearly-independent nontrivial solutions, the conjugate statical problem of the membrane theory *has* a solution (perhaps nonunique) if, *and only if*, J conditions (4) are satisfied. (Italicized words are those inserted into the original formulation of the theorem.)

In terms of the statical-kinematic analysis, the theorem simply means that a membrane, as any other structural system, can support only equilibrium loads, i.e., those orthogonal to the virtual displacements; and only in the absence of virtual displacements is the virtual work (4) always a zero, so that every load is an equilibrium load and the statical problem always has a solution.

Infinitesimal Mobility of a Toroidal Membrane

Analytical Signs of Infinitesimal Mobility. The fact that a toroidal membrane is incapable of balancing an arbitrary smooth load is a statical sign of its being underconstrained. Taking into account the uniqueness of configuration, proved in geometry, this means that the membrane must possess infinitesimal mobility. This feature can be detected by means of a linear analysis establishing either: (i) the existence of a nontrivial solution to the linear homogeneous geometric (strain-displacement) equations; or (ii) the possibility of unbounded displacements at the expense of small strains. These two equivalent signs of a degenerate tangent operator exhibit the inadequacy of the linear model. The linear strain-displacement equations for an axisymmetric membrane are

$$\varepsilon_1 = u'/A - \sigma_1 w, \quad \varepsilon_2 = B'u/AB - \sigma_2 w, \qquad (8.5)$$

where ε_1, ε_2 are, respectively, the meridional and hoop strains and u, w are the tangential and normal displacements.

Choosing as an independent variable the angle θ between the normal to the surface and the axis of revolution (Fig. 1) enables the metric coefficients and the principal curvatures of the torus to be taken as

$$A = R, \quad B = r = R_0 + R\sin\theta, \quad \sigma_1 = R, \quad \sigma_2 = r/\sin\theta. \qquad (8.6)$$

Exploring first the alternative (i), a nontrivial solution to homogeneous equations (5) with $\varepsilon_1 = \varepsilon_2 = 0$ is obtained:

$$u = \Delta\sin\theta, \quad w = \begin{cases} \Delta\cos\theta, & 0 \le \theta < \pi, \\ -\Delta\cos\theta, & \pi \le \theta < 2\pi, \end{cases} \qquad (8.7)$$

where Δ is arbitrary. Being physically inconsistent because of discontinuous w, solution (7), nevertheless, formally satisfies the homogeneous geometric equations (note that $\sigma_2 = 0$ at the crowns, where $\theta = 0$, π).

The alternative check (ii) confirms the outcome. Assuming, for example, constant strains and symmetry with respect to the equatorial plane leads to the following solution to equations (5):

$$u/R = (\varepsilon_1 - \varepsilon_2)\sin\theta\ln\tan\theta/2 + \varepsilon_2(R_0/R)\cos\theta,$$
$$w/R = (\varepsilon_1 - \varepsilon_2)\cos\theta\ln\tan\theta/2 - \varepsilon_2[1 + (R_0/R)\sin\theta]. \qquad (8.8)$$

Almost everywhere in the membrane, the normalized displacements are of the same order as the strains, but in the vicinity of the crown w/R can exceed the strains by an arbitrarily large factor if $\varepsilon_1 \ne \varepsilon_2$.

Infinitesimal Bending Pattern. Solutions (7) and (8) to the linearized geometric equations indicate infinitesimal first-order mobility requiring only second-order strains. Moreover, the solutions provide enough clues to the actual deformation for its geometric possibility to be demonstrated explicitly (Fig. 2). First, the inner and outer segments of the torus undergo an infinitesimal, rigid-body mutual axial shift Δ accompanied by an inversion of the narrow annuli 2–3 and 2'–3' (Fig. 2a). So far, the deformation is a piecewise smooth isometric bending with $\theta = \theta^*$ being the line of angular discontinuity. However, in the obtained configuration, the membrane still cannot support the load shown in Fig. 1. This requires an infinitesimal tilt, ϕ, of the horizontal tangents at points 2 and 2' resulting in an additional axial displacement (Fig. 2b)

$$2d/R \approx 2\phi\theta^*. \qquad (8.9)$$

The displacement d is possible at the expense of strains of the order ϕ^2, whereas the total normalized axial shift

$$2(\Delta + d)/R \approx \theta^*(\theta^* + 2\phi) \qquad (8.10)$$

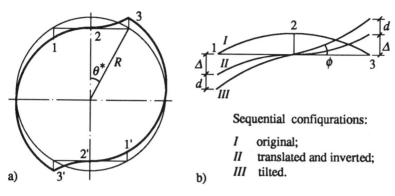

Sequential confiqurations:

I original;

II translated and inverted;

III tilted.

Figure 8.2. Infinitesimal deformation of toroidal membrane: a) geometry of bending accompanied by second-order strain; b) sequential stages of deformation.

is still of the first order, since

$$\theta^* \approx \sqrt{2\Delta/R}. \tag{8.11}$$

The magnitude of the angular discontinuity at points 3 and 3' is

$$2\theta^* + \phi \approx 2\theta^*. \tag{8.12}$$

The foregoing analysis demonstrates the possibility and reveals the pattern of nonsmooth infinitesimal bending in a toroidal membrane shell subjected to smooth surface loads. It is the presence of asymptotic lines that makes this bending deformation possible. Consistent with the behavior of discrete quasi-invariant systems, the deformed configuration of the torus (with two annular ridges) is geometrically invariant within the class of smooth loads. Note, however, that for a toroidal membrane this is predicated upon the ridges being capable of supporting a concentrated tension or compression hoop force (in reality, a stress concentration). Only then any perturbation load of the original configuration can be equilibrated in the deformed configuration. As usual, the problem can be linearized if and only if the applied load contains an equilibrium component, i.e., is not a pure perturbation load.

The response of an elastic toroidal shell with nonvanishing bending stiffness should exhibit the basic features of the described behavior. Steele (1964) found an almost step-wise jump in the displacements near the crowns indicating the formation of the ridges 3 and 3'. The ridges, originating at the crowns, will gradually steepen and shift in the course of loading. This might eventually lead to a development of plastic strains along the ridges followed by formation and propagation of a plastic zone.

Toroidal Membrane of a General Profile. The surface load acting on a toroidal membrane of a circular profile was intentionally chosen to demonstrate the membrane's inability to support it. Even more impressive are examples of

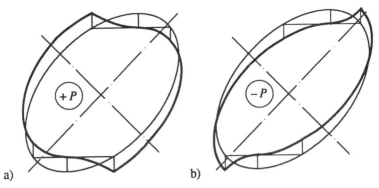

Figure 8.3. Infinitesimal bending of a toroidal membrane with a noncircular profile under a uniform pressure: a) internal; b) external.

the opposite kind, showing that toroidal membranes of many common noncircular profiles cannot balance the simplest surface load—a uniform pneumatic pressure. Libai and Simmonds (1988) raise a question, "What happens in a very thin, pressurized toroidal shell whose undeformed cross section does *not* have points of horizontal tangency lying on the same vertical line?... We conjecture that wrinkling must occur."

To analyze this problem, consider a toroidal membrane of the described type with a profile in the form of a tilted ellipse (Fig. 3). Obviously, for this membrane, a uniform pressure is not an equilibrium load, as it would produce a transverse shearing force along the crowns (parallels with horizontal tangency), making equilibrium in the original configuration impossible. This situation and its resolution are all too familiar in the statical-kinematic analysis of underconstrained systems: as with any quasi-invariant system, the membrane must deform in a certain way in order to balance a uniform pressure. The deformation is a combination of piecewise smooth bending and second-order infinitesimal stretching. The deformed profiles of such a membrane shell subjected to an internal or external pressure are shown in Fig. 3a and b. Geometric invariance of the final configurations is confirmed in that the parallels of horizontal tangency vanished altogether. Instead, two annular ridges emerged which ensure that no smooth surface load can produce a transverse shear in the deformed membrane configuration.

As to the above conjecture on wrinkling, it requires just one qualification. Wrinkling usually implies the formation of a uniaxial stress zone and a resulting tension field in a true membrane. As shown in Chapter 6, in the presence of only a normal surface load, a geodesic tension field must develop. However, this is not the case with the two circular ridges representing the key feature of the above deformation pattern. These formations are neither geodesic nor necessarily in tension; therefore, in the context of a memrane shell, a connotation with wrinkling is undesirable. However, for a true membrane, the conjecture is quite correct, as is demonstrated below in Section 4.

Observations made and features revealed in the foregoing statical-kinematic analysis are in every way typical of the behavior of an underconstrained structural system. In particular, uniqueness of configuration (kinematic determinacy) of a membrane does not necessarily entail virtual determinacy and geometric invariance, even in the context of smooth loading. As has been mentioned, smooth loading implies the absence not only of line and point loads, but also of support reactions of this type. If such a reaction evolves, it represents perturbation loading and the membrane may come to equilibrium only by changing its configuration: it undergoes displacements accompanied by second- or higher-order strains. Generally, the configuration change is confined to the immediate vicinity of the perturbation reaction and then the deformation is called a nonlinear edge effect. As is the case with any underconstrained system, in the presence of an equilibrium load or prestress, the problem can be linearized. Note that this nonlinear, momentless edge effect may coexist with, but is in no way related to, the classical, linear, *moment* edge effect in shells with bending stiffness.

8.2   Net of Principal Force Trajectories

At $\omega = \pi/2$ and with load terms restored, equilibrium equations (6.24) become

$$(N_{u,i} - N_u R_i) u^i - N_v \kappa_v = P_i u^i,$$
$$(N_{v,i} - N_v R_i) v^i + N_u \kappa_u = P_i v^i, \tag{8.13}$$
$$N_u \sigma_u + N_v \sigma_v = P.$$

According to (6.12), for an orthogonal net in natural coordinates,

$$\kappa_u = \lambda_u = -A_2/AB, \qquad \kappa_v = \lambda_v = B_1/AB. \tag{8.14}$$

Introducing (14), and going over to the natural coordinates, allows the above equilibrium equations to be rewritten

$$BN_{u,1} + B_1(N_u - N_v) = BP_1,$$
$$AN_{v,2} - A_2(N_u - N_v) = AP_2, \tag{8.15}$$
$$N_u \sigma_u + N_v \sigma_v = P.$$

This system coincides with the equilibrium equations of a membrane in coordinate lines of principal force trajectories. It establishes formally the physically apparent analogy between a material orthogonal net and the net of principal force trajectories for a membrane under the same load. Therefore, the results obtained in the statics of orthogonal nets can be transferred into the theory of membranes and vice versa. In doing so, the requirement of both of the net forces being tensile is unnecessary for a momentless shell capable of supporting compression; however, for a true membrane, the requirement stands.

Note that equations (15) describe equilibrium of the deformed state of a net or membrane, and are applicable to the original geometry only when it can be considered identical to the final one.

Since any orthogonal fiber net can balance numerous equilibrium loads, any orthogonal net of a surface may become a net of principal force trajectories of a membrane. In what follows, the trajectory nets are explored in the context of particular surface geometries, loads, and support conditions.

Homogeneous Problem

Net of Principal Force Trajectories of a Membrane as a Singular Orthogonal Net. It is obvious that in a membrane supporting only a tangential load at the edge, the net of the principal force trajectories is a singular orthogonal net with a virtual self-stress pattern. Conversely, each singular orthogonal net of a surface is the trajectory net for some tangential load acting at the membrane edge.

For example, in a membrane of revolution under an axisymmetric edge load, the meridional, hoop, and shear forces are

$$N_1 = C_1 \sigma_2 / \sin^2 \theta, \quad N_2 = - C_1 \sigma_1 / \sin^2 \theta, \quad S = C_2 \sigma_2^2 / \sin^2 \theta, \qquad (8.16)$$

where θ, as before, is the meridional slope. Using the known formula for the orientation of the principal forces gives

$$\tan \gamma = (\sqrt{H^2 + C^2 \sigma_2^4} - H)/C\sigma_2^2, \quad C = C_2/C_1. \qquad (8.17)$$

Here H is the mean curvature and γ is the angle between meridian and the first trajectory. The value

$$\tan 2\gamma = C \sigma_2^2 / H$$

found from (17), coincides with (7.44), evaluated in an entirely different way.

It is not difficult to show that a simply connected membrane of positive total curvature (a convex cap, $K > 0$) does not admit an edge equilibrium load normal to the edge (in its tangential plane). For proof, assume that such a load does exist. Then the edge, being free of a tangential shearing force, is one of the principal force trajectories, hence, one of the lines (say, a v-line) of a singular orthogonal net. The forces $N_u = N\sigma_v$ [cf. (6.25)] are of the same sign everywhere along the edge (and, for that matter, over the entire membrane) since $N > 0$ and σ_v cannot change its sign on a surface with $K > 0$. But then the edge force resultant is different from zero and, due to the absence of other loads or reactions, the membrane cannot be in equilibrium as a free body. This contradicts the above assumption on the existence of the edge equilibrium load, thereby proving the following.

Theorem. A simply connected membrane of positive total curvature does not have an equilibrium load normal to the edge.

The theorem provides a somewhat unexpected answer to the question: What happens when a *self-balanced* normal edge load is applied along the edge of a convex membrane cap? In this case, the overall (free-body) equilibrium is assured. Still, since the membrane cannot be in equilibrium, it must change its configuration and undergoes isometric bending. However, Gaussian curvature K is a bending invariant, so that equilibrium cannot be attained in any configuration, in spite of the load being self-equilibrated. This apparent paradox is resolved by noticing that the edge load must be normal to the edge while acting in the tangent plane. The latter changes its orientation in the membrane bending, so that the load can be realized only as a follower load. Such a load is known to be nonconservative. Under it, the membrane will start bending and twisting and will continue to do so, without coming to equilibrium, as long as the follower load is maintained.

Thus, equilibrium edge loads for a membrane, even when they are self-balanced, are far from being arbitrary. Every such load must be an edge equilibrium load of some singular orthogonal net of the surface. Obviously, any combination of such loads for several singular nets of the surface also represents an edge equilibrium load for the membrane and for some new singular orthogonal net of the surface.

Closed-Surface Membranes.   The following corollary is immediately obtainable from the above theorem.

Corollary.   A convex closed surface (an ovaloid) does not allow virtual self-stress; it is statically determinate.

Indeed, if self-stress was possible, the membrane could be cut along a principal force trajectory and both parts would be in equilibrium under a normal edge load replacing the forces in the cut trajectories; this, according to the theorem, is impossible.

Impossibility of self-stress in an ovaloid is known in the membrane theory where it is derived (using the statical-geometric analogy) from the known proprerty of infinitesimal rigidity of an ovaloid. As follows from the statical-kinematic analysis of Chapter 1, a topologically adequate and statically determinate system is geometrically invariant. Hence, the absence of virtual self-stress may be considered as a statical criterion of the infinitesimal rigidity (geometric invariance) of an ovaloid.

In this light, it is interesting to observe that a closed membrane in the form of a multiply connected surface is statically indeterminate and, therefore, cannot be geometrically invariant (recall the preservation property (1.44) for the difference $V - S$). The best example is a toroidal membrane, where the possibility of self-stress is not difficult to demonstrate. Initial forces can be induced by cutting the membrane along any meridian or parallel circle, applying equal and opposite tangential shearing forces (with a torque moment resultant) to both edges of the cut, and reconnecting the surface. Thus, in this case, the classification as quasi-invariant, hence not infinitesimally rigid and possessing infinitesimal mobility, could be deduced from the existence of virtual self-stress. The degree of statical indeterminacy of an n-connected membrane is $S = n - 1$, which entails an equal degree of virtual indeterminacy, with the resulting $V = n - 1$ linearly independent modess of infinitesimal displacements.

Inhomogeneous Problem

Singular Versus Nonsingular Nets. Consider a problem of identifying all loads such that some orthogonal net on a given membrane surface becomes the net of principal force trajectories. This problem is relevant, for instance, in the optimization of reinforcement layout for composite membranes. For an edge load, the problem, in fact, has already been solved: only singular orthogonal nets can be trajectory nets, and their edge equilibrium loads are found as a by-product in specifying their geometry. If the net happens to contain a pole, an appropriate concentrated force might be required for the overall equilibrium.

Going over to the inhomogeneous problem, note that for a given orthogonal net of a membrane to become the net of principal forces, it is necessary that both a surface equilibrium load of the net and the corresponding edge load are acting simultaneously. In this regard, a singular net differs from a general orthogonal net in that it allows an independent equilibrium load at the edge. Superimposing this load preserves the singular net as the net of principal forces of the membrane.

Example 8.1 (A Catenoid of Revolution). As an example of a singular net of principal force trajectories, consider an orthogonal net obtained by rotating the net of principal curvatures of a catenoid of revolution through an angle $\beta = \beta(z)$. The resulting net has

$$\tan \alpha = - \operatorname{ctn} \beta, \quad \omega = \beta - \alpha = \pi/2, \quad \sigma_u = - \sigma_v, \qquad (a)$$

and is singular as an orthogonal net of a minimal surface. Taking into account the characteristic property of minimal surfaces, $\sigma_1 = \sigma_2$, it is easily found from equations (7.17) that

$$\sin \theta = r_e/r, \qquad (b)$$

where r_e is the radius of the equator.

Introducing (b) into the statical equation (7.85) of an orthogonal net gives

$$d(T_z r)/dr - T_z + 2 M_z r_e \operatorname{ctn} 2\beta/r^2 = 0.$$

From here, using (7.57), the normal equilibrium load of the net is found:

$$P = - M_z r_e \operatorname{ctn} 2\beta/r^4. \qquad (c)$$

Note that at any parallel where $\beta = \pi/4$, the pressure P turns into zero. The reason is that at such locations the net lines have asymptotic directions and the net cannot support any normal surface load. Taking into account relations (a) and (b), the net forces given by (7.56) become

$$2\pi T_u = T_z \sin\beta/\sin\theta - (M_z/r)\cos\beta,$$
$$2\pi T_v = T_z \cos\beta/\sin\theta + (M_z/r)\sin\beta. \qquad (d)$$

The obtained results amount to a solution of an axisymmetric statical problem for a membrane in the form of a catenoid subjected to an arbitrary normal surface pressure, axial force, and torque moment. The principal force trajectory is inclined to a meridian under an angle β determined by

$$\text{ctn}\,2\beta = -Pr^4/M_z r_e. \qquad (e)$$

To evaluate the principal forces in the membrane, the Lamé parameters in the natural coordinates of the net are needed. They are found from (7.19):

$$A_n = r\cos\beta, \quad B_n = r\sin\beta. \qquad (f)$$

The principal forces in the membrane are

$$N_u = T_u/B_n = T_z/2\pi r_e - (M_z/r^2)\,\text{ctn}\,\beta,$$
$$N_v = T_z/2\pi r_e + (M_z/r^2)\tan\beta. \qquad (g)$$

Since the net is singular, it will remain in equilibrium (but with different forces) if, in addition to the existing normal surface pressure, the equilibrium edge load is introduced. For the membrane this means that adding a uniform meridional force of any magnitude at the edge parallels affects the magnitudes of the principal forces but not the orientation of their trajectories.

Let a membrane in the form of a convex cap be supported such that only normal membrane forces are possible at the edge. This edge condition is modeled as an array of support bars located in the tangent plane and perpendicular to the membrane contour. In the absence of a tangential load component, the contour must be a principal force trajectory. As follows from the consideration of a homogeneous problem, any orthogonal net involving the edge of a convex cap does not allow self-stress. Therefore, if any of the equilibrium loads of such a net is applied to the membrane, the resulting stress state is unique, with the principal forces given, respectively, by N_u and N_v evaluated for the net.

For example, when a membrane of revolution is supported in a described way and the meridian is known to be the funicular curve for the surface load, the geographical net is the trajectory net, and the membrane hoop force is zero. This is the case with a shallow suspended membrane having a meridian in the form of a cubic parabola which is a funicular curve for a linearly varying load produced by a uniform surface pressure. The situation is easily recognizable as a particular case of a semigeodesic net with a force-free nongeodesic array, considered in Chapter 6.

The observations made for a convex cap are valid for an ovaloid as well: in both cases, the uniqueness of a statically possible state is a straightforward consequence of statical determinacy.

Trajectory Nets of a Uniformly Stressed Membrane. One simple particular case of equilibrium of an orthogonal net described in Chapter 6 deserves special consideration. This was the case of a normal surface load proportionate to the mean curvature of the surface, H. Under this load, forces N_u and N_v in any orthogonal net of the surface were found to be constant and equal to each other. For the membrane, this means simply that the stress state is a uniform biaxial tension or compression. In this state, the net of principal force trajectories is undefined and, generally speaking, any orthogonal net (singular or not) can be considered as the trajectory net. For a convex cap or an ovaloid, by virtue of their statical determinacy, the uniform force field is the only one possible under this load, but for a multiply connected membrane, say, a toroidal one, this is not so.

Somewhat surprisingly, the described, almost trivial, case for a membrane is much more interesting and diverse in application to nets. The reason is that, when going over to the net forces, the constant membrane forces must be multiplied by the coefficients of the first quadratic form of the surface. This is seen even in the simplest case of a uniform plane stress, when the net of the principal stress trajectories is undefined. As a result, every plane orthogonal net can be considered as the net of the principal stresses. Consider, for instance, plane isothermic nets which are obtainable by the conformal mapping of a Cartesian net with the aid of analytical functions

$$w = w(z), \quad w = u + iv, \quad z = x + iy, \quad \bar{z} = x - iy. \qquad (8.18)$$

With an isothermic coordinate net, the square of the linear element (the first quadratic form) is given by

$$ds^2 = dx^2 + dy^2 = dz \, d\bar{z} = \lambda(u, v) \, (du^2 + dv^2) \qquad (8.19)$$

and the net forces, according to (6.22), are

$$T_u = T_v = C\sqrt{\lambda}. \qquad (8.20)$$

For an orthogonal net formed by two arrays of confocal hyperbolas,

$$w = z^2, \quad \lambda = 1/\sqrt{u^2 + v^2}; \qquad (8.21)$$

for a net of confocal parabolas,

$$w = \sqrt{z}, \quad \lambda = u^2 + v^2; \qquad (8.22)$$

for a plane radial-hoop net,

$$w = \ln z, \quad \lambda = e^{2u}; \qquad (8.23)$$

and so on. It must be borne in mind that a plane net, as a particular case of an asymptotic net, is infinitely statically indeterminate; accordingly, the force pattern (20) with any appropriate λ is but one of the statically possible states for each net under an edge load. On the other hand, the respective force patterns of the orthogonal nets on uniformly stressed membranes are unique, as long as the net is nonsingular.

One Unique Property of Trajectory Nets. As has been mentioned, an orthogonal fiber net does not preserve its characteristic property in kinematic and, the more so, elastic deformations. A similar question posed with respect to orthogonal nets of a membrane is more interesting. It is obvious that for a net drawn on the membrane surface, the net angle generally is not preserved in the membrane deformations. Hovever, there is one exeption. If, in an arbitrarily deformed isotropic elastic membrane, the principal force trajectories are marked and then the load is removed, the marked lines on the membrane in its unloaded, natural state, again form an orthogonal net. Thus, the net of principal forces of an isotropic elastic membrane possesses the following property.

Preservation of Orthogonality. There exists one, and only one, net that is orthogonal on an arbitrarily deformed membrane surface and on the initial, stress-free membrane surface; it is the net of the principal force trajectories.

For proof, note that the strain tensor e_{ij} is the difference of the metric tensors in the deformed and initial states of the membrane:

$$e_{ij} = G_{ij} - g_{ij}. \tag{8.24}$$

In an isotropic material, the stress and strain tensors are coaxial. If the principal force trajectories are taken as coordinates, the strain tensor is diagonal, and such is, due to the coordinate orthogonality, the metric tensor G_{ij} of the deformed surface. But then, by virtue of (24), the tensor g_{ij} must also be diagonal, meaning that the coordinate net is orthogonal on the stress-free surface as well.

A few remarks are in order. First, the considered net, generally, is orthogonal only in the original and final states of the membrane deformation; in transition, the net angle may vary depending on the loading path, although a path preserving the orthogonality must exist. Second, in the obvious (and trivial) case of a uniform biaxial deformation, all nets, not just orthogonal ones, preserve the net angle. And, finally, the described observation is valid for the principal stress trajectories of three-dimensional isotropic elastic solids.

8.3 Wrinkling Membranes: Fundamentals

Wrinkling is a post-buckling equilibrium configuration and load transmission mode in membranes, thin-wall shear panels, flexible sheathing of sandwich plates and shells, and other similar structural components. Assuming the post-buckling compression stress sufficiently small, wrinkling can be defined as an equilibrium configuration of a two-dimensional continuum with unilateral constraints in tension disengaged in one direction. Wrinkling may occur under static and dynamic loads; as a mode of free or forced vibrations; be reversible or irreversible; and be an intended feature of structural response (as in restraint air bags) or a mode of failure (as in a high-precision membrane reflector).

Structural components susceptible to wrinkling are usually modeled analytically as a membrane whose zero bending stiffness implies no resistance to buckling as well. When equilibrium equations of a membrane require one of the principal forces to be compressive, wrinkling occurs and the principal force remains zero. To attain an equilibrium configuration, the membrane undergoes displacements such that the array of nonzero stress trajectories forms a uniaxial stress zone—a tension field. Geometrically, a tension field is represented by a smooth surface (called in the literature a pseudosurface) approximating the continuous distribution of infinitely fine and closely spaced wrinkles.

The tension field theory has evolved within the context of elastic membranes, culminating in more recent studies by Pipkin (1986), Steigmann and Pipkin (1989b) and Steigmann (1990). The latter work contains an exhaustive bibliography on the subject. At the same time, the statical-geometric aspect of the theory has always been distinct and in many cases predominant. Moreover, works on inextensible membranes (Wu, 1974; Zak, 1982) are, by implication, confined to this aspect alone. Aside from producing a comprehensive solution for an inextensible membrane, statical-geomtric analysis provides a good entry point for taking into account constitutive relations.

In keeping with the overall character of this book, the investigation here will be confined to the statical-geometric aspect of the problem, with an emphasis on analytical solutions.

Preservation of Geodesics

Continuous Model and Kinematics of Wrinkling. According to the basic property of a net with one force-free array described in Chapter 6, the net of the principal forces in a uniaxially stressed membrane is semigeodesic, with the nonzero force trajectories forming a geodesic tension field on the resulting pseudosurface. In particular, when only edge loads are present (no surface pressure), the geodesic trajectories of the tension field are straight lines, giving rise to a ruled pseudosurface.

There exists only one link between the intrinsic geometry of the original membrane surface and that of the tension field pseudosurface. It reflects the fact that, as a mode of deformation, wrinkling of an inextensible membrane is a special case of isometric bending. Along with all metric properties of the surface, this deformation preserves the surface geodesics which are the shortest and "straightest" (zero geodesic curvature) lines on the surface. The isometric character of wrinkling is preserved even in the limit, as the wrinkles become infinitely fine and continuously distributed. At this stage, the original surface geodesics become infinitesimal ripples about their respective mean paths located on the resulting pseudosurface. Generally, these mean paths are not geodesic lines of the pseudosurface, but there is one exception; it is the path following the direction of wrinkling.

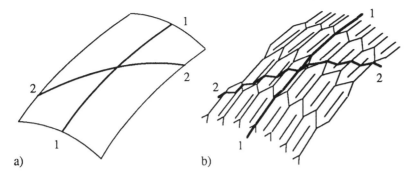

Figure 8.4. Preservation of geodesics: a) original surface geodesics; b) only one of the original geodesics is preserved on the pseudosurface.

Shown in Fig. 4 are two geodesics of the original membrane surface before and after wrinkling. The microstructure of the wrinkled surface represents a stretched honeycomb pattern which is a nonsmooth isometric deformation of the original surface. With the length of an elementary hexagon becoming vanishingly small, its width approaches a second-order infinitesimal. As a result, the length of the original surface geodesic 1–1, directed along the wrinkle, coincides in the limit with the length of the tension field trajectory, leading to the following conclusion.

Preservation of Geodesics in Wrinkling. The tension field trajectory, being geodesic on the pseudosurface, is also a geodesic of the same length on the original membrane surface.

The described preservation of a geodesic array in wrinkling is a condition leading to a determination of the sought pseudosurface and the key to the solution of the entire problem. This preservation property links the intrinsic geometries of the original membrane surface and of the pseudosurface. As to the extrinsic geometry of the original surface, it affects only the extent and, thereby, the overall shape of the tension field.

Turning to the statical aspect of the problem, note that, as is always the case in the statical-kinematic analysis, the pseudosurface geometry is determined solely by the pattern, but not the intensity, of the applied load. This includes surface loads, such as pneumatic or hydrostatic pressure, and the accompanying traction loads at the membrane edge. As a result, the assortment of pseudosurface shapes is relatively small and for the most important types of load can be established once and for all. However, solving this problem analytically requires explicit solutions for geodesics. Such solutions are known only for a few special classes of surfaces, of which surfaces of revolution are the most important. Accordingly, in what follows, selected pseudosurface shapes are obtained for pressurized axisymmetric membranes with and without an external axial force or torque moment.

Pressurized Axisymmetric Membranes with Axial Forces

Euler's Elastica Revisited. Meridians are geodesic lines of surfaces of revolution. Accordingly, in a wrinkling pressurized axisymmetric membrane, meridians of the original surface become the meridians, of the same length but different shape, of the tension field pseudosurface. The pseudosurface shape in this case is identical to the shape of the zero-hoop-stress axisymmetric membrane (Steele, 1964) or, which is the same, the shape of a continuously distributed array of meridional fibers supporting a normal surface pressure. The equilibrium equation for such an array is

$$\sigma_1 T_1 = Pr, \tag{8.25}$$

where σ_1 is the first principal curvature of the pseudosurface (i.e., the curvature of its meridian); T_1 is the constant meridional force per unit polar angle; and P is the normal surface pressure.

This equation, like equation (5.5) for a cable under a hydrostatic load, happens to coincide with the equation of column buckling, which once again provides an avalanche of readily available results transferred from Euler's elastica theory. A concise statement of the observed analogy is as follows.

Axisymmetric Tension Field with Uniform Pressure. Equilibrium configuration of a pressurized axisymmetric wrinkling membrane (or of any segment of such a membrane) is a surface of revolution of Euler's elastica, regardless of the original membrane shape.

This analogy, like the one established in Chapter 5 with a hydrostatically loaded cable, is not a pure coincidence. Indeed, tension induced by internal pressure is the source of the statical-kinematic stiffness of the uniaxially stressed membrane with respect to an external axial force, either tension or compression.

Combining the familiar formulas for the principal curvatures of a surface of revolution,

$$\sigma_1 = 1/R_1 = d(\sin \theta)/dr, \quad \sigma_2 = 1/R_2 = \sin \theta/r, \tag{8.26}$$

with equation (25) produces a first-order differential equation

$$d(\sin \theta)/dr = a^2 r, \quad a^2 = P/T_1.$$

Integration gives

$$\sin \theta - a^2 r^2/2 = C. \tag{8.27}$$

The integration constant C can be expressed in terms of some convenient geometric parameter, such as the meridian slope, θ_a, at the axis of revolution (the pseudosurface apex); the radius, r_e, of the equator (where $\theta = \pi/2$); or the radius, r_c, of the pseudosurface crown (the parallel circle in the plane of horizontal tangency, where $\theta = 0$):

$$C = \sin \theta_a = 1 - a^2 r_e^2/2 = a^2 r_c^2/2. \tag{8.28}$$

From here, the first principal curvature and the Gaussian curvature of the pseudosurface are found as algebraic functions of the meridian slope θ

$$\sigma_1 = a^2 r = a\sqrt{2(\sin\theta - \sin\theta_a)}, \quad K = \sigma_1\sigma_2 = a^2\sin\theta.$$

The meridian arclength, s, the axial coordinate, z, and the radius of revolution, r, of the sought pseudosurface are evaluated by integrating their respective differentials expressed in terms of θ

$$ds = -d\theta/\sigma_1 = -d\theta/a\sqrt{2(\sin\theta - C)}, \quad dz = ds\sin\theta, \quad dr = ds\cos\theta. \quad (8.29)$$

Both s and z are measured from the equator, so that the arclength s increases with the decreasing slope. Using the substitution

$$k\sin\varphi = \sin[(\pi/2 - \theta)/2], \quad k^2 = (1 - C)/2, \quad (8.30)$$

the solution is obtained in terms of elliptic integrals:

$$as = F(\varphi, k), \quad az = 2E(\varphi, k) - as, \quad ar = 2k\cos\varphi = 2k\,\mathrm{cn}\,as. \quad (8.31)$$

Here $F(\varphi, k)$ and $K(\varphi, k)$ are, respectively, incomplete elliptic integrals of the first and second kind, with amplitude φ and modulus k.

Geometry of the Tension Field Pseudosurface. The meridian slope at the membrane apex is determined from the condition of axial equilibrium

$$\sin\theta_a = T_{za}/2\pi T_1 = T_{za}a^2/2\pi P. \quad (8.32)$$

The radii of the pseudosurface equator and of the crown (the latter being present in the case of axial compression) are found from (28) and (30) by setting θ, respectively, to $\pi/2$ and to zero:

$$ar_e = \sqrt{2(1 - \sin\theta_a)} = 2k, \quad ar_c = \sqrt{-2\sin\theta_a}. \quad (8.33)$$

A useful by-product of the foregoing development is a solution to the problem of compression of an inflated membrane by two parallel rigid plates. The contact zone of the membrane, from the apex to the parallel of horizontal tangency, undergoes nonsmooth isometric bending. With the exception of the border parallel, this zone is stress-free, hence, is not a tension field. As to the border parallel, theoretically it has an infinite hoop stress but, for a membrane made of a real material, the stress is finite, with a narrow transition annulus over which the stress decays to zero.

The obtained equations, along with the condition of preservation of geodesics, suffice for an analysis of pressurized axisymmetric membranes, either partially or completely wrinkled. The pseudosurface meridian arclength between a reference parallel, φ_0, and a terminal one, φ^*, is found from $(31)_1$ and must be equal to the length of the original surface meridian:

$$[F(\varphi^*, k) - F(\varphi_0, k)]/a = L = \int_{\theta_0}^{\theta^*} R_1\,d\theta, \quad (8.34)$$

where R_1 is the radius of curvature of the original meridian. For a convex membrane cap bordered by the equator, $\varphi_0 = 0$, $\varphi^* = \pi/2$, and, in view of $(33)_1$, the meridian arclength is

$$L = K(k)/a = r_e K(k)/2k,\tag{8.35}$$

where $K(k)$ is the complete elliptic integral of the first kind. Assuming the meridian arclength L known, this expression, combined with (32), leads to a transcendental equation in the unknown modulus, k:

$$\sqrt{1-2k^2}/K(k) = \sqrt{T_{za}/2\pi PL^2}.\tag{8.36}$$

The function in the left-hand side has a maximum of $2/\pi$ (at $k = 0$) and a minimum of 0.351 (at $k = 0.923$). These values are the bounds of the feasible range for the nondimensional input ratio at the right. The upper bound represents the limit of the very concept of tension field pseudosurface as a model for an axisymmetric wrinkling membrane. The reason is that, with the maximum radius r_e approaching zero [cf.(33)], the idea of continuously distributed infinitely fine meridional wrinkling does not hold. As to the lower bound, it just indicates a transition to an alternative solution range, which will be considered in the next section. For the time being, it should be noted that, in addition, there exists yet another, physical, restriction: at a certain negative value of T_{za}, the axial length of the pseudosurface segment turns into zero, making further deformation impossible without the membrane self-intersection. As follows from $(31)_2$, this happens at

$$aZ = 2E(k) - K(k) = 0,$$

leading to

$$k = 0.909, \quad K(k) = 2.318, \quad a = K(k)/L, \quad r_e = 0.784L,$$
$$r_c = 0.493L, \quad \sin\theta_a = -0.653, \quad \theta_a = -40.7°.$$

The obtained magnitude of the modulus k provides a refined lower bound for the ratio in (36). Thus, a physically meaningful input requires this ratio to be

$$2/\pi > \sqrt{T_{za}/2\pi PL^2} > 0.348 \quad (0.351),$$

with the previously established value (in parentheses) still valid in the case of a membrane compressed by two rigid plates.

The graph in Fig. 5 represents the solution of equation (36). The sign rule for the input parameter (plotted along the horizontal axis) reflects the fact that the two radicands in the equation change sign simultaneously. With the sought value of k available, the chain of calculations employing the foregoing equations unwinds easily: parameters $\sin\theta_a$, a, r_e, and r_c are evaluated successively from equations (30)–(33) and the axial length Z from $(29)_2$, thereby providing an exhaustive description of the tension field pseudosurface. The sequence of calculations differs slightly if the membrane is subjected to axial displacement loading instead of force loading. According to (31),

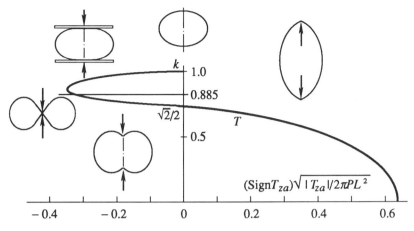

Figure 8.5.   Solution graph of equation (36) and the evolution of shapes of pressurized wrinkling membranes with opposing axial forces.

$$Z/L = [2E(k) - K(k)]/K(k). \tag{8.37}$$

Given the axial length of the pseudosurface, the ratio Z/L is known and equation (37) replaces its counterpart (36) used in the force loading problem.

Note that none of the above equations contains any parameters describing the original geometry of the membrane surface except for the length of the preserved geodesic (in this case, meridian). Thus, when wrinkling comprises the entire membrane, this length is the only required item of information on the original surface geometry. The assumption on comprehensive wrinkling can be verified by checking that the pseudosurface radius given by $(31)_3$ nowhere exceeds the radius r_0 of the same material point on the meridian of the original membrane surface. This amounts to

$$\min (r_0 - r) = \min_s \; [r_0 (s) - r_e \; \mathrm{cn} \; as] \geq 0. \tag{8.38}$$

A location along the meridian where this minimum is attained is identified by equating to zero the first derivative of $(r_0 - r)$ with respect to s. This condition yields

$$\cos \theta_0 = \cos \theta. \tag{8.39}$$

Thus, in verifying that wrinkling is comprehensive, the only points that are to be checked are those where the slope of the pseudosurface meridian equals the slope of the original surface meridian. If condition (38) is satisfied at all such points, wrinkling comprises the entire membrane.

Example 8.2. Axial forces are applied at the poles of a spherical membrane of radius $R = 1$ (Fig. 6) and are directed either toward or away from each other. The forces induce localized wrinkling in the polar zones over the as-yet unknown length of the meridian, $R\theta^*$, where θ^* designates the transition parallel

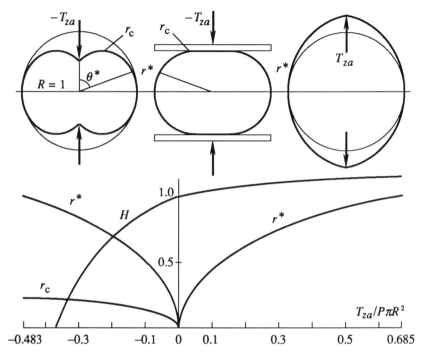

Figure 8.6. Wrinkled pressurized spherical membrane compressed by concentrated axial forces; compressed between two rigid plates; subjected to concentrated axial tension forces; and comprehensive graphs decribing the deformation.

between the uniaxial and biaxial stress zones. This length must equal the arclength of the pseudosurface meridian found from (31). Recall that the latter is measured from the equator, so that the formula has to be adjusted accordingly, resulting in the following equation:

$$a\theta^* = K(k) - F(\varphi^*, k). \qquad (a)$$

The remaining simultaneous equations are: equation (31)$_3$; a relation for the radius r^* of the transition parallel; and equation (30) for two locations:

$$ar^* = 2k \cos \varphi^*, \quad k = \sin[(\pi/2 - \theta_a)/2],$$
$$r^* = \sin \theta^*, \quad k \sin \varphi^* = \sin[(\pi/2 - \theta^*)/2]. \qquad (b)$$

This system of equations has been solved iteratively and the results are presented graphically in Fig. 6. The normalized radius, r^*/R, of the transition parallel, the radius, r_c/R, of the crown (in the case of compression), and the height, H/R, of the deformed membrane, are plotted against the normalized axial force, $T_{za}/P\pi R^2$. As expected, the axial forces, either tension or compression, produce local wrinkling in the vicinity of the poles at the onset of loading. With an increasing magnitude of the forces, the uniaxial stress zones propogate towards the equator and reach it when the transition radius becomes $r^* = r_e = 1$, so that the slope $\theta^* = \pi/2$, and $\varphi^* = 0$. With this, equations (a) and $(b)_1$ yield

$$K(k)/k = \pi. \tag{c}$$

This transcendental equation has two roots (one for compression, the other for tension) leading, with the aid of (31)–(33), to the following two solutions:

Root	a	$\sin \theta_a$	$T_{za}/P\pi R^2$	r_c/R	H/R
0.983	1.966	−0.933	−0.241	0.694	−0.508
0.545	1.089	0.406	0.342	n/a	1.085

In further compression or tension loading, the membrane meridians are elasticas whose geometric attributes depend only on the normalized load ratio, $T_{za}/P\pi R^2$.

Pressurized Membranes Without Axial Forces

Geometric Properties of the Tension Field. For the simplest, yet important, case of uniform pressure, the condition of axial equilibrium yields

$$T_1 = Pr R_2/2.$$

Comparing this to (25) yields a characteristic geometric property of the resulting pseudosurface—a fixed ratio of the principal curvatures:

$$\sigma_1/\sigma_2 = R_2/R_1 = 2. \tag{8.40}$$

Furthermore, it also reveals the geometric meaning of the statical parameter a:

$$a^2 = \sigma_1/r = 2\sigma_2/r = 2 \sin \theta/r^2. \tag{8.41}$$

Thereby yet another geometric property of this tension field pseudosurface is established—the constancy of the ratios figuring in (41).

The above equations are referred to cylindrical coordinates with the origin in the equatorial plane. This is the most convenient reference system, even though the radius r_e of the equator is as-yet unknown and, moreover, the sought pseudosurface segment may not even contain an equator. For a pressurized membrane without an external axial force, $\theta_a = 0$ and, as follows from (27), (30), and (41),

$$k = 1/\sqrt{2}, \quad a = \sqrt{2}/r_e, \quad r/r_e = \sqrt{\sin \theta} = \cos \varphi. \tag{8.42}$$

As a result, expressions (31) for the meridian arclength and the axial length of the pseudosurface section measured from the equatorial plane become

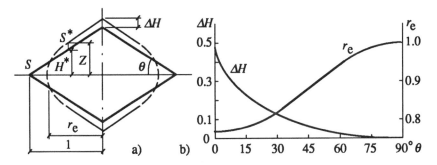

Figure 8.7. A pressurized double conical membrane: a) geometric parameters of the membrane; b) graphs of axial expansion and lateral contraction.

$$L = (r_e/\sqrt{2})\, F(\varphi^*,\, 1/\sqrt{2}), \quad Z = r_e\sqrt{2}\, E(\varphi^*,\, 1/\sqrt{2}) - L. \quad (8.43)$$

Here the asterisk indicates the terminal parallel of the tension field pseudosurface; it is either the edge of the membrane or the transition parallel between the uniaxial and biaxial stress zones. The meridian slope θ^* at the transition parallel is, generally, unknown and must be determined from the condition of the slope continuity. Note that this condition entails continuity of the second principal curvature, but not of the first one.

Example 8.3. A closed membrane is assembled of two identical conical membranes joined at the edges (Fig. 7). For a cone, $\theta = $ const. and the length s of the generator replaces the slope as the meridional coordinate. Relation $(42)_3$ yields

$$r_e = r^*/\sqrt{\sin\theta} = S^*\cos\theta/\sqrt{\sin\theta} \qquad (a)$$

and equation $(43)_1$ becomes

$$L = S - S^* = S^*\cos\theta\; F(\varphi^*,\, \sqrt{1/2})/\sqrt{2}\sin\theta. \qquad (b)$$

Upon evaluating from (b) the parameter S^* of the transition parallel, the radius of the equator is found from (a). The axial expansion of the membrane is now evaluated with the aid of $(43)_2$:

$$\Delta H = Z - H^* = r_e\sqrt{2}\, E(\varphi^*,\, \sqrt{1/2}) - (1 + \sin\theta)(S - S^*). \qquad (c)$$

The obtained solution provides a complete description of the membrane deformation. Shown in Fig. 7 are plots of the normalized axial expansion and lateral contraction as functions of the base angle of the cone. The limiting case $\theta = 0$ is a membrane formed by two flat circular membranes joined at the edge (like a mylar ballroom balloon).

Case Study: Pressurized Ellipsoid of Revolution. Consider a membrane whose surface is generated by rotating an ellipse with a semimajor

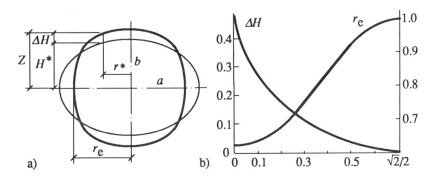

Figure 8.8. A pressurized ellipsoid of revolution: a) geometric parameters of the membrane; b) graphs of axial expansion and lateral contraction.

axis $a = 1$ and a semiminor axis $b < 1$ about the minor axis (Fig. 8). The parameter b represents in this case a nondimensional aspect ratio of the obtained ellipsoid. Expressed in terms of the meridian slope, θ, the principal curvatures of an ellipsoid of revolution are

$$\sigma_1 = (1 + d^2 \sin^2 \theta)^{3/2}, \quad \sigma_2 = (1 + d^2 \sin^2 \theta)^{1/2},$$

with

$$d^2 = e^2/(1 - e^2) = (1 - b^2)/b^2, \quad e^2 = 1 - b^2,$$

where e is the eccentricity of the ellipse.

According to (40), a pressurized membrane of this shape will wrinkle if at some location the principal curvature ratio is

$$\sigma_1/\sigma_2 = 1 + d^2 \sin^2 \theta > 2. \tag{8.44}$$

From here it follows that wrinkling occurs in membranes with $b^2 < 1/2$ and comprises a belt containing the surface equator.

The main difficulty of the problem is evaluating the extent of the wrinkled zone. It turns out that the transition parallel, θ^*, separating the uniaxial and biaxial stress zones, is not the parallel where the principal curvature ratio attains the value of 2. Instead, θ^* is farther from the equator and the uniaxial zone extends into the part of the membrane with $\sigma_1/\sigma_2 < 2$. The reason is that the membrane statics requires continuity of the meridian slope at the transition parallel, whereas the principal curvature σ_1 can (and generally does) have a discontinuity at θ^*.

The arclength of the ellipsoid meridian measured from the equator equals

$$L_m = E(e) - E(\psi, e). \tag{8.45}$$

The amplitude ψ is related to the radius and the meridian slope by

$$r = \sin \psi, \quad b \tan \psi = \tan \theta, \tag{8.46}$$

so that

$$\sin^2\theta = b^2 r^2/(1 - e^2 r^2). \tag{8.47}$$

Equating the arclength and the slope of the meridian at the transition point to their respective counterparts for the tension field [cf. (42) and (43)] gives

$$E(e) - E(\psi^*, e) = (r_e/\sqrt{2})\, F(\varphi^*, 1/\sqrt{2}), \tag{8.48}$$

$$b^2 r^{*2}/(1 - e^2 r^{*2}) = \cos^4\varphi^*. \tag{8.49}$$

In the obtained closed system of equations, some of the unknowns appear as arguments in the elliptic integrals (48) necessitating an iterative solution. Fortunately, the remaining equations are rather simple and so is the employed iterative loop. After assuming an initial approximation for the transition radius r^*, the unknown radius r_e and the amplitudes ψ^* and φ^* are found from equations (42), (46), and (49). Their substitution into equation (48) produces an error which is used to refine the initial approximation, thus closing the loop.

The process converges rapidly, except for ellipsoids with $b^2 \to 1/2$, when wrinkling is confined to a narrow belt at the pseudosurface equator and the equations become ill-conditioned. The difficulty is overcome by devising an asymptotic solution based on power expansions in φ. As follows from (42) and (47), within a narrow equatorial belt

$$\sin\theta \approx 1 - \varphi^2, \quad r \approx 1 - \varphi^2/2, \quad r_e \approx 1 - \varphi^2/4. \tag{8.50}$$

The expansions of arclengths (43) and (45) are

$$L \approx r_e\,\varphi(1 + \varphi^2/12)/\sqrt{2}, \tag{8.51}$$

$$L_m \approx t\,[b + (1 - b^2 t^2)/6b]. \tag{8.52}$$

Here $t = \pi/2 - \psi$ is the parametric angle of the ellipse related to the slope θ by

$$\sin^2\theta = 1 - (\tan^2 t)/b^2.$$

This relation, combined with (50), allows L_m in (52) to be expressed in terms of φ and equated to L in (51). The resulting equation in the unknown value of the transition angle φ^* is

$$(1 - \varphi^{*2}/4)\,\varphi^*\,(1 + \varphi^{*2}/12)/\sqrt{2} = b^2\sqrt{2}\,\varphi^*\,(1 + \varphi^{*2}/6),$$

which yields

$$\varphi^{*2} \approx 3\,(1 - 2b^2). \tag{8.53}$$

The axial extension of the membrane is found with the aid of (43)

$$\Delta H = Z - H^* \approx (2 - \sqrt{2})\varphi^{*3}/4, \tag{8.54}$$

where $H^* = b\sin t$ is the axial coordinate of the transition parallel on the original ellipsoid.

The obtained results are presented graphically in Fig. 8b. As expected, the maximum lateral contraction and axial expansion occur in the limiting case of an initially "flat" ellipsoid—a ballroom balloon. At the opposite extreme is an ellipsoid with $b^2 = 1/2$ where a uniaxial zone does not form at all, since the maximum value of the principal curvature ratio nowhere exceeds 2 (it just attains this magnitude at the equator).

8.4 Wrinkling Membranes: Advanced Problems

The analogy with Euler's elastica, although providing a ready solution for a certain class of axisymmetric wrinkling membranes, does not cover all of these problems. However, with the analytical basis—the condition of preservation of geodesics—in place, all that needs to be done for solving other problems is to accommodate the particular types of membrane geometry and loading conditions. Three such problems, all of them axisymmetric, are solved in this section. The first one is a pressurized membrane compressed axially through two concentric rigid discs of a given radius; the second is a pressurized toroidal membrane of a general profile; and the third is a pressurized membrane with axial torque.

Membrane Compressed or Indented by Rigid Disks

Alternative Solutions and Their Domains. The problem of a pressurized membrane with axial forces at the poles can easily be adapted for the case of axial compression or indentation by rigid disks of a given radius r_i. In fact, the solution remains intact for the membrane segment $r > r_i$ as long as the condition of axial equilibrium is satisfied, so that [cf. (32)]

$$\sin \theta_a = \sin \theta_a + a^2 r_i^2/2 = a^2(T_{za}/2\pi P + r_i^2)/2. \qquad (8.55)$$

In the case of axial tension (assuming the two discs are attached to the membrane, replacing its polar caps), the solution is always valid. However, in the case of compression, two qualifications are necessary. One is straightforward: in a contact problem with the disks unattached to the membrane, there must be $r_i \leq r_c$, since the part of the disk outside the crown is irrelevant for the problem. The other, more substantial, qualification affects the form of the above solution.

Upon incorporating the input parameter r_i into equations (31), the first two of them are rewritten as

$$2k \cos \varphi_i/F(\varphi_i, k) = r_i/L,$$
$$2E(\varphi_i, k)/F(\varphi_i, k) = Z/L + 1, \qquad (8.56)$$

where L and Z are, respectively, the meridian arclength and the axial length of the membrane segment between r_e and r_i. In the absence of the surface apex, the constant of integration C in solution (27) has to be expressed in terms of geometric parameters other than θ_a, for example,

$$C = 1 - a^2 r_e^2/2 = - a^2 r_c^2/2 = \sin \theta_i - a^2 r_i^2/2. \qquad (8.57)$$

As is readily seen, C is always negative and it is possible that $C < -1$. Within this new range, the solution of differential equations (29) in terms of elliptic integrals requires a replacement of variables different from (30), namely,

$$\phi = (\pi/2 - \theta)/2, \quad c^2 = 2/(1 - C). \qquad (8.58)$$

This leads to an alternative solution

$$as = cF(\phi, c), \quad caz = 2E(\phi, c) - (2 - c^2)F(\phi, c), \quad car = 2\sqrt{1 - c^2 \sin^2 \phi}, \qquad (8.59)$$

which describes an axisymmetric tension field pseudosurface different from that in (31). Its characteristic radii, found from (57) and (59), are given by

$$car_e = 2, \quad car_c = \sqrt{2(2 - c^2)}, \quad car_i = 2\sqrt{1 - c^2 \sin^2 \phi_i}, \qquad (8.60)$$

[cf. (33)]. The counterpart of the system of equations (56) is easily obtainable upon specifying $r = r_i$ in (59):

$$2\sqrt{1 - c^2 \sin^2 \phi_i}/F(\phi_i, c) = c^2 r_i/L,$$
$$2E(\phi_i, c)/F(\phi_i, c) - 2 = c^2(Z/L - 1). \qquad (8.61)$$

Note that the feasible ranges of φ and ϕ in equations (56) and (61) coincide: the lower bound corresponds to $c = k = 1$, when the two systems of equations become identical, and the upper bound of $\pi/2$ is determined by $\theta_i = -\pi/2$ according to both transformations (30) and (58).

When solving a particular problem, it is not known in advance which one of the two alternative solutions to follow. Fortunately, this is an occasion when equations "think," making this aspect of the analysis error-proof. If the wrong alternative is chosen, a value of the modulus k or c greater than 1 results, thereby signaling that the other solution must be followed. Furthermore, the respective elliptic integrals of the two solutions are obtainable from each other by the reciprocal modulus transformation. For the case $k > 1$, this transformation is given by

$$F(\varphi, k) = cF(\phi, c),$$
$$E(\varphi, k) = [E(\phi, c) - (1 - c^2)F(\phi, c)]/c,$$
$$c = 1/k, \quad \sin \phi = k \sin \varphi \le 1.$$

In terms of the elastica theory, the range $C < -1$ corresponds to large deflections of a prismatic bar subjected to end forces and moments such that the mutual rotation of the bar ends exceeds 2π. Note that this deformation cannot be caused by compression forces alone, without a bending moment.

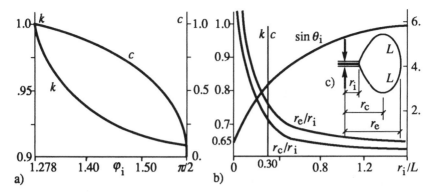

Figure 8.9. A pressurized membrane indented by rigid disks: a) moduli k and c for the alternative solution domains; b) graphs of normalized geometric parameters; c) geometric parameters of the membrane.

Example 8.4. A pressurized axisymmetric membrane is indented axially by two rigid disks such that the disks are pressing against each other (Fig. 9c shows one half of the symmetric profile of the membrane). Assume that the entire membrane is in a uniaxial stress state, i.e., completely wrinkled. This assumption, making the problem independent of the original membrane geometry, must be verified with the aid of condition (38) after the pseudosurface geometry is established.

Equations $(56)_2$ and $(61)_2$ with $Z = 0$ in the right-hand sides were solved numerically and the solution graphs $k = k(\varphi)$ and $c = c(\phi)$ are presented in Fig. 9a. At $Z = 0$, the common lower bound for φ and ϕ was found to be 1.278. For every matching pair of values of the amplitude and the modulus, substitution in the appropriate one of the equations $(56)_1$ or $(61)_1$ gives the corresponding ratio of r_i/L. After that, the geometric parameters of interest are obtainable from (57) and other available equations.

The graphs in Fig. 9b illustrate the geometric evolution of the pseudosurface with reference to the only input parameter of the problem—the ratio r_i/L of the inner radius and meridional arclength. One interesting feature is the lower bound on the meridian slope at the inner radius: $\sin\theta_i = 0.653$. The vertical line marks the transition value of $r_i/L = 0.301$ (produced by $C = -1$) between the domains of the two alternative solutions.

Pressurized Toroidal Membrane

Toroidal Tension Field. As demonstrated in the beginning of this chapter, a toroidal membrane is underconstrained even within the class of smooth bending. Moreover, for this membrane, a uniform normal pressure, generally, is not an equilibrium load. For it to be an equilibrium load, it is necessary (but not sufficient) that the two crowns of the toroid are of the same radius. If the undeformed surface does not satisfy this requirement, a uniform pressure, either internal or external, will produce piecewise smooth isometric bending in a membrane shell and wrinkling in a true membrane.

An internally pressurized toroidal membrane is different in principle from a simply connected closed membrane in that it does not allow comprehensive wrinkling. Indeed, wrinkling in a pressurized axisymmetric membrane is a result of constraint disengagement in the hoop direction, which is the direction of the nongeodesic principal force trajectories. However, an internal pressure P produces a resultant hoop tension force $T_\varphi = PA$, where A is the area of the meridional section of the deformed toroid. This force can be supported by the membrane only by developing a tensile hoop stress at some location, thereby creating at least one biaxial stress zone. As shown below, such a zone may be confined to just one singular parallel. Nevertheless, the conclusion stands: *a closed and smooth toroidal tension field is impossible.*

A few geometric characteristics of a toroidal tension field pseudosurface are obtainable from simple statical considerations. Because of the presence of two crowns of the same radius r_c, there is no interaction in the axial direction between the inner $(r \leq r_c)$ and outer $(r > r_c)$ segments of the membrane. As a result, the conditions of axial equilibrium for the two segments are independent, although they are of one and the same form:

$$2\pi T_1 \sin \theta = \pi(r^2 - r_c^2)P. \tag{8.62}$$

From here, using (25) and (26), the principal curvature ratio for a toroidal pseudosurface is found:

$$\sigma_1/\sigma_2 = 2r^2/(r^2 - r_c^2). \tag{8.63}$$

Solution (27), along with several equivalent expressions (57) for the integration constant, is valid for a wrinkled section of a toroidal membrane. However, the unavoidable presence of a biaxial stress zone makes this membrane more difficult to analyze in comparison with a simply connected membrane. To further complicate the matter, a pattern of alternating uniaxial and biaxial stress belts is quite common in wrinkling toroidal membranes.

Case Study: A Conical Toroid. The utter geometric simplicity of a conical toroid (a double-walled cone) is due to a constant meridian slope. This allows a variety of typical situations to be examined unobscured by tedious calculations. The geometric parameters chosen for the study are the cone base angle θ_0 and the ratio, L/r_i, of the length of the generator to the smaller base radius (Fig. 10a).

A good starting point is $\theta_0 = 0$, representing a membrane whose original shape is an annular toroid. A conspicuous geometric feature of the inflated membrane is a salient point at the inner equator (Fig. 10b). The slope of the meridian at this point, as well as all other geometric parameters of the pseudosurface, are those of the preceding example; in fact, the two problems are

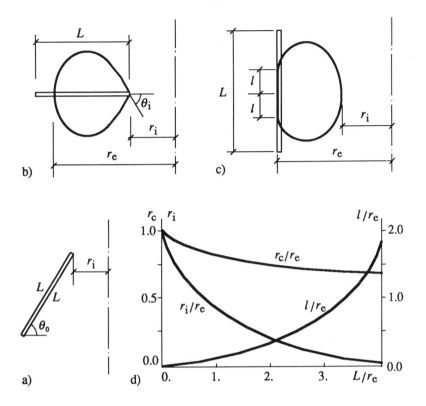

Figure 8.10. A pressurized conical toroid: a) initial geometry; b) annular toroid; c) cylindrical toroid; d) graphs of normalized parameters of a cylindrical toroid.

identical. Note that an implicit requirement for this identity to hold is a comprehensive wrinkling of the membrane, a condition apparently met in this case [cf. (38)]. Indeed, the unavoidable biaxial stress zone supporting the entire hoop tension force T_φ is confined in this membrane to a single parallel—the inner equator. Were the membrane extensible, this zone would spread over an equatorial belt of a finite width, accommodating a rapidly decaying stress concentration.

A singular inner equator is not an inevitable attribute of a wrinkling toroidal membrane. This is readily seen from the behavior of a membrane representing the other extreme shape, $\theta_0 = \pi/2$, i.e., a cylindrical toroid (Fig. 10c). In this case, it is obvious that the biaxial stress zone must be located at the outer equator and be confined to a cylindrical belt of a certain height $2l$. The slope of the meridian at the transition point to the uniaxial stress zone must be continuous, hence $\theta^* = \pi/2$. According to (58), $\theta_i = -\pi/2$ yields $\phi_i = \pi/2$, and the arclength of the pseudosurface meridian found from $(59)_1$ is

$$a(L - l) = c K(c).$$

The only known parameter in this equation is the length L. The second equation, derived from $(59)_2$, implements the condition for the combined height of the biaxial and uniaxial stress zones of the membrane to be zero:

$$cal + 2E(c) - (2 - c^2)\,K(c) = 0.$$

Since the radius of the equator in this problem is known, it can be used to eliminate, with the aid of $(60)_1$, the parameter a from the above two equations. Adding them together, yields one solving equation in one unknown, c:

$$K(c) - E(c) = L/r_e. \tag{8.64}$$

The geometric parameters of interest are expressed in terms of c as follows:

$$r_c/r_e = \sqrt{1 - c^2/2}, \quad r_i/r_e = \sqrt{1 - c^2}, \quad l/r_e = L/r_e - c^2 K(c)/2. \tag{8.65}$$

Their graphs are shown in Fig. 10d as functions of the input ratio L/r_e.

The obtained solutions for the two extreme shapes (the annular and cylindrical toroids) shed light on the deformation patterns of membranes within the entire range $0 < \theta_o < \pi/2$. As long as θ_o is small (a shallow cone), condition (38) is satisfied and the membrane has the shape shown in Fig. 10b. The biaxial stress zone remains confined to the inner equator and the slope θ_i is independent of the cone base angle θ_o. This continues until a second biaxial stress zone forms at the outer segment of the membrane which can be detected with the aid of (38) and (39).

Let s_o be the pseudosurface meridian arclength from the outer equator to the point where the meridian slope is $\theta = \theta_o$. According to (39), at this point the difference between the radii of the original surface and the pseudosurface is a minimum. The radius of the original cone at this point is

$$r_o = r_i + (L - s_o)\cos\theta_o = L[r_i/L + (1 - s_o/L)\cos\theta_o].$$

The lengths s_o and L figuring in this expression, and the pseudosurface radius to be compared with $r_o(s_o)$, are given by the alternative solutions (31) and (59)

$$as_o = F(\varphi_o, k), \quad aL = F(\varphi_i, k), \quad ar(s_o) = 2k\cos\varphi_o,$$

$$as_o = cF(\phi_o, c), \quad aL = cF(\phi_i, c), \quad ar(s_o) = 2\sqrt{1 - c^2\sin^2\phi_o}/c.$$

Requiring the difference between the radii to be positive (the condition for wrinkling to occur) yields two alternative inequalities

$$(r_i/L + \cos\theta_o)\,F(\varphi_i, k) - F(\varphi_o, k)\cos\theta_o - 2k\cos\varphi_o \geq 0,$$

$$(r_i/L + \cos\theta_o)\,F(\phi_i, k) - F(\phi_o, k)\cos\theta_o - 2\sqrt{1 - c^2\sin^2\phi_o}/c^2 \geq 0. \tag{8.66}$$

These inequalities assure wrinkling and can be used to identify the maximum base angle θ_o of a conical toroid with only one biaxial stress zone at the inner equator (point 1 in Fig. 11a). The angle, evaluated as a function of the given ratio r_i/L, was found to be changing very slowly and within a rather narrow

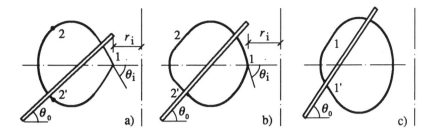

Figure 8.11. A general conical toroid: a) biaxial stress zone is confined to the inner equator; b) two additional biaxial zones evolve; c) wrinkling comprises the inner equator, leaving two biaxial zones that eventually merge.

range, from $\theta_0 = 32°$ for $r_i/L \to 0$ to $\theta_0 = 43°$ for $r_i/L \to \infty$. When θ_0 exceeds the above values, two new biaxial stress zones develop in the membrane (2 and 2' in Fig. 11a and b) comprising a part of the original conical surfaces. With a further increase in θ_0 the tensile hoop stress in these zones takes up a gradually increasing portion of the overall hoop force produced by the internal pressure. In the process, the meridian slope at the inner equator increases and eventually becomes vertical (Fig. 11c). From this point on, only the two outer biaxial zones exist, supporting the entire resultant hoop force. In the limit, at $\theta_0 = \pi/2$, these zones merge, producing the already familiar profile of an inflated cylindrical toroid shown in Fig. 10c.

Axisymmetric Membrane with Torque

Twisted Axisymmetric Pseudosurface. Let an axisymmetric membrane supporting a normal surface pressure, P, be subjected to a gradually increasing axial torque moment, M_z. At a certain magnitude of M_z, one of the principal forces in the membrane approaches zero and the net of principal force trajectories becomes semigeodesic. When the force in the nongeodesic array is zero, it follows from (7.74) that

$$M_z \sigma_2 / T_z = \tan \beta. \tag{8.67}$$

Introducing this into equation (7.71), and taking into account that the axial force induced by the internal pressure is $T_z = \pi r^2 P$, gives

$$\sigma_1 \sigma_2 = 2 - \tan^2 \beta, \tag{8.68}$$

which is a geometric characteristic of a twisted axisymmetric pseudosurface.

Since the tension field trajectory is geodesic, it obeys the Clairaut formula (7.21), so that

$$\tan^2 \beta = b^2/(r^2 - b^2), \quad b = r_e \sin \beta_e.$$

When this relation, along with expressions (26), is introduced into (67), a first-order differential equation results:

$$d(\sin \theta)/dr = 2 \sin \theta/r - b^2 \sin \theta/r(r^2-b^2).$$

Its solution is the first integral of the twisted axisymmetric pseudosurface:

$$\sin \theta \cos \beta/r^2 = C. \tag{8.69}$$

An explicit description of the pseudosurface, as in all previous cases, is obtainable by forward integration from the following differential equations:

$$ds_t = dr/\cos \theta \cos \beta, \quad dz = dr \tan \theta, \quad d\varphi_t = dr \tan \beta/r \cos \theta. \tag{8.70}$$

Here s_t is the arclength of the tension field (or principal force) trajectory, z is the axial coordinate measured from a reference plane perpendicular to the axis of revolution, and φ_t is the polar angle increment along the trajectory.

By using a substitution

$$x = r^2/r_e^2, \quad Q = \tan \beta_e = M_z/T_{ze}r_e, \tag{8.71}$$

the above differentials can be transformed as follows:

$$ds_t = r_e \, dx/2G(x), \quad dz = r_e \, x \, dx/2G(x), \quad d\varphi_t = Q \, dx/2x \, G(x), \tag{8.72}$$

where

$$G(x) = \sqrt{- x^3 + (1 + Q^2)x - Q^2}.$$

The roots of the radicand polynomial are

$$x_1 = 1, \quad x_{2,3} = - (1 \pm q)/2, \quad q = \sqrt{1 + 4Q^2},$$

so that the problem is once again reducible to elliptic integrals. However, in this problem, their computational utility is questionable and, at least for practical purposes, a numerical forward integration of (70) is preferable.

The above equations provide some qualitative insight into the evolution of the shape of a pressurized wrinkled membrane with an increasing torque moment. Initially, the pseudosurface is convex and has a constant principal curvature ratio of 2 everywhere. Under an applied torque, $\beta \neq 0$ and, according to (68), the principal curvature ratio starts to decrease, causing the pseudosurface meridian to flatten. When $\tan^2 \beta_e$ reaches the value of 2, the curvature σ_1 becomes zero resulting in a cylindrical pseudosurface. In further loading, an hourglass-shaped pseudosurface evolves.

If the membrane edge parallels are of unequal radii, a cylindrical pseudosurface is precluded. In this case, $\tan^2 \beta$ first reaches the magnitude of 2 at the smaller edge; as a result, at this location σ_1 becomes zero and an inflection point appears on the meridian. In further loading, the inflection point advances toward the larger edge, leaving behind a segment with negative meridional and total curvature. At this stage, the pseudosurface may, but not necessarily does, have an equator (a parallel with $\theta = \pi/2$). However, sooner or later, an equator

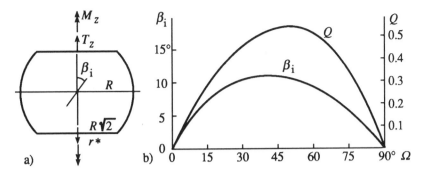

Figure 8.12. A spherical segment of two bases under axial tension and torque: a) geometric parameters; b) solution graph for the slope of the limiting geodesic and torque-rotation diagram.

forms at the smaller edge and then moves away from it, but never reaches the greater edge. Instead, a striction parallel will form within the pseudosurface, dividing it into two segments.

Example 8.5. A membrane in the form of a spherical segment of two bases (Fig. 12) is subjected to a combination of axial tension and torque moment. In the absence of a surface pressure, the tension field pseudosurface is ruled (its geodesic trajectories are straight lines). The only ruled surface of revolution is a one-sheet hyperboloid. By virtue of the preservation of geodesics, the linear generators of the hyperboloid (which are, of course, geodesics) must have been geodesic lines on the original sphere as well. For each load combination, only one of the original surface geodesics becomes taut, thereby controlling the axial length of the wrinkled membrane. This observation underlies the following sequence of calculations.

A geodesic line on a sphere (an arc of the great circle) can be identified by the angle β it forms with a meridian at the equator. The length of the geodesic between the segment bases is evaluated easily:

$$S = 2R \tan^{-1}(\sqrt{\sec 2\beta}),$$

where R is the radius of the sphere. Next, the increment, 2Φ, of the polar angle along the geodesic is found by integrating $(70)_3$; for a sphere the result is obtainable in a closed form:

$$\sin 2\Phi = \sin \beta \, [\sqrt{5 - 9 \sin^2\beta} - (1 - 3 \sin^2\beta / \cos \beta].$$

Suppose, the torque moment causes a mutual axial rotation of the membrane edges through an angle 2Ω. When one of the original sphere geodesics becomes a generator of the wrinkled hyperboloid, the axial length of the latter is given by

$$Z^2 = S^2 - [\sqrt{2} \, R \sin (\Phi + \Omega)]^2.$$

For a given magnitude of rotation 2Ω, the preserved and, therefore, controlling geodesic is the one that corresponds to the minimal length Z. Hence, minimizing the above expression with respect to β (which is present in both S and Φ) identifies the sought geodesic for any given Ω. One of the curves in the graph of Fig. 11b represents the parameter β of the controlling geodesic as a function of the angle of rotation Ω. It turns out that the range of potentially controlling geodesics is very narrow: it is limited to $0 < \beta < 11°$. The other curve represents the normalized torque as a function of Ω and is, in fact, the torque–rotation diagram for a fixed axial tension level. The moment peaks shortly after the angle of rotation attains the value $\Omega = 45°$. Further deformation is accompanied by a rapidly developing striction at the equatorial plane; at $\Omega = 90°$ the striction is complete.

9
Other Underconstrained Systems and Contact Problems

This chapter comprises a selection of underconstrained structural systems of which some are new and others are combinations of the already familiar components such as cables, nets, and membranes. The subject of the first section is a band or a cable-band assembly. These assemblies involve one or more arrays of structural members in the form of thin narrow flexible bands, which represent a relatively new addition to the list of generic structural components. When used in place of one (or, possibly, both) of the cable arrays of a net, bands advantageously combine load-carrying and surface-cladding functions. The kinematic properties of a band, together with the arrangement of band or cable-band intersections, account for some interesting features in the system statics and kinematics. At the same time, the properties of a band as a structural component impose certain restrictions on the feasible geometry of a band or cable-band system as compared with a similar cable net.

The remaining two sections of the chapter are concerned with structural systems whose analysis requires solving certain contact problems. Here underconstrained components are interacting with one another or with highly elastic components. Axisymmetric 3-nets analyzed in the second section represent assemblies involving three arrays of structural members: one array directed along the surface meridians and two inclined arrays. These systems found some interesting applications in structural engineering (e.g., tensile roofs, large natural draft cooling towers). Their modifications with an array of flexible bands replacing one of the cable arrays have also been discussed in the literature.

The last section deals with contact problems where highly elastic components are interacting with an extensible or inextensible fiber net. Structural behavior of such systems is characterized by a blend of kinematic and elastic mobility of the components. One of the considered systems is an elastic, initially cylindrical membrane stiffened by a geodesic or Chebyshev net. Analysis of this system gives rise to a nonlinear contact problem which is solved using the results of Chapter 7. An analysis of a tensile structural member comprised of a highly elastic cylindrical core and a woven braid leads to yet another contact problem. Observing the behavior of this member proved useful in the synthesis of composite structural systems with a desired type of response. This is illustrated by an example of a highly elastic spring with a stiffening load-displacement diagram.

9.1   Band and Cable-Band Systems

The Band as a Structural Component

Statical and Kinematic Properties. The band is a distinct item in the nomenclature of generic structural components such as a bar, a beam, or a cable. Compared to other components, the statical and kinematic properties of a band are closest to those of a cable and a thin-walled bar. A band is similar to a cable in that it has no torsional stiffness and no bending stiffness in one plane; hence, like a cable, it cannot support compression. On the other hand, a band resembles a thin-walled bar of a narrow rectangular cross section in having a similar relation among its length, width, and thickness, namely, $L \gg w \gg t$. However, the band is different from a thin-walled bar in that it is completely devoid of torsional and out-of-plane bending stiffness, quite like a membrane. In statical terms, a band is a structural component that can be defined as a narrow membrane strip capable of resisting only tension and in-surface bending and shear. It does not resist compression, torsion, and out-of-surface bending and shear.

Because of the small width, the geometry of a band is determined by the geometric parameters of the centerline. Moreover, in an analytical model of a band or cable-band assembly, a band is completely represented by its centerline. Amending the above statical definition with a kinematic one, a band is idealized as inextensible and preserving the geodesic (in-plane) curvature of its centerline. For example, a narrow conical membrane belt, when cut, applies to a plane as a narrow open annulus of a certain curvature. A narrow geodesic strip cut out of any surface develops on a plane as a straight band and, conversely, a flat rectilinear band can be applied to any curved surface such that its centerline follows one of the surface geodesics. In both cases, the absence of torsional stiffness (due to the small width and very small thickness) ensures a close compliance. However, continuous covering of a surface, generally, requires either varying widths (both along and among the bands) or an overlap. The presence of overlaps or gaps, as well as the widths of individual bands, are irrelevant for an analytical model of a band assembly since the bands are represented in it by their centerlines.

Implications for Structural Layout and Design. In the band and cable-band systems considered below the bands are assumed fabricated flat and straight (zero geodesic curvature). It is further assumed that the band end support fixtures are such that in assembling the system (and, if this is the case, in prestressing it), in-surface bending of bands is avoided. Then a continuous model of a band or cable-band system is a geodesic or semigeodesic network. Under certain conditions, described below, these systems can be analyzed using the methods developed earlier for geodesic and semigeodesic cable nets. If the system is prestressed, its original shape is that of a singular geodesic or

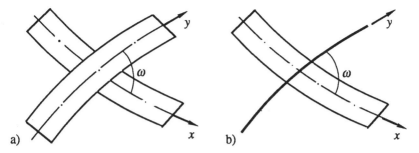

Figure 9.1. Generic elements of an analytical model: a) of a band system; b) of a cable-band system.

semigeodesic net of negative total curvature. The equations and solutions developed in the preceding chapters for singular geodesic and semigeodesic nets, in particular, for axisymmetric ones, are valid for the respective cable and cable-band systems.

Whereas the band centerlines in a band or cable-band system are always geodesic, the cable trajectories may or may not be geodesic, depending on the design of the member intersections. Most interesting for applications is an assembly where member slippage at the intersections is not prevented (Fig. 1), at least, in one direction (Fig. 2). In what follows, only systems of this type are considered. Accordingly, for a band system with two (or more) band arrays, the intersections are always assumed not fastened, so that only a normal force interaction (contact pressure) between the arrays is possible. In such systems, an external normal surface load and the array contact pressure do not induce bending in the bands. The only potential source of bending lies in the band end support conditions.

When the band ends are free to rotate about the normal, there are no end moments, and the bands remain entirely moment-free. Under these conditions, a band or a cable-band system behaves like a geodesic net, and it preserves this property in both kinematic and elastic deformations. When cables in a cable-band system are free to slip at fixed locations along the bands (Fig. 2), a semigeodesic system results with the slipping array nongeodesic. Such is, for example, a net of meridians and parallels on a surface of revolution where the hoop members (nongeodesic) are allowed to slip across, but not along, the meridians. In a system of this type, the nongeodesic array is always a cable array, for two reasons. First, initially straight bands cannot acquire geodesic curvature kinematically, while cables can. Second, cables are more difficult to restrain against slipping; although technically feasible, this requires special fixtures with tight grips.

In general, geodesic lines of a surface are space (three-dimensional) curves. An initially straight and flat band acquires the third dimension without resistance, due to the absence of bending and torsional stiffness. Geodesic lines can also be plane and then they are the lines of principal curvature of the surface. This

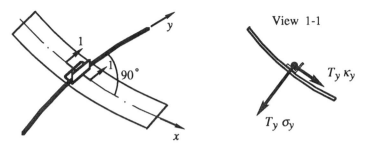

Figure 9.2. Generic element of a semigeodesic cable-band system with a cable slipping across a band.

transformation of initially straight bands requires only bending, but no torsion: the lines of principal curvatures are known to have zero geodesic torsion. Finally, geodesic lines can be rectilinear, like, for instance, the linear generators of ruled surfaces. In this case, the bands must undergo torsion but no bending. Furthermore, even torsion is unnecessary when the surface is developable; then the generators are straight lines with zero normal and geodesic curvatures and zero geodesic torsion.

When selecting a particular shape of a band or a cable-band system, it makes sense to maximize the normal cuvature of load-supporting (as opposed to prestressing) elements, thereby reducing their tension level. Maximum curvature is the property of lines of principal curvature, so it is advantageous to choose one of the principal curvature arrays for the band trajectories. However, to avoid in-plane bending, bands must remain geodesic. A geodesic line of principal curvature is plane (but not straight): as a geodesic, it has zero geodesic curvature, and as a line of principal curvature, it has zero geodesic torsion. Thus, a surface with one array of plane geodesic lines of principal curvature is, from this point of view, the best choice for the system. The most prominent example of such surfaces is a surface of revolution, where meridians are plane lines of principal curvature and, therefore, are geodesic. Unfortunately, if such a system is to be prestressed, the second array must be nongeodesic, resulting in a semigeodesic system; otherwise, the axial symmetry is lost. (Recall that prestressable axisymmetric geodesic nets are skew and do not include the meridional array.)

Kinematics of Systems with Unrestrained or Partially Restrained Intersections

Networks with Unrestrained Intersections. A kinematic model of a two-array band, cable-band, or cable system with the member intersections not fastened involves two separate, although related, displacement fields. Ruling out the possibility of delamination, the two member arrays must always be located in one surface but otherwise are independent. The resulting displacement pattern may be considered as a compound displacement field described by five independent displacements: two in-surface displacements for each array and one common normal displacement.

With the two member arrays designated by x and y, their respective displacement vectors, Δ_x and Δ_y, are expressed in terms of the longitudinal, U, transverse, V, and normal, W, displacement components:

$$\Delta_x = U_x \mathbf{u}_x + V_x \widetilde{\mathbf{u}}_x + W\mathbf{n}, \quad \Delta_y = U_y \mathbf{u}_y + V_y \widetilde{\mathbf{u}}_y + W\mathbf{n}, \quad (9.1)$$

where $\mathbf{u}_{x(y)}$ are unit direction vectors of the two arrays and $\widetilde{\mathbf{u}}_{x(y)}$ are their orthogonal unit vectors in the reference configuration. The coordinate derivatives of the displacement vectors (designated by subscripts 1 and 2) are evaluated with the aid of Gauss formulas (6.17). The formulas simplify considerably due to the zero geodesic curvatures of the members in the original configuration. As a result,

$$\Delta_{x,1} = U_{x,1} \mathbf{u}_x + U_x A \sigma_x \mathbf{n}$$
$$+ V_{x,1} \widetilde{\mathbf{u}}_x + V_x A \tau_x \mathbf{n} + W_1 \mathbf{n} - WA(\sigma_x \mathbf{u}_x + \tau_x \widetilde{\mathbf{u}}_x), \quad (9.2)$$

$$\Delta_{y,2} = U_{y,2} \mathbf{u}_y + U_y B \sigma_y \mathbf{n}$$
$$+ V_{y,2} \widetilde{\mathbf{u}}_y + V_y B \tau_y \mathbf{n} + W_2 \mathbf{n} - WB(\sigma_y \mathbf{u}_y + \tau_y \widetilde{\mathbf{u}}_y).$$

Recall that the geodesic torsions of the band centerlines, τ_x and τ_y, are among the known geometric parameters of the original configuration of the system.

It is convenient to consider the deformed configuration as a two-layer surface which can be defined by two radius vectors:

$$\mathbf{r}_x^* = \mathbf{r} + \Delta_x, \quad \mathbf{r}_y^* = \mathbf{r} + \Delta_y, \quad (9.3)$$

where $\mathbf{r}$ is the radius vector of the original surface. The unit direction vector of the first array on the deformed surface is

$$\mathbf{u}_x^* = \mathbf{r}_{x,1}^* / A^* = (\mathbf{r} + \Delta_x)_1 / A (1 + \varepsilon_x). \quad (9.4)$$

Assuming strains negligibly small in comparison with unity and repeating the calculation for the second array gives

$$\mathbf{u}_x^* \approx \mathbf{u}_x + \Delta_{x,1}/A, \quad \mathbf{u}_y^* \approx \mathbf{u}_y + \Delta_{y,2}/B. \quad (9.5)$$

When introducing here the derivatives (2), rotations $V_{x,1}$ and $V_{y,2}$ of the binormals to the member trajectories, as well as rotations W_1 and W_2 of the normal, are assumed moderate and, unlike strains, are not disregarded in the presence of unity:

$$\mathbf{u}_x^* = \mathbf{u}_x + (V_{x,1}/A)\widetilde{\mathbf{u}}_x + W_1/A)\mathbf{n},$$
$$\mathbf{u}_y^* = \mathbf{u}_y + (V_{y,2}/B)\widetilde{\mathbf{u}}_y + W_2/B)\mathbf{n}. \quad (9.6)$$

This level of accuracy is maintained in evaluating the derivatives of the unit direction vectors for the deformed system:

$$\mathbf{u}_{x,1}^{\bullet} = [A\sigma_x + (W_1/A)_1]\mathbf{n} + (V_{x,1}/A)_1\tilde{\mathbf{u}}_x,$$

$$\mathbf{u}_{y,2}^{\bullet} = [B\sigma_y + (W_2/A)_2]\mathbf{n} + (V_{y,2}/B)_2\tilde{\mathbf{u}}_y. \tag{9.7}$$

Within the assumption of small strains, the relations between displacements and longitudinal strains in a two-array system are similar to (6.67):

$$\varepsilon_x \approx (\mathbf{u}_x + \Delta_{x,1}/2A)(\Delta_{x,1}/A), \qquad \varepsilon_y \approx (\mathbf{u}_y + \Delta_{y,2}/2B)(\Delta_{y,2}/B). \tag{9.8}$$

Substituting all of the necessary vectors and their derivatives gives

$$\varepsilon_x = U_{x,1}/A - W\sigma_x + (V_{x,1}^2 + W_1^2)/2A^2,$$

$$\varepsilon_y = U_{y,2}/B - W\sigma_y + (V_{y,2}^2 + W_2^2)/2B^2. \tag{9.9}$$

The member displacements impart geodesic curvatures absent in the initial configuration:

$$\kappa_x = (V_{x,1}/A)_1 + (W\tau_x)_1, \qquad \kappa_y = (V_{y,2}/B)_2 - (W\tau_y)_2. \tag{9.10}$$

These curvatures appear, together with the respective bending moments, in the constitutive relations of the problem.

Orthogonal Semigeodesic Network.   A similar development can be undertaken for a semigeodesic two-array system. In this case, it is assumed that the arrays are orthogonal, the x-array (either bands or cables) is geodesic, and the nongeodesic y-array is an array of cables (Fig. 2). Orthogonality between the two arrays and the presence of the constraints at the intersections entail, respectively,

$$\tilde{\mathbf{u}}_y = -\mathbf{u}_x, \qquad V_y = -U_x. \tag{9.11}$$

The displacement Δ_x in (1) preserves while Δ_y becomes

$$\Delta_y = U_y\tilde{\mathbf{u}}_x + U_x\mathbf{u}_x + W\mathbf{n}. \tag{9.12}$$

The derivative $\Delta_{y,2}$ must be evaluated anew, since the y-array (cables) is not geodesic. This leads to a restoration in $(2)_2$ of the terms containing the previously absent initial geodesic curvature κ_y,

$$- U_yB\kappa_y\,\tilde{\mathbf{u}}_y \quad \text{and} \quad V_yB\kappa_y\mathbf{u}_y.$$

With these terms, and taking into account relations (11), the more complete expression for the derivative of the second displacement vector is obtained [cf. the counterpart expression $(2)_2$]:

$$\Delta_{y,2} = U_{y,2}\mathbf{u}_y + U_yB(\sigma_y\mathbf{n} + k_y\mathbf{u}_x)$$

$$+ U_{x,2}\mathbf{u}_x - U_xB(\tau_y\mathbf{n} + k_y\mathbf{u}_y) + W_2\mathbf{n} - WB(\sigma_y\mathbf{u}_y - \tau_y\mathbf{u}_x). \tag{9.13}$$

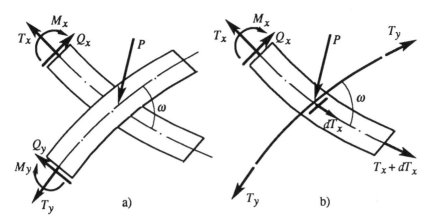

Figure 9.3. External and internal forces acting on generic elements: a) of a band system; b) of a cable-band system.

Among the remaining kinematic relations, $(6)_2$ has to be replaced by

$$\mathbf{u}_y^* = \mathbf{u}_y + U_{x,2}\mathbf{u}_x + W_2\mathbf{n}/B, \qquad (9.14)$$

and $(7)_2$ by

$$\mathbf{u}_{y,2}^* = [B\sigma_y + (W_2/B)_2]\mathbf{n} + [B\,\kappa_y + (U_{x,2}/B)_2]\mathbf{u}_x. \qquad (9.15)$$

The resulting new expression for the longitudinal strain in the cables is found from $(8)_2$:

$$\varepsilon_y = U_{y,2}/B - U_x\kappa_y - W\sigma_y + (U_{x,2}^2 + W_2^2)/2B^2. \qquad (9.16)$$

It differs from $(9)_2$ (which it replaces) in two ways: first, the presence of a term containing the nonzero geodesic curvature of the y-array; and second, the absence of displacement V_y which is no longer independent due to constraints imposed at the intersections.

Equilibrium of Two-Array Systems

Band Systems. The vectorial equation of equilibrium for an infinitesimal element of a band system (Fig. 3a) has the same form as the corresponding equation for a net:

$$\mathbf{F}_{x,1} + \mathbf{F}_{y,2} = \mathbf{P}\,AB\,\sin\omega. \qquad (9.17)$$

Here $\mathbf{P}$ is the external load per unit surface area, A and B are the Lamé parameters of the surface, and ω is the net angle. The internal forces $\mathbf{F}_x$ and $\mathbf{F}_y$ are acting in a certain constant number of bands contained in the respective coordinate strips $dy = 1$ and $dx = 1$.

The forces **F** are the resultants of two band force components—longitudinal force T and in-surface shear Q—in the final, deformed configuration of the system. Taking into account the dominating role of the longitudinal forces, it makes sense to account for the changes in the array directions only in the factors accompanying these forces:

$$\mathbf{F}_x = T_x \mathbf{u}_x^* + Q_x \widetilde{\mathbf{u}}_x, \quad \mathbf{F}_y = T_y \mathbf{u}_y^* + Q_y \widetilde{\mathbf{u}}_y. \tag{9.18}$$

Introducing these expressions, along with the unit direction vectors and their derivatives (5) into the equilibrium equation (17) yields

$$T_{x,1} \mathbf{u}_x + [(T_x V_{x,1}/A)_1 + Q_{x,1}] \widetilde{\mathbf{u}}_x + [T_x A \sigma_x + (T_x W_1/A)_1 + Q_x A \tau_x] \mathbf{n}$$

$$+ T_{y,2} \mathbf{u}_y + [(T_y V_{y,2}/B)_2 + Q_{y,2}] \widetilde{\mathbf{u}}_y + [T_y B \sigma_y + (T_y W_2/B)_2 - Q_y B \tau_y] \mathbf{n}$$

$$= \mathbf{P} \, AB \sin \omega. \tag{9.19}$$

In most cases external loads do not contain tangential traction components; then $\mathbf{P} = P\mathbf{n}$ and the analysis simplifies. The number of translational degrees of freedom of a generic element of a band system is five. The corresponding five simultaneous equations of equilibrium are obtained by scalar multiplication of (19) successively by $\mathbf{u}_x$, $\widetilde{\mathbf{u}}_x$, $\mathbf{u}_y$, $\widetilde{\mathbf{u}}_y$, and $\mathbf{n}$ (reflecting the absence of in-surface constraints, these vectors are treated as independent):

$$T_{x,1} = 0, \qquad\qquad\qquad T_{y,2} = 0,$$

$$T_x(V_{x,1}/A)_1 + Q_{x,1} = 0, \qquad T_y(V_{y,2}/B)_2 + Q_{y,2} = 0, \tag{9.20}$$

$$T_x[A\sigma_x + (W_1/A)_1] + Q_x A \tau_x + T_y[B\sigma_y + (W_2/B)_2] - Q_y B \tau_y = P.$$

The first two equations indicate that the bands are isotensoid—the tension forces do not vary along their lengths. This is the result of the absence of in-surface traction loads and of tangential interactions between the two arrays (there are no fasteners and friction is disregarded). In the process of their development, the last three equations are simplified by taking into consideration the constancy of the band tension forces.

Two conditions of in-surface moment equilibrium of the band elements produce conventional relations between the bending moments and shear forces found in the theory of beams:

$$M_{x,1} = AQ_x, \quad M_{y,2} = BQ_y. \tag{9.21}$$

The kinematic equations (9)–(10) and equilibrium equations (20)–(21) must be complemented by constitutive equations relating strains (9) and geodesic curvatures (10) to the band tension forces and bending moments:

$$T_x = EA_x \, \varepsilon_x, \quad M_x = EI_x \, \kappa_x,$$

and similar equations for the second array. These equations complete a closed system of equations of statics for a geodesic band assembly. If one (or both) of the band arrays is replaced by cables, the necessary modifications (in fact, simplifications) of the equations are obvious and straightforward: the related shear forces and bending moments must be set to zero.

As was the case with cable nets, the initial configuration of a band or cable-band system serves as a reference for the displacements, but not always for strains. In a prestressed system, initial strains exist before the onset of loading and must be subtracted from the left-hand sides in relations (9) and (16). (Bending moments and shear forces were assumed absent in the prestressed state.) As with any quasi-variant system, it is the presence of initial forces that assures linearizability of the problem at the first and, perhaps, the only step of the incremental statical analysis.

Cable-Band Systems.   For a semigeodesic cable-band or cable system (Fig. 3b), equilibrium equation (17) stays intact but the direction vectors and displacements of the nongeodesic array must be represented by (11) and (15). Their introduction into (18) and then into (17) reduces the vectorial equation of equilibrium to

$$T_{x,1}\mathbf{u}_x + [(T_x V_{x,1}/A)_1 + Q_{x,1}]\widetilde{\mathbf{u}}_x + [T_x A \sigma_x + (T_x W_1/A)_1 + Q_x A \tau_x]\mathbf{n}$$

$$+ T_{y,2}\mathbf{u}_y + [(T_y V_{y,2}/B)_2 + T_y B \kappa_y]\widetilde{\mathbf{u}}_y + [T_y B \sigma_y + (T_y W_2/B)_2]\mathbf{n}$$

$$= \mathbf{P} \, AB \sin \omega. \quad (9.22)$$

Comparing this equation to (19) note the new term with geodesic curvature κ_y and the absence of the shear force Q_y—both resulting from the y-array being now a nongeodesic array of cables.

In a semigeodesic system with partially restrained intersections, the number of translational degrees of freedom of a generic element is four.   Scalar multiplication of (22) by $\mathbf{u}_x$, $\widetilde{\mathbf{u}}_x$, $\mathbf{u}_y$, and $\mathbf{n}$ (conditionally treating these vectors as independent) produces four simultaneous scalar equations of equilibrium:

$$T_{y,2} = 0, \qquad T_{x,1} + T_y[B \kappa_y + (U_{x,2}/B)_2] = 0,$$

$$(T_x V_{x,1}/A)_1 + Q_{x,1} = 0, \qquad\qquad\qquad\qquad (9.23)$$

$$T_x A \sigma_x + (T_x W_1/A)_1 + Q_x A \tau_x + T_y[B \sigma_y + (W_2/B)_2] = P.$$

As is seen from the first equation, in this system only one of the two tension forces is constant—the force T_y in the nongeodesic (cable) array. The bending aspect of the problem is still described by equations $(10)_1$ and $(21)_1$. These equations, combined with kinematic equations $(9)_1$ and (16), equations of equilibrium (23), and constitutive relations represent a closed system of equations for semigeodesic cable-band assemblies.

Some applications (e.g., large-span roofs) require shallow systems. In this case, the intrinsic geometry of the system can be approximated by the plane geometry.   Then the Lamé parameters A and B can be considered constant, producing a chain of simplifications.   For an (approximately) orthogonal shallow band assembly, the solving system of equations becomes

$$T_x V_{x,11} = - M_{x,11}, \quad T_y V_{y,22} = - M_{y,22},$$

$$T_x(\sigma_x + W_{11}) + T_y(\sigma_y + W_{22}) = P,$$

$$U_{x,1} - W\sigma_x + (V_{x,1}^2 + W_1^2) = (T_x - T_x^0)/EA_x, \qquad (9.24)$$

$$U_{y,2} - W\sigma_y + (V_{y,2}^2 + W_2^2) = (T_y - T_y^0)/EA_y,$$

$$V_{x,11} = M_x/EI_x, \qquad V_{y,22} = M_y/EI_y.$$

where $EA_{x(y)}$ and $EI_{x(y)}$ are, respectively, the extensional and bending stiffnesses of the bands.

9.2   Axisymmetric 3-Nets

A 3-net is formed by three intersecting arrays (Fig. 4) of linear structural members with each intersection involving one member of each array. The subject of this section is an axisymmetric 3-net where one array is meridional while the other two are inclined to the meridian (generally, under different angles) and form an axisymmetric net. If members of all three arrays are fastened at the intersections the system becomes a discrete triangulated membrane: like its continuous counterpart, it possesses kinematic mobility in the form of surface bending, whereas the intrinsic geometry is invariant. However, if the meridional members are allowed to slip at the intersections, the system becomes intrinsically underconstrained. Obviously, this is also the case when the intersections are not constrained in any way so that members of all three arrays can mutually slip.

Only intrinsically underconstrained 3-nets are studied here (Kuznetsov, 1984b). Such a system is considered as an assembly of meridional members supported only at their ends and a net of fibers or cables whose intersections may be either free or fastened. The meridional members can be made either of flexible bands or of not-so-flexible narrow strips enveloping the volume within the net. The strips are assumed capable of supporting both tension and compression. In a 3-net subjected to a normal surface load and, perhaps, prestressed, the meridional members, being geodesic, are isotensoid (uniformly stressed along the length). The net forces, generally, vary; they are constant only when the net is geodesic. As long as the net is in tension it exerts contact pressure on the meridional array, acting as a continuous lateral constraint and preventing the system delamination. In the case of compressed meridional members, the net is assumed sufficiently fine to prevent their local buckling between two adjacent net intersections.

It is readily seen that an entirely tensile prestressed 3-net must be of negative total curvature and the axial resultant of its prestressing forces is tension. It is not obvious in advance what the member forces can be for a prestressed system

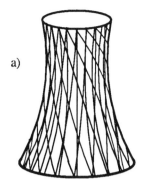
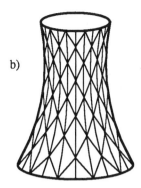

Figure 9.4. Axisymmetric 3-nets: a) general (skew); b) reflection-symmetric.

involving compressed meridional strips. In particular, it is interesting whether a self-contained (tensegrity) prestressed system is feasible, that is, whether the axial resultants of the compression in the meridional array of members and the tension in the net can cancel each other. This question will be answered, along with a few others, in the process of establishing general statical-geometric interrelations for underconstrained 3-nets.

General Statical-Geometric Relations

Singular Versus Nonsingular 3-Nets. For the sake of simplicity, the governing equations are developed for axisymmetric 3-nets where torque moment is either absent or uncouples (the latter is the case of a reflection-symmetric 3-net). A 3-net is considered as involving two subsystems: one is the meridional array, the other an axisymmetric net, generally, a skew one. The first integral (7.64) and statical-geometric relation (7.65) obtained earlier for torque-free nets present a convenient point of departure:

$$T_n = T_{n0} f/f_0, \tag{9.25}$$

$$T_{n0} (\sigma_1 - \sigma_2 \tan \alpha \tan \beta) f/rf_0 = P_n. \tag{9.26}$$

Here T_n is the meridional force (Fig. 5a) in the net per unit polar angle and the subscript 0 designates the reference plane $z = z_0$. The above two relations are valid for a net under any normal surface load, P_n; in particular, this load may be a contact pressure exerted on the net by the meridional array in a 3-net.

Let F_m be the force per unit polar angle in the meridional members. This force does not vary along the meridian and is related to the normal load, P_m, of the meridional array by

$$F_m \sigma_1/r = P_m. \tag{9.27}$$

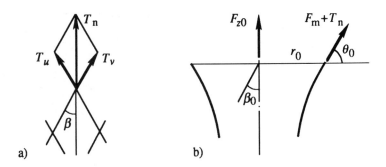

Figure 9.5. Forces in a 3-net: a) forces in the net; b) force resultants.

Adding together relations (26) and (27) yields the sought equation interrelating the normal equilibrium loads and geometric configurations of underconstrained axisymmetric 3-nets:

$$F_m \sigma_1 / r + T_{n0} (\sigma_1 - \sigma_2 \tan \alpha \tan \beta) f / r f_0 = P_m + P_n = P. \qquad (9.28)$$

Note that the two components of the total equilibrium load, P_m and P_n, supported by each of the subsystems, generally, do not depend on the patricular way the external load P is acting on the system. The load can be applied to the subsystems in any proportion or act entirely on one of them, yet each subsystem will isolate and support only its own equilibrium load, provided a delamination does not occur. This is the familiar, observed in Chapter 2, "load-filtering" feature of a nonsingular underconstrained structural system with an in-series arrangement of subsystems. However, the contact pressure between the two component subsystems does depend on the actual way the external pressure is applied. If the portions of the total pressure acting on the meridional members and the net are, respectively, P_{ma} and P_{na}, the mutual contact pressure, P_c, exerted by the two component subsystems is given by

$$P_c = P_{ma} - P_m = P_n - P_{na}. \qquad (9.29)$$

A complete geometric description of a 3-net requires two independent inputs: the shape of the meridian determining the two principal curvatures of the surface; and the intrinsic geometry of the net determining the function $f(r)$ in relation (25). As soon as the geometry of an axisymmetric 3-net is specified, its equilibrium normal loads are found as linear combinations of the two known functions—those multiplied by the parameters F_m and T_{n0} in (28). The accompanying edge loads are uniformly distributed meridional forces applied at the edge parallels and producing either a tension or compression axial resultant force. In addition, if the 3-net is reflection-symmetric, uniformly distributed membrane shearing forces (with two equal and opposite torque moment resultants) may be acting at the edge parallels. The torque produces forces (7.48) in the net while not affecting the equilibrium configuration of the system and the meridional member forces.

Equation (28) reveals a profound difference between the respective statical-geometric properties of singular and nonsingular 3-nets. For a singular one, the existence of virtual self-stress means that the mentioned two functions in (28) are linearly dependent and the system has only one normal equilibrium load rather than two. This is the usual trade-off between the degree of singularity (statical indeterminacy) and the number of independent equilibrium loads. It is impossible to evaluate by purely statical means the way the sole equilibrium load is shared by the two subsystems, since for a singular system this depends on the material properties and constitutive relations.

Equilibrium Configurations of 3-Nets. Adding together the forces F_m and T_n given, respectively, by (27) and (25) and projecting the sum on the z-axis (Fig. 5b), allows the axial resultant, F_z, of internal forces of the system to be evaluated. It must be equal to the axial force produced by external loads, in this case, the normal surface pressure P:

$$2\pi \, (F_m + T_n) \sin \theta = F_z = F_{z0} + 2\pi \int_{r_0}^{r} P \, r \, dr, \qquad (9.30)$$

where F_{z0} is the axial force at the reference parallel:

$$F_{z0} = 2\pi \, (F_m + T_{n0}) \sin \theta_0. \qquad (9.31)$$

The equilibrium configuration of a 3-net for a given load (F_{n0} and P) is obtainable from equation (30) by forward integration after the intrincic geometry of the net is specified and the function $f(r)$ is evaluated. This function is available in closed form (7.69) for a geodesic net and (7.76) for a Chebyshev net. Upon introducing it into (30), the equilibrium shape is evaluated by expressing $\sin \theta$ in terms of r' and integrating the resulting first-order equation. It has the familiar structure of equation (7.61):

$$\text{ctn} \, \theta \equiv dr/dz = \sqrt{[2\pi \, (F_m + T_n)/F_z]^2 - 1}. \qquad (9.32)$$

As before, this equation is integrated together with (7.57) or (7.62) depending, respectively, on whether pressure P is given as a function of r or z.

Singular Axisymmetric 3-Nets

Feasible Shapes and Their Prestress Patterns. After setting $P = 0$, equation (30) can be modified with the aid of (25) as follows:

$$(C - 1 + f/f_0) \sin \theta = C \sin \theta_0, \quad C = (F_m + T_{n0})/T_{n0}. \qquad (9.33)$$

where f_0 is the value of the function $f(r)$ at the reference plane z_0. Singular configurations of 3-nets are evaluated from here by forward integration of a first-order equation similar to (32). Note that in the corresponding self-stress patterns, the axial components of the meridional member force F_m and of the net force T_n vary along the z-axis, while the sum of the two components preserves.

The geometry of singular 3-nets can be assessed qualitatively by evaluating the principal curvature ratio at the surface equator, assuming for the time being that $z = z_0$ is the equatorial plane. With $P = 0$, equation (28) yields for this location

$$C\,(\sigma_1/\sigma_2)_{|0} = \tan\alpha_0 \tan\beta_0. \qquad (9.34)$$

The obtained relation between the statical parameter C and the curvature ratio enables a few observations to be made on the singular configurations. For an entirely tensile torque-free 3-net, $C > 0$ while α and β are of the opposite signs. As a result, the surface has negative Gaussian curvature (it is hourglass shaped). Such is also the shape of a 3-net where the meridional array is force-free $(C = 1)$. In this system, the net is singular, so that the only axisymmetric normal equilibrium load is that of the meridional array. Any other load of this type would cause the net to rotate and deflect or, if the net is reflection-symmetric, hence, asymptotic, just to deflect.

The case $C = 0$ would correspond to an axially self-equilibrated prestress. According to (33), this is precluded, thereby ruling out a self-contained prestressed system and indicating that a prestressed 3-net cannot exist without external supports balancing the nonzero axial force resultant. As follows from (34), systems with a compression axial resultant must be convex (barrel- or spindle-shaped). Finally, both of the limiting cases, $C \to \pm\infty$, produce cylindrical shapes.

In the remainder of this section only reflection-symmetric $(\alpha = -\beta)$ singular 3-nets are considered. Further progress in their investigation is possible upon specifying the relation between r and β and evaluating the resulting function $f(r)$. This, in turn, requires knowing the intrinsic geometric properties of the net. As before, the two most important particular types of net are considered in detail—geodesic and Chebyshev.

Singular Geodesic 3-Nets. In this system all three arrays are geodesic. The function $f(r)$ for a reflection-symmetric geodesic net is given by (7.73); inserting it into (33) gives

$$(C - 1 + \cos\beta/\cos\beta_0)\sin\theta = C\sin\theta_0. \qquad (9.35)$$

The shapes of singular geodesic 3-nets can be found from here explicitly by forward integration after expressing $\sin\theta$ in terms of r'. Parameters to be specified are C, β_0, and r_0; however, the latter, as in all previous cases, is just a scale factor that determines only the size of the system and does not contribute to the variety of feasible shapes. The foregoing qualitative survey of singular 3-nets can now be specialized and elaborated on using the additional geometric information.

In Fig. 6, the graph of equation (34) (a hyperbola relating the statical parameter C and the principal curvature ratio of the surface) is used as a

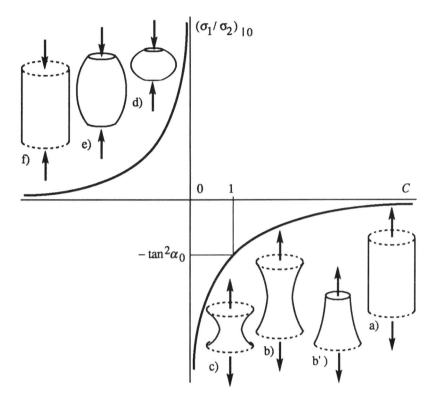

Figure 9.6. Feasible shapes of prestressed geodesic 3-nets.

reference framework for the set of the evolving singular shapes of 3-nets. The limiting case $C \to \infty$ corresponds to a cylindrical 3-net (a) with a force-free net and the meridional members supporting the entire axial tension force, F. With a decreasing C, the originally force-free net takes on a gradually increasing portion of the total axial force, and at $C = 1$ the net supports the entire tension force. In this configuration the meridional members are force-free and, therefore, do not exert any normal contact pressure on the net. As a result, the inclined cables have zero normal curvature in addition to zero geodesic curvature. This means that the cables are straight and the surface is ruled. Since the only axisymmetric ruled surface of double curvature is a one-sheet hyperboloid of revolution, this must be the shape of a singular geodesic 3-net at $C = 1$ [(b) in Fig. 6].

The succession of hourglass shapes between (a) and (b) includes configurations that do not contain an equator. Whether or not this is the case can be verified by evaluating $\cos \beta$ at the equator $(\theta = \pi/2)$ from equation (35); $\cos \beta$ being positive requires that

$$\sin \theta_0 > C_t \equiv (C - 1)/C. \qquad (9.36)$$

When this inequality is not satisfied, the 3-net surface truncates at the axial location with $\beta = \pi/2$. Here the meridian slope, θ_t, and the surface radius, r_t, are determined, respectively, by

$$\sin \theta_t = \sin \theta_0/C_t, \quad r_t = r_0 \sin \beta_0. \tag{9.37}$$

The corresponding terminal (truncation) parallels are shown in Fig. 6 by solid lines. Singular geodesic 3-nets of negative total curvature do not have a maximum radius and extend in the axial direction indefinitely with increasing radius, which is indicated in the figure by broken lines.

As is seen from (33) and (36), tensile systems with compressed meridional members reside in the range $0 < C < 1$ and do not truncate. When approaching the lower limit of this range the system flattens (c) and degenerates to emerge in a pill shape (d) as a compressed system. Its further evolution with decreasing C leads to a barrel shape (e) and with $C \rightarrow -\infty$ it approaches the limiting cylindrical shape (f).

Note that for a compressed system, $C < 0$, condition (36) cannot be satisfied and the corresponding feasible shapes are always truncated. Indeed, since each of these shapes contains an equator, the equatorial plane can be designated as the reference plane, $z = z_0$ then $\sin \theta_0 = 1$ and, in accordance with equations (36) and (37),

$$\sin \theta_t = C/(C - 1) < 1, \tag{9.38}$$

meaning that the surface truncates at the location where the meridian slope is given by (36).

Truncation of geodesic nets is well known in wound shell design where the shell shapes can be nearly arbitrary and both r_t and θ_t for a given shape depend on the chosen angle of winding, β_0, at the reference parallel. The situation is quite different for singular geodesic 3-nets. Here the feasible shapes are governed by statics, and the slope θ_t which is determined by the parameter C does not depend on the chosen β_0.

Singular Chebyshev 3-Nets. This name is retained for a reflection-symmetric 3-net where the meridional array is diagonal to the Chebyshev net formed by the two inclined arrays with fastened intersections. The function $f(r)$ for a Chebyshev net is given by (7.80); its introduction into (33) gives

$$(C - 1 + \cos \beta_0/\cos \beta) \sin \theta = C \sin \theta_0. \tag{9.39}$$

This equation is very similar to its counterpart for a singular geodesic 3-net, which is the reason for a certain parallelism in the evolution of the shapes of the two systems. The hyperbola of equation (34) is once again used as a display framework for the succession of feasible shapes (Fig. 7).

The cylindrical Chebyshev 3-net (a) corresponding to $C \rightarrow \infty$ is identical with its geodesic counterpart: in both cases the inclined arrays are two counterwound helices forming a net which is simultaneously geodesic and Chebyshev. As C decreases, a succession of entirely tensile, hourglass shapes

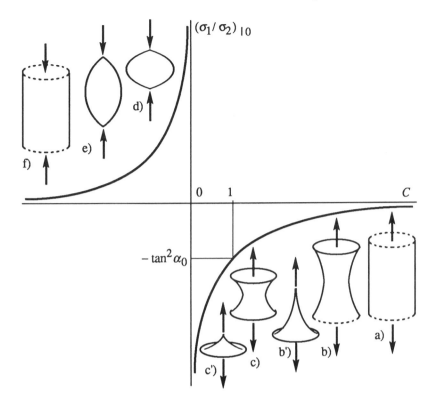

Figure 9.7. Feasible shapes of prestressed Chebyshev 3-nets.

evolves, with a noteworthy case (b) where $C = 1$. In this configuration, the meridional members are idle ($F_m = 0$, in transition from tension to compression) while the net supports the entire axial force. Being rotation- and reflection-symmetric, the net must be asymptotic; therefore, it has the shape of one of the three pseudospheric surfaces described in Chapter 7 and depicted in Fig. 7.5.

To the left from the point $C = 1$ lies the domain of singular 3-nets with the meridional array in compression. However, the total axial force F remains tensile until another station, $C = 0$, is reached. Here $F = 0$ and the system should have become self-equilibrated; instead, as was the case with a geodesic 3-net, with $C \to 0$, the system flattens (c) and degenerates.

All of the Chebyshev 3-nets within the investigated range (except for the two limiting cases) have negative Gaussian curvature, but not all of them contain an equator. An equator exists when the minimum radius and, by virtue of (7.34), the minimum net angle occur at $\theta = \pi/2$ and are positive. According to (39) and (31), this condition is satisfied when

$$\sin \theta_0 > C_a \equiv (C - 1 + \cos \beta_0)/C. \tag{9.40}$$

If this condition is not satisfied, the 3-net has an apex where $\beta = 0$ and the meridian slope, θ_a, is given by

$$\sin \theta_a = \sin \theta_0 / C_a. \tag{9.41}$$

The maximum radius of the net is the radius of its regression circle. At this paralel $\beta = \pi/2$, $\theta = 0$ and, as is seen from (39), a prestressed system cannot contain a regression circle.

After disappearing at $C = 0$, the equilibrium shape emerges in a completely different form as a compression system (d). From here on, the system always contains an equator and two apexes where the meridian slope is determined by (41). With a decreasing value of C, the equilibrium shapes tend to become more oblong (e) until, finally, the limiting cylindrical configuration (f) with compressed generators is reached. It closes the assortment of feasible shapes of singular Chebyshev 3-nets.

9.3 Elastic Systems Stiffened by Nets

Many structural systems involve components very dissimilar in their elastic properties—very rigid and very flexible in comparison to each other. It is sometimes advantageous to model such a system as a combination of underconstrained and elastic subsystems. Then the system behavior is determined not only by the elastic properties but also by the current geometric configuration, and the magnitude of the member forces induced by the equilibrium load in the underconstrained subsystem. By combining the geometric and elasic features of the subsystems in a meaningful way it is possible to achieve a desired type of structural response. Following are two examples of such composite systems where highly elastic components are stiffened by either inextensible or extensible fiber nets.

Axisymmetric Elastic Membrane
Stiffened by an Inextensible Net

In this system, an elastic membrane of revolution is stiffened on the inside or the outside by an enveloping inextensible net which is assumed to slip without friction over the membrane surface. Both the membrane and the net are attached to two edge rings of the same radius and the initial shape of the system is cylindrical. The system is subjected to a normal surface pressure (external or internal) giving rise to an interesting contact problem: the two subsystems have independent tangential displacements but a common normal displacement. Thus, elastic displacements of the membrane are matched in the normal direction by kinematic displacements of the net.

The two edge rings may be fixed at a given axial distance from each other, in which case the membrame or the entire system can be prestressed. Alternatively, if the rings are not fixed, an external axial force (tension or compression) can be applied to the system simultaneously with a normal surface load. Recall, however, that the net subjected to an axial force (either as an external load or as prestress) cannot be cylindrical.

Membrane Analysis. A solution to the problem will be sought capitalizing on the fact that, barring delamination, the deformed configurations of the net and the membrane are identical. Equations (25) and (26) for an axisymmetric net provide the necessary statical-geometric relations between a normal surface load and the corresponding equilibrium configuration of the net. As it turns out, an initially cylindrical elastic membrane also allows an explicit determination of the normal surface pressure that gives rise to a given axisymmetric deformed shape. The calculation is made possible by an exquisite solution (Pipkin, 1968) that relates the principal extension ratios, λ_1 and λ_2, of such a membrane:

$$w - \lambda_1 (\partial w / \partial \lambda_1) = C_0. \tag{9.42}$$

Here w is the elastic strain energy density (per unit initial area of the membrane) which is assumed to be a known function of the principal extension ratios; C_0 is a constant of integration evaluated, for instance, in terms of the extension ratio λ_{10} at a reference parallel z_0 with a known axial force T_{z0}.

The possibility of a closed-form solution is due to the fact that the hoop extension ratio is simply $\lambda_2 = r/r_0$ where r_0 is the constant radius of the undeformed surface. Solution (42) is valid for an arbitrary homogeneous elastic material, even an anisotropic one, if the pattern of elastic properties is reflection-symmetric. The significance of this solution for the problem in consideration is in that it enables the meridional extension ratio λ_1 to be expressed as an explicit function of λ_2, that is, of the radius r, which represents the membrane geometry. Thus, both extension ratios are known explicit functions of r and some constants:

$$\lambda_2 = r/r_0, \quad \lambda_1 = \lambda_1 (r, T_{z0}). \tag{9.43}$$

In the absence of a meridional traction load, the equilibrium equation of an axisymmetric membrane in the meridional direction has the form [cf., for example, equation $(8.15)_1$ with $B = r$]:

$$r (dt_1 / ds_1) + (dr / ds_1) (t_1 - t_2) = 0, \tag{9.44}$$

where t_1 and t_2 are, respectively, the meridional and hoop tensions in the deformed membrane. By virtue of $dr = ds_1 \sin \theta$ and assuming $\sin \theta \neq 0$, this equation reduces to

$$d (t_1 r) / dr - t_2 = 0. \tag{9.45}$$

The products $\lambda_2 t_1$ and $\lambda_1 t_2$, as the membrane forces per unit initial length, are expressed in terms of the elastic potential energy:

$$\lambda_2 t_1 = \partial w / \partial \lambda_1, \quad \lambda_1 t_2 = \partial w / \partial \lambda_2. \tag{9.46}$$

Introducing this into (45) along with relations (43) verifies solution (42).

Membrane–Net Interaction. The analysis of the net-stiffened membrane can now be completed within the already familiar framework developed for the analysis of axisymmetric 3-nets. Going over to the previously used notation F_m for the meridional (this time, membrane) force per unit polar angle, equation $(46)_1$, in view of $(43)_1$, can be rewritten

$$F_m = (\partial w / \partial \lambda_1) r_0. \tag{9.47}$$

Substituting this expression into (28) and (30) yields, respectively,

$$(\partial w / \partial \lambda_1) r_0 \sigma_1 / r + T_{n0} (\sigma_1 - \sigma_2 \tan \alpha \tan \beta) f / r f_0 = P, \tag{9.48}$$

$$2\pi [(\partial w / \partial \lambda_1) r_0 + T_{n0} f / f_0] \sin \theta = F_z = F_{z0} + 2\pi \int_{r_0}^{r} P \, r \, dr. \tag{9.49}$$

A conventional problem statement calls for establishing the deformed configuration of the system under a given load. In this sense, equation (48) provides a solution to a reversed problem: it establishes equilibrium normal loads for a net-stiffened membrane in a given geometric configuration. These loads are linear combinations of the two known functions of the system geometry in the left-hand side of the equation. If the functions happen to be linearly dependent, the configuration is singular: it has only one normal equilibrium load and is statically indeterminate, i.e., allows virtual self-stress. To be prestressable, a singular configuration must have $\sigma_1 < 0$, that is, to be of negative total curvature (hourglass-shaped).

Equation (49) is the solving equation of the problem. In it, the two functions in the left-hand side become known and explicit functions of r as soon as the membrane material properties and the intrinsic geometry of the net are specified. Evaluation of the first function (the derivative of the strain energy density) for a Mooney material can be found, for example, in Kydoniefs (1969). The second function, $f(r)$, evolved in (7.60), has been evaluated for particular types of nets in Chapter 7. With this information available, the deformed configuration of the system under a given load is determined by forward integration of an equation of the form (32) where F_m is replaced by (47).

The above equations clearly reveal the respective contributions of the two subsystems with their dissimilar types of stiffness in supporting the external load. The conventional, elastic, stiffness of the membrane depends on the material properties, and the corresponding elastic moduli are implicitly present in an expression for the strain energy function w. On the other hand, the stiffening net, assumed inextensible, possesses stiffness of a purely statical-kinematic nature. It is proportionate to the parameter T_{n0} quantifying the tension level in the net. The amalgamated response of the net-stiffened membrane can be finely tailored to satisfy a variety of requirements.

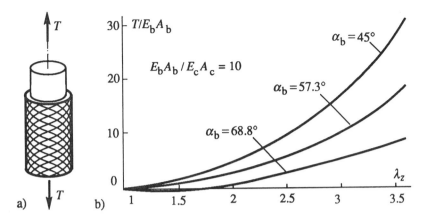

Figure 9.8. A composite elastic cord: a) structure of the cord; b) normalized force-extension diagram for three braid weave angles.

A Composite Elastic Cord

Statics and Kinematics of a Composite Cord.   This structural item consists of an elastic cylindrical core (usually rubber) and an outer woven braid made of fibers or strands with high strength and stiffness (Fig. 8a). The braid represents two symmetrically counterwound helices forming an angle α with the generator of the cylinder. Frictionless mutual slippage of the braid and the core is assumed possible at the interface. Properly selecting the mechanical properties of the component materials and the braid geometry makes it possible to design a cord with a desired type of stiffening nonlinear response. The following analysis of the composite cord illustrates the basic idea of a structural design capitalizing on the interplay of kinematic and elastic mobility of subsystems involved.

The axial force in the composite cord is the sum of the component forces, the braid force, T_b, and the core force, T_c:

$$T = T_b + T_c = 4\pi T_\alpha \cos \alpha + T_c. \tag{9.50}$$

Here T_α is the force per unit polar angle in each of the two braid strands. The hoop component of this force per unit axial length is

$$T_\varphi = T_\alpha \sin \alpha \tan \alpha / r. \tag{9.51}$$

The interaction between the braid and the core is the interface pressure representing the radial resultant of the hoop tension in the braid. The two of the braid hoop forces produce a contact pressure

$$P = 2T_\varphi/r = T_b \tan^2 \alpha / 2\pi r^2 = T_b \tan^2 \alpha / 2A, \tag{9.52}$$

where α, r, and A are, respectively, the braid angle, braid/core interface radius, and the core cross-sectional area in the current configuration of the cord.

Let α_b be the initial value of the braid angle. In a uniform axial extension of the cord, the braid trajectory remains helical, and the following kinematic relations describing the braid deformation can be written:

$$\lambda_r/\lambda_\alpha = \sin\alpha/\sin\alpha_b, \quad \lambda_z/\lambda_\alpha = \cos\alpha/\cos\alpha_b. \tag{9.53}$$

Here λ_α, λ_r, and λ_z are, respectively, the longitudinal, radial, and axial extension ratios. The braid is considered linear elastic with a small strain

$$\varepsilon_\alpha = \lambda_\alpha - 1 = 2\pi T_\alpha/E_b A_b, \tag{9.54}$$

where E_b is the modulus of elasticity and A_b is the total cross-sectional area of the braid. The core material is assumed neo-Hookean, obeying the following constitutive equations:

$$t_r = E_c\lambda_r^2/3 - p, \quad t_z = E_c\lambda_z^2/3 - p, \quad \lambda_r^2\lambda_z = 1, \tag{9.55}$$

where $t_r = t_\varphi$ and t_z are the principal tensions, p is the hydrostatic stress component, and the third equation is the condition of material incompressibility.

In loading, the braid and the core undergo the same axial extension. Their respective radial contractions must be such as to preclude delamination; otherwise, the braid would not engage, i.e., would be force-free. The engagement is assured if an interface contact pressure exists between the braid and the core. This is verified by comparing the rates of the kinematic radial contractions of the braid and the core caused by a unit axial extension of the cord at the onset of loading. A contact pressure develops if the rate of the radial contraction of the braid, taken as inextensible, is greater than the rate of the radial contraction of the core due to incompressibility of the core material.

Setting $\lambda_\alpha = 1$ in (53) as the condition of braid inextensibility gives

$$\lambda_r^2 = (1 - \lambda_z^2\cos^2\alpha_b)/\sin^2\alpha_b. \tag{9.56}$$

The corresponding radial contraction of the core is found from $(55)_3$:

$$\lambda_r^2 = 1/\lambda_z. \tag{9.57}$$

Evaluating the derivatives of the above two expressions with respect to λ_z at $\lambda_z = 1$ (the onset of loading) yields the desired condition of the immediate engagement of the braid:

$$2\,\mathrm{ctn}^2\alpha_b \ge 1, \quad \alpha_b \le \tan^{-1}\sqrt{2} = 54.7°. \tag{9.58}$$

The obtained magnitude of α_b is all too familiar: it is the geometric parameter describing the equilibrium configuration of a cylindrical Chebyshev net under a uniform pressure [cf. (7.72)]. Comparing the right-hand sides in (56) and (57) shows that the rate of the kinematic radial contraction of an inextensible braid is much faster than that of the incompressible core. This observation leads to two interesting conclusions.

First, with an initial braid angle greater than indicated in (58), the braid engagement, although not immediate, will occur later in the process of loading, as illustrated in the numerical example below. This delayed engagement can, in fact, be used advantageously as an additional means in synthesizing a composite member with a desired response.

The second, and somewhat less expected conclusion is that *the two kinematic models—an inextensible braid and an incompressible core—are incompatible*: the former does not preserve the volume of the core as required by the latter. Hence, for an analytical model of the composite structure to be consistent, at least one of the components must be modeled differently. In the following statical analysis the braid is considered linearly elastic but very rigid (a high modulus of elasticity as compared to the core). The core material is assumed incompressible neo-Hookean which is one of the simplest nonlinear constitutive models.

Cord–Braid Interaction. The force-extension relation for the composite member in consideration is not difficult to obtain. From equations (55) it follows that

$$t_z = E(\lambda_z^2 - 1/\lambda_z)/3 + t_r. \tag{9.59}$$

The radial stress t_r (compression) is given by (52):

$$t_r = -P = -T_b \tan^2 \alpha / 2A. \tag{9.60}$$

By virtue of the incompressibility condition $(55)_3$, the current cross-sectional area of the core is related to the initial area, A_c, by

$$A \lambda_z = A_c. \tag{9.61}$$

The current value of $\tan \alpha$ is obtained from (53) and $(55)_3$:

$$\tan^2 \alpha = \tan^2 \alpha_b / \lambda_z^3. \tag{9.62}$$

With the aid of the last four equations, the tension force in the core is evaluated:

$$T_c = t_z A = E_c A_c (\lambda_z^3 - 1)/3\lambda_z^2 - T_b \tan^2 \alpha_b / 2\lambda_z^3. \tag{9.63}$$

The tension in the braid is found from (53) and (54):

$$T_b = E_b A_b \varepsilon_\alpha \cos \alpha = E_b A_b (\lambda_z \cos \alpha_b - \cos \alpha), \tag{9.64}$$

which, with the aid of (62), reduces to

$$T_b = E_b A_b \lambda_z [\cos \alpha_b - \sqrt{\lambda_z/(\lambda_z^3 + \tan^2 \alpha_b)}]. \tag{9.65}$$

Adding up expressions (63) and (65) yields

$$T = T_b + T_c = T_b(1 - \tan^2 \alpha_b / 2\lambda_z^3) + E_c A_c (\lambda_z^3 - 1)/3\lambda_z^2. \tag{9.66}$$

The term representing the contribution of the braid is positive only if condition (58) is met (recall that $\lambda_z \geq 1$). If this term is negative, it must be omitted since, in the absence of contact pressure, the braid does not engage and is statically irrelevant. On the other hand, when this term is positive, it increases rapidly with the increasing extension λ_z, thereby contributing to the stiffening of the nonlinear response.

The load-extension relation for the composite cord can be presented in a normalized form:

$$T = E_c A_c \, (\lambda_z^3 - 1)/3\lambda_z^2$$
$$+ E_b A_b \, [\cos \alpha_b - \sqrt{\lambda_z/(\lambda_z^3 + \tan^2 \alpha_b)}] \, (2\lambda_z^3 - \tan^2 \alpha_b)/2\lambda_z^2. \tag{9.67}$$

The two extreme cases corresponding, respectively, to the beginning of the loading process and to a well-developed elastic deformation are described by

$$\lambda_z = 1 + \varepsilon_z, \quad T \approx E_c A_c \varepsilon_z,$$
$$\lambda_z \gg 1, \quad T \approx E_b A_b \varepsilon_\alpha \cos \alpha, \tag{9.68}$$

where ε denotes a small strain and ε_α is given in terms of extension λ_z by (64). Thus, the nonlinear stiffness of the composite cord exhibits a gradual transition between the elastic stiffness of the core and that of the braid.

Three load-extension diagrams are presented in Fig. 8b. The upper curve corresponds to $\alpha_b = 45°$ which, according to (58), guarantees an immediate braid engagement. This geometry produces a smooth and rapidly stiffening response. The second curve describes a composite member with $\alpha_b = 57.3°$ (1 rad), which is only slightly greater than the border value in (58). After a relatively short transition, this curve approximates the behavior of the first one.

The third curve ($\alpha_b = 68.8°$) is peculiar in that it dips below zero. The explanation to this aberration is simple: instead of disregarding the braid for as long as it is not engaged, equation (66) describes a physically meaningless effect of a radially shrinking core pulling on the braid and inducing a compression stress in it. Obviously, this part of the diagram must be disregarded; more precisely, the first term in (66) must be taken as zero for as long as it stays negative.

10
Related Applications

Several different applications presented in this chapter deal with some aspects of structural behavior typical of underconstrained systems. The first section is concerned with systems that exhibit the type of behavior called kinematic adaptivity: the system kinematically adapts to an applied load by acquiring a configuration that is, in a sense, most suitable for supporting the load. This feature is very useful in actively controlled systems, in particular, those implementing the concept of statically controlled geometry.

The second section is devoted to a scantily explored phenomenon—bending instability of plates and shallow shells. In contrast to membranes, plates and shells with bending resistance are not underconstrained. Yet, the kinematic hypothesis on inextensibility of the middle surface entails an interesting interplay of kinematic and elastic mobility. The geometric theory of surfaces provides an analytical relation between surface stretching and bending. Since the relation is nonlinear, it eludes and, thereby, may debilitate a linear problem formulation. For example, surface deformation without stretching (isometric bending), if at all possible, must preserve the total curvature. One difficulty in the conventional linear theory of plates is a paradoxical edge effect in the cylindrical bending of inextensible plates: nonpreservation of the total curvature within a narrow zone adjacent to the plate edges. In this case, the linear model can be salvaged only by taking into account the transverse shearing deformation of the plate (Kuznetsov, 1982).

Normal and shearing strains at the middle surface of a shell are formally of the second order of smallness (in terms of nondimensional normal displacements) and cannot be accounted for in a linear theory. However, the corresponding terms in the nonlinear equations are accompanied by large coefficients describing the extensional stiffness of the shell. With small but finite displacements, these terms can be comparable to, or even exceed, the first-order terms. Because of the absence of such nonlinear terms, the linear theory does not reflect the fact that a plate or shell in bending tends toward an isometric deformation mode. In loading and unloading, a more or less abrupt switch between nonisometric and nearly isometric modes may occur. This represents the phenomenon of bending instability, to which the linear theory is inherently blind.

Shape optimization of a membrane shell of revolution is the topic of the third section. Statical determinacy of this problem allows it to be formulated so as to satisfy the condition of separability of variables. This makes it possible to solve the problem by a modified dynamic programming method.

10.1 Kinematic Adaptivity and Statically Controlled Geometry

Concept Description and Applications

Kinematic Adaptivity. When an underconstrained structural system subjected to a perturbation load changes its geometric configuration, it does so in a very interesting, not to say meaningful, way. To the extent possible, the system first deforms by adapting to the load kinematically and then, in addition, develops elastic deformations as well. Both of the deformations, and especially the kinematic ones, modify the geometric configuration so as to enable the system to support the applied load in the best possible way, that is, with the lowest possible member forces. This familiar phenomenon of absorbing a perturbation load of the initial configuration and transforming it into an equilibrium load of the final configuration has been described earlier as geometric hardening. It is displayed graphically by the stiffening load-displacement and load-stress diagram.

Kinematic adaptivity as a type of structural response, in a way amounts to a spontaneous optimization of the geometric configuration of the system. Recall, for example, the remarkable force-leveling effect in radial cable systems under an antisymmetric (cosine) surface load and in skew axisymmetric nets under symmetric loads. Kinematic adaptivity is observed in many other familiar structural systems, such as a hammock, a fishing net, a string bag, or a fabric sack, which conform to the shapes they are containing. These examples demonstrate, in particular, the enhanced kinematic mobility due to the absence of shearing resistance in nets and fabrics as compared to membranes and solid films (liquid films also lack sheering resistance). An important technical application of this higher kinematic mobility is in the manufacture of doubly curved composite laminate shells and membranes from preimpregnated fabrics.

Example 10.1. An interesting example of the advantageous use of kinematic adaptivity is a pipeline design for hot or cryogenic products. The problem with a conventional, metal pipeline is thermal expansion and contraction, which necessitates special devices, such as thermal loops, for lowering the thermal stresses. A thermal loop significantly increases the pumping resistance and this is especially detrimental in cryogenic pipelines: the added friction contributes to the parasitic heat generation, necessitating an increase in the required cooling capacity.

The alternative concept employs the idea of separating the load-supporting function, on the one hand, and space enveloping and sealing functions on the other, as in the case of a tire and an inner tube. The sealing component is a thin flexible liner, impermeable to the product and capable of withstanding the extreme temperatures required, but not possessing the necessary strength. The load-supporting component is, structurally, a woven hose with two high-strength counterwound helical strands. When a fixed length section of such a pipeline is subjected to a thermal action, the two strands expand or shrink radially producing only an insignificant change in the helix angle but practically no axial force:

$$\varepsilon_r = \Delta R / R = cT / \sin 2\alpha, \quad \Delta\alpha = cT/2.$$

Here R is the pipe radius, α is the helix angle, c is the thermal expansion coefficient, T is the temperature change, and Δ denotes a small increment.

Shape Control in Conventional and Underconstrained Structures. Taking the concept of kinematic adaptivity in underconstrained structural systems a step further leads to a method of obtaining precise geometric forms by statical means, i.e., by utilizing suitable field and boundary loads. Upon obtaining a desired configuration, the underconstrained system can be either immobilized (by introducing additional discrete or continuous constraints) or actively controlled (by monitoring the external loads, support reactions, or forces of prestress) using some kind of feedback.

From the shape control viewpoint, a difference of principle exists between a geometrically invariant system, however flexible, and a variant one. A variant system is more than flexible, it is inherently shapeless, thus promising ultimate ease of control. Indeed, geometric reconfiguration of such a system can be purely kinematic, in which case it requires only minute control loads. The following comparison presents some key differences between geometrically invariant and variant systems, relevant for their control:

Invariant System	Variant System
Possesses a unique natural (stress-free) configuration.	Is inherently shapeless, can admit a variety of configurations.
Allows only elastic displacements (small or large).	Allows both kinematic and elastic displacements.
Linear response is described by a (tangent) *elastic* stiffness matrix.	Response is described by a tangent *statical-kinematic* stiffness matrix.
System resistance is determined by member elastic stiffnesses.	Elastic stiffnesses are irrelevant for kinematic displacements.

Even from this incomplete list, it is clear that a variant system as an object of shape control is drastically different from an invariant one in its basic properties, type of response, and applicable analytical tools. In the following case study, a particular but rather typical practical problem is addressed—one exemplifying the concept of statically controlled geometry.

Case Study: Kinematic Reconfiguration of a Mesh Reflector

Problem Description. The available means of maintaining the reflecting surface precision for large, conventional structure, space antennas include stiffening the system, reducing its temperature sensitivity, and employing passive or active damping of maneuver-induced vibrations. An alternative and, in a sense, opposite approach involves a very flexible, low-mass reflecting membrane actively controlled by an electrostatic force field (Lang and Staelin, 1982). Being very thin and flexible, a metal-coated elastic polymer membrane seems to be easily controllable. At the same time, surface smoothness and continuity suggest no apparent practical limit to operational wavelength. However, as discussed earlier (Chapter 8), a convex membrane attached to a rigid contour is geometrically invariant within the class of smooth loads: it does not allow any smooth deformations without either surface stretching or wrinkling. Since wrinkling as a means of shape control is unacceptable, the membrane shape can be changed only at the expense of elastic, plastic, thermal, or other material deformations. Of these, only elastic deformations are reversible, hence, at least theoretically, suitable for shape control.

Some serious difficulties arise at an attempt to implement this approach. Obviously, a membrane must be made sufficiently flexible for the feasible electrostatic forces to control reflector shape through stretching. For all practical purposes, this limits the membrane material selection to some kind of a highly elastic polymer. On the other hand, this same membrane must have an acceptable lifetime in an environment of extreme temperatures and intense radiation. These formidable difficulties necessitate a consideration of alternative implementations of the concept. One of them is the use of an underconstrained, variant system in the form of a two-array wire mesh (Janiszewski and Kuznetsov, 1988).

A Formal Statement of the Problem. A wire mesh with rhombic cells, being a Chebyshev net, can be applied without folds or ripples to any smooth surface. In particular, the mesh can be given the shape of a paraboloid of revolution. However, for a mesh with thermally distorted wire lengths, such an ideal compliance may become infeasible. Yet, the mesh is still an underconstrained system and allows an infinite number of kinematically possible configurations (i.e., those compatible with the changed wire lengths and the thermally distorted contour ring). The problem of kinematic reconfiguration calls for establishing the best among all kinematically feasible configurations of the mesh. This configuration must be, in some sense, closest to the ideal, reference paraboloid. Assuming the mean-square surface deviation as the measure of closeness, the problem can be formulated in terms of mathematical programming:

$$\text{Minimize} \sum_{1}^{N} W_n^2 \tag{10.1}$$

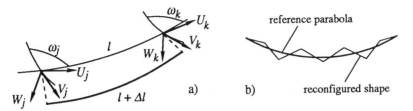

Figure 10.1. Illustrations of constraint equations for a wire mesh: a) generic wire section; b) the meaning of convexity requirement (3).

subject to

$$F^i(U_n, V_n, W_n) = 0, \quad i = 1,\ldots, C; \quad n = 1,\ldots, N, \qquad (10.2)$$

and

$$F^k(W_n) > 0, \quad k = 1,\ldots, 2N. \qquad (10.3)$$

Here U_n, V_n, and W_n are nodal displacements referred to a curvilinear coordinate system formed by the two arrays of wire trajectories and the normal to the reference paraboloid (Fig. 1a). Note that in-surface displacements U_n and V_n are not present in the objective function (1) since they do not affect the reflector shape.

The C equality constraints (2) assure displacements consistent with known (thermally distorted) wire lengths. For a wire section connecting nodes j and k (Fig. 1a), the constraint equation is a discrete version of the kinematic differential equation developed for a continuous model of a net in Chapter 6; it is simply the condition of wire inextensibility:

$$\Delta l = \Delta U + \Delta V (\cos \omega_j + \cos \omega_k)/2 + (\sigma_j W_j + \sigma_k W_k)/2$$

$$+ [(\Delta U)^2 + (\Delta V)^2 + (\Delta W)^2 + 2\Delta U \, \Delta V]/2l = 0, \qquad (10.4)$$

where $\Delta U = U_k - U_j$, etc., are the nodal displacement increments and σ_j is the normal curvature of the wire at the node:

$$\sigma_j^x = H/A^2, \quad \sigma_j^y = H/B^2, \quad H = 1/2FAB \sin \omega_j. \qquad (10.5)$$

Here A and B are the discretized values of the Lamé parameters of the net

$$A^2 = 1 + (x_j/2F)^2, \quad B^2 = 1 + (y_j/2F)^2,$$

expressed in terms of Cartesian coordinates x_j and y_j of node j and the focal length F. With the z-axis directed along the axis of revolution, the equation of the reference paraboloid is

$$z = (x^2 + y^2)/4F. \qquad (10.6)$$

Inequality constraints (3) are convexity conditions for the optimal mesh configuration. Without these constraints, in the case of a thermally expanded mesh, the chosen objective function is likely to result in a rippled surface (Fig. 1b), which would be functionally unsatisfactory as a reflecting surface. The convexity conditions are imposed twice at each node (x and y directions) and have the form

$$\sigma_j - \frac{l_2 W_{j-1} - (l_1 + l_2)W_j + l_1 W_{j+1}}{l_1 l_2 (l_1 + l_2)} > 0. \tag{10.7}$$

The second term in this expression is the incremental curvature resulting from the nodal displacements and evaluated as a central second difference. For nodes at the contour, forward and backward second differences are employed.

Numerical Realization. The stated optimization problem (1)–(3) is one of nonlinear programming, with a large number of variables and constraints. Since thermal patterns on the reflector surface result mostly from self-shadowing, a plane of symmetry is present which allows the problem size to be nearly halved. Still, the main resort in rendering the problem size manageable is an aggregation of the actual mesh in a coarser mesh of the analytical model. However undesirable, this compromise is the only available approach to the problem.

The displacements of the contour ring due to its thermal distortion have been evaluated prior to the optimization. This has been done using a conventional technique based on the Mohr integral and taking advantage of the temperature pattern symmetry. In the process of optimal reconfiguration, the deformed ring is allowed to shift, as a free body, along the axis of symmetry. The magnitude of the free-body shift, X, figures in the expressions for the $3N_c$ contour node displacements:

$$U_c = (X + x) \cos \varphi, \quad V_c = y \cos \psi,$$
$$W_c = (X + x) \sin \varphi - y \sin \psi. \tag{10.8}$$

Here x and y are thermal distorsion displacements of the ring, and φ and ψ are the respective slopes of the wire tangents at the contour nodes.

Further reduction of the problem size is more painful and consists of satisfying the nonlinear equality constraints (2) outside of the optimization loop proper. This is achieved by iterative linearization of the constraint equations with subsequent elimination of dependent displacements from both the objective function and the inequality constraints. As a result, the original general problem of nonlinear programming has been reduced to a recursive sequence of quadratic programming problems of a substantially smaller size. Specifically, estimating the number of wire sections as

$$C \approx 2(N + \sqrt{N}), \tag{10.9}$$

where N is the number of internal nodes in the mesh, there should be

$$V \approx 3N - C \approx N - 2\sqrt{N} \tag{10.10}$$

variables (independent displacements) in the recursive optimization cycle.

In designating the independent variables, preference was given to normal displacements W_n since these alone figure in the objective function and in the inequality constraints. The total number of normal displacements is more than sufficient for this purpose since it is greater than the number V of virtual degrees of freedom. However, not any V of the normal displacements can be designated as independent. The reason is that the column space of the Jacobian matrix of equations (2) must remain of full rank upon transferring the selected V columns to the right-hand side. A selection rule for normal displacements satisfying this requirement has been formulated and implemented. As a result, the recurrent optimization problem has a quadratic objective function in V variables and $2N$ linear inequality constraints (3).

The elimination of dependent displacements from nonlinear constraint equations (2) has been carried out using the modified Newton–Raphson iterative procedure. Its first step was different from a typical step due to an accidental singularity of the original configuration of the system. As shown in Chapter 6, this mesh, being a net of translation, is singular, with $S = D = 1$. Although the singularity (along with the resulting statical indeterminacy) is irrelevant for this particular application, it still has certain computational implications. Specifically, it necessitated an increase by one in the number of independent displacements at the first step of the analysis.

After obtaining an optimizing set of independent displacements, the dependent ones (all of the U_n, V_n, and some of W_n) were found from the nonlinear constraint equations using the same Newton-Raphson method. A frequently encountered complication was lack of convergence due to an overshoot in the evaluation of the independent displacements. It is not difficult to visualize an exaggerated normal displacement such that it cannot be incorporated into any compatible geometric configuration of the system. In the implemented algorithm, the lack of convergence triggers a scaling down of the independent displacements, followed by a new attempt to evaluate the dependent ones.

Upon convergence (within an assigned tolerance), the obtained configuration is kinematically feasible, since the nonlinear constraint equations are satisfied, but not optimal, since the optimization was performed subject to linearized constraint equations. In the next cycle, the original constraint equations are linearized at the recently obtained configuration and the entire procedure repeated. The process continues until the attained reduction in the value of the objective function becomes too small to justify yet another cycle of optimization.

Discussion of Results. The single most important parameter characterizing the reconfigured reflecting surface is antenna gain. It is usually estimated as being inversely proportionate to the mean square normalized deviation of the actual surface from the ideal reference paraboloid. In terms of the objective function employed, this measure is

$$\varepsilon^2 = (\sum_{1}^{N} W_n^2)/N. \tag{10.11}$$

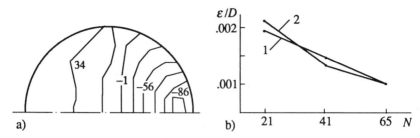

Figure 10.2.   Kinematic reconfiguration: a) isotherms of temperature pattern produced by self-shadowing; b) normalized root-mean-square surface error.

Three thermal patterns were investigated: uniform cooling, uniform heating and a transition temperature field resulting from reflector self-shadowing. For each case, the contour ring was either subjected to a corresponding thermal distortion or, for comparison, was assumed unaffected.

The cases of uniform cooling and heating turned out to be very different in terms of their respective optimal reconfigurations. In the case of cooling, a contracted mesh, geometrically similar to the ideal one, minimizes the objective function, leaving the inequality (convexity) constraints inactive. The surface deviation practically does not depend on the parameter N (the number of nodes) characterizing the mesh fineness, which is the consequence of the geometric similarity between the original and optimized configurations.

The optimal reconfiguration in the case of a uniformly heated mesh is more sophisticated and more efficient due to the presence of appreciable tangential displacements, partially "absorbing" the thermal expansions. Compared on a temperature degree-per-degree basis, the mean square surface error in the case of heating is smaller than in the case of cooling. In addition, the attained surface error under heating reduces, however slightly, with increasing N. In both the cooling and heating cases, the respective uniform contraction or expansion of the contour ring improves the situation, as could be expected.

Most interesting is the optimization outcome for the third temperature pattern (Fig. 2a), produced by self-shadowing. The results of the optimal kinematic reconfiguraion are presented graphically in Fig. 2b. The coefficient of thermal expansion in this study was 10^{-5} in/in °F for both the mesh and the contour ring materials. The two lines in the graph depict the normalized root-mean-square surface error in relation to the mesh fineness as measured by the number of nodes. Line 1 describes the case when the thermal distortion of the ring is disregarded and line 2 the case when it is accounted for. As is seen from the graph, the role of the ring distortion is insignificant and rather ambiguous. In contrast, a strong positive correlation was established between the quality of the reconfigured surface and the mesh fineness. This was traced back to an increased kinematic mobility (a larger number of degrees of freedom) of a finer mesh.

The optimal kinematic reconfiguration is but the first stage of the problem. The second one is the synthesis of a control load producing the desired configuration. With the intent of using an electrostatic force field as the control load, it is necessary to devise a system of electrostatic charges on a command surface located behind and roughly parallel to the reflecting surface. For a given geometric pattern of charges, this calls for the evaluation of the charge magnitudes approximating one of the equilibrium loads of the optimal configuration. A uniform electrostatic force field is an equilibrium load for the original surface—an ideal paraboloid of revolution. For the optimally reconfigured distorted mesh, the most suitable equilibrium load must be evaluated, taking into account the number and the pattern of independently controlled electrostatic charges.

10.2 Bending Instability of Improperly Supported Plates and Shallow Shells

Problem Formulation and Analysis

Limitation of the Linear Theory and Nonlinear Formulation. One deficiency of the linear theory of thin plates was first demonstrated by Kelvin and Tait (1890). They noticed that the experimentally observed deformation of a square plate under a self-balanced system of equal corner loads (Fig. 3a) is cylindrical instead of saddle-shaped, as predicted by the theory. Later studies, most recently by Lee and Hsu (1971), revealed that the saddle-shaped deflection develops at the onset of loading but at some point becomes unstable and transforms into cylindrical.

Bending instability occurs only in plates and shells whose support conditions do not preclude isometric bending. In this case, geometric invariance of the plate or shell is due to its bending resistance, whereas the corresponding membrane model, whether extensional or inextensional, is underconstrained. In typical conventional applications, such support conditions are not too common (e.g., certain point or line supports, an elastic foundation, all kinds of unilateral supports, and contact problems). It is the outer space applications employing support-free plates and shells that rekindled interest in the problem of bending instability. The following analysis of the problem expands the results of Kuznetsov and Hall (1986).

Bending instability in plates and shells can be investigated using the simplest nonlinear formulation of the problem (small displacements and strains but finite rotations). This model is represented by the von Karman equations for thin flexible shells:

$$D\Delta^4 W = [W, F] + \Delta_\sigma^2 F, \tag{10.12}$$

$$(1/B)\Delta^4 F = -(1/2)[W, W] - \Delta_\sigma^2 W. \tag{10.13}$$

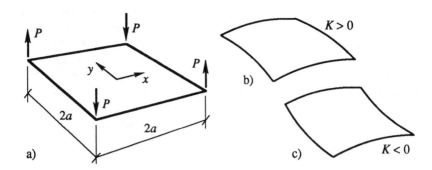

Figure 10.3. A plate and a shallow shell of square planform.

Here

$$[W, F] = W_{xx}F_{yy} - 2W_{xy}F_{xy} + W_{yy}F_{xx}, \tag{10.14}$$

and

$$\Delta_\sigma^2 F = \sigma_1 F_{yy} + \sigma_2 F_{xx}, \tag{10.15}$$

with subscripts x and y denoting the corresponding partial derivatives; $\Delta^4(\ )$ is the biharmonic operator; W is the normal deflection of the shell (positive when directed towards the center of curvature of the x-line); F is the Airy function; σ_1 and σ_2 are the principal curvatures of the middle surface which are assumed to be constant (as one of the consequences of the shell being shallow); finally, B and D are, respectively, the extensional and bending stiffnesses of the shell. Retaining the two independent stiffness parameters makes the solution applicable to truss-like, sandwich-type, and other layered plates and shells with isotropic layers, in which case B and D represent the reduced properties of the structure.

In a problem of the type considered here, any general deformation mode of a plate or shell has a threshold beyond which some nearly-inextensional deformation becomes, energy-wise, preferred to the original combination of bending and stretching. The existence of this critical deformation, which can be attained in a variety of loadings, is a manifestation of the principle due to Gerard (1962), stating that bifurcation is a deformation-controlled, rather than load-controlled, phenomenon. Seeking the critical deformation allows bifurcation under different load combinations and other actions (such as temperature or creep) to be studied efficiently in one analytical procedure. It also provides a link to an analysis of nonlinear dynamic behavior, when loads as such may not even figure, but the deflection amplitude is always present since it is necessary for describing the plate or shell motion.

Solution Method and Analysis. To render this investigation analytical to the utmost, a combination of the Galerkin method and numerical techniques is employed. The shell deflection is sought in the form of an approximating series with undetermined coefficients. The membrane forces produced by the approximated deflection of the shell are evaluated by integrating numerically the compatibility equation (13). The obtained values are then substituted into the equilibrium equation (12) and the Galerkin procedure reduces the problem to a system of simultaneous cubic equations. It is crucial for the analysis that the deflection mode corresponding to isometric bending is present among (or obtainable as a linear combination of) the base functions chosen. It turns out that for a shallow shell of a square planform (Fig. 3b and c), surprisingly comprehensive and accurate solutions can be obtained with just two base functions

$$W = d_1 W_1 + d_2 W_2 = d_1 xy/a^2 + d_2(x^2 + y^2)/2a^2. \tag{10.16}$$

Here d_1 and d_2 are the respective magnitudes of the corner deflections for the anticlastic (saddle-shaped) and synclastic (cap-shaped) displacement modes with reference to the coordinate system shown in the figure.

Carrying out the operations prescribed by (14) and (15) gives

$$[W, W] = 2(d_2^2 - d_1^2)/a^4, \quad \Delta_\sigma^2 W = 2H_0 d_2/a^2, \tag{10.17}$$

where $H_0 = (\sigma_1 + \sigma_2)/2$ is the mean curvature of the undeformed midsurface. Substituting (17) into (13) yields

$$\Delta^4 F = B(d_1^2 - d_2^2 - 2H_0 d_2 a^2)/a^4 \equiv BG(d_1, d_2)/a^2. \tag{10.18}$$

The introduced nondimensional parameter $G(d_1, d_2)$ will later be shown to be proportionate to an incremental change in the total curvature of the surface due to deformation. Since total curvature preserves in isometric bending, the corresponding deflection of the shell is characterized by the relation between d_1 and d_2 turning G into zero.

Next, the biharmonic equation (18) along with the boundary conditions corresponding to the stress-free edges was to be integrated. Instead, a substitute plane thermoelasticity problem,

$$\Delta^4 F = -E\alpha \Delta^2 T, \tag{10.19}$$

with the same free-edge boundary conditions has been solved, since only the second derivatives of F (i.e., the stress components) were needed for subsequent use in equation (12). Because the right-hand side in (18) is a constant, any temperature distribution $T(x, y)$ that obeys $\Delta^2 T = $ const. is suitable and (it may come as a surprise) leads to one and the same stress distribution. A finite element analysis with 400 plane stress elements was performed utilizing the system symmetry. This model was deemed (and later proved) to be consistent with, or exceeding, the perceived overall level of accuracy, taking into account the approximations inherent in the Galerkin method.

Governing Equations for the Discretized Model. The solving equations are obtained by evaluating and setting to zero the work done by all external and internal forces over each of the two modal virtual displacements.

These are unit magnitude displacements described by the base functions in (16). Both functions are utterly simple, resulting in $\Delta^4 W = 0$. However, neither of them satisfies the statical boundary conditions at the force-free edges of the shell. Instead, unbalanced bending and twisting moments occur at all four edges:

$$M_1 = -D(1 + \mu)W_{xx} = -D(1 + \mu)d_2/a^2 = M_2, \qquad (10.20)$$

$$M_{12} = -D(1 - \mu)W_{xy} = -D(1 - \mu)d_1/a^2. \qquad (10.21)$$

Here both the bending stiffness D and the Poisson ratio μ represent the reduced properties of the layered structure.

According to the modified Galerkin method used in such cases, work done by the unbalanced forces is taken into account when evaluating the total work. The edge moments perform work over the respective rotations

$$\phi = \partial W/\partial n, \qquad \psi = \partial W/\partial t, \qquad (10.22)$$

where n and t are the directions normal and tangent to the shell contour. The corresponding work is

$$Q = \int_L (M_n\phi + M_t\psi)\, ds \qquad (10.23)$$

or, evaluated separately for the antisymmetric, 1, and symmetric, 2, modes,

$$Q_1' = 8D\,(1 - \mu)\,d_1/a^2, \quad Q_2' = 8D\,(1 + \mu)\,d_2/a^2. \qquad (10.24)$$

In checking the dimension of this work it should be borne in mind that it is performed over virtual displacements of unit magnitude.

Upon the introduction of (16) the nonlinear term in equation (12) becomes

$$[W, F] = (d_2 N + 2d_1 S)/a^2, \qquad (10.25)$$

where

$$N = N_1 + N_2 = F_{yy} + F_{xx} = \Delta^2 F, \quad S = -F_{xy}. \qquad (10.26)$$

With the membrane forces N_1, N_2, and S available from the numerical (finite element) solution of equation (15), their work over each of the modal virtual displacements can be calculated:

$$Q_1'' = \iint_A [W, F]\, W_1\, dA = 0.051\, BGd_1, \quad Q_1''' = \iint_A \Delta_\sigma^2 F\, W_1\, dA = 0, \quad (10.27)$$

$$Q_2'' = \iint_A [W, F]\, W_2\, dA = 0.051\, BGd_2, \quad Q_2''' = \iint_A \Delta_\sigma^2 F\, W_2\, dA = 0.051\, BGH_0 a^2. \qquad (10.28)$$

The coincidence of the numerical factors in (27) and (28) is not an accident. After being evaluated independently, their values were found conspicuously close to each other (within a few per cent). An explanation, found later, stems from the curious fact that

$$\iint_A N(x^2 + y^2)dA = -4 \iint_A S\,xy\,dA$$

for any self-stressed square plate with normal stresses symmetric relative to the x and y axes. This relation can be verified by integrating by parts the identity

$$\iint_A (\Delta^2 S + N_{xy})(x^2 + y^2)xy\,dA = 0,$$

while taking into account the equilibrium equations of plane stress and the boundary conditions at free edges. Thus, the numerical factors in (27) and (28) must be equal and the small discrepancy noted above is a confirmation of the adequate accuracy of the finite element solution of equation (18).

For the following, it is helpful to introduce the normalized displacements

$$\delta_1 = d_1/t, \quad \delta_2 = \delta_{20} + d_2/t, \quad \delta_{20} = H_0 a^2/t = (\sigma_1 + \sigma_2)a^2/2t, \quad (10.29)$$

where

$$t = \sqrt{D/0.051B} \qquad (10.30)$$

is a convenient parameter that can be interpreted as the effective shell thickness.

The governing equations describing the behavior of a shallow shell with isotropic layers as a system with two degrees of freedom can now be assembled. This is done by equating the work of external loads over the two modal virtual displacements to the combined work given by (24), (27), and (28):

$$[C_1^2 + G(\delta_1, \delta_2)]\,\delta_1 = Q_1, \qquad (10.31)$$

$$[C_2^2 + G(\delta_1, \delta_2)]\,\delta_2 - C_2^2\delta_{20} = Q_2. \qquad (10.32)$$

Here

$$C_1^2 = 8(1 - \mu), \quad C_2^2 = 8(1 + \mu), \quad G(\delta_1, \delta_2) = \delta_1^2 - \delta_2^2 + \delta_{20}^2. \qquad (10.33)$$

The right-hand side in (31)–(32),

$$Q_{1(2)} = Q_{1(2)}^0\, a^2/Dt, \qquad (10.34)$$

is the normalized virtual work of the external load, which can be considered as the load parameter. (Q^0 is the virtual work before normalization.)

Bending Bifurcation of Plates

Critical Deformation for a Symmetric Mode. A solution for a square plate is obtained from the system of simultaneous equations (31)–(32), by setting $\delta_{20} = 0$ and considering symmetric and antisymmetric modes separately. For a symmetric mode, $Q_1 = 0$, and equation (31) is satisfied either by $\delta_1 = 0$ or by $C_1^2 = -G(\delta_1, \delta_2)$. In the first case, equation (32) becomes

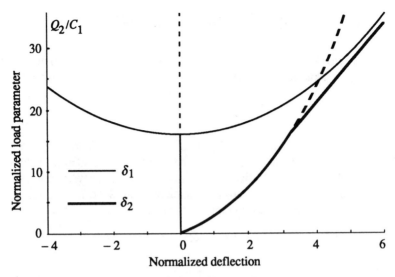

Figure 10.4. Load-displacement diagram with bifurcation for a square plate.

$$(C_2^2 + \delta_2^2)\delta_2 = Q_2 \tag{10.35}$$

and has only one real root. The solution of the system in this case is

$$\delta_1 = 0, \quad \delta_2 = (Q_2/2 + R)^{1/3} + (Q_2/2 - R)^{1/3}, \quad R = (Q_2^2/4 + C_2^6/27)^{1/2}. \tag{10.36}$$

In the second case, taking into account (33),

$$\delta_1 = \sqrt{\delta_2^2 - C_1^2}, \quad \delta_2 = Q_2/16. \tag{10.37}$$

The solution is meaningful when δ_1 is real, that is, at

$$|\delta_2| \geq C_1 = \delta_2^{cr}, \quad Q_2 \geq 16C_1 = Q_2^{cr}. \tag{10.38}$$

Accordingly, at a certain point in the course of loading, there are two solutions to simultaneous equations (31) and (32). The bifurcation occurs at the point $\delta_1 = 0$, $\delta_2 = C_1$, $Q_2 = 16C_1$ (Fig. 4) where the original symmetric deflection mode (37) becomes unstable (broken lines). In the process of loading beyond this point, the antisymmetric component of deflection δ_1 increases rapidly and asymptotically approaches δ_2, producing a nearly isometric deformation. The plate gradually acquires cylindrical shape with a generator parallel to one of the square diagonals.

The symmetric deformation of the plate at the bifurcation point is described by the critical values of curvature, σ_{cr}, and total curvature, K_{cr}:

$$\sigma_{cr} = d_2^{cr}/a^2 = t\delta_2^{cr}/a^2 = C_1 t/a^2, \quad K_{cr} = \sigma_{cr}^2 = C_1^2 t^2/a^4. \tag{10.39}$$

Upon attaining the critical deformation, energy required by further plate stretching in the symmetric mode becomes too high, causing bifurcation. The elastic energy associated with the already present extensions of the middle plane is absorbed in the new, predominantly bending, mode of deformation. Curiously, the critical magnitude of the corner deflection,

$$d_2^{cr} = t \delta_2^{cr} = C_1 t, \tag{10.40}$$

is independent of the plate size and is determined solely by the effective thickness of the plate defined in (30).

As has been mentioned, bifurcation occurs whenever critical deflection or curvature is reached, regardless of the physical cause. For example, the critical value of a uniform transverse thermal gradient in the plate is

$$\tau_{cr} = \sigma_{cr}/c, \tag{10.41}$$

where c is the thermal expansion factor. As to critical loads, they can be found from the expression (38) of the critical value of work produced by any such load. For instance, in the case of pure bending by a uniform moment M acting along the plate perimeter, this work is

$$Q_2^0 = 8M, \tag{10.42}$$

so that, according to (35) and (39),

$$M_{cr} = 2C_1 Dt/a^2. \tag{10.43}$$

For a corner-supported plate under an arbitrary normal surface load w,

$$Q_2^0 = \iint_A w(x,y) \, [(x^2 + y^2)/2a^2] \, dA. \tag{10.44}$$

From here, the critical magnitude for any given load pattern is easily evaluated. For a uniformly distributed load it is found to be

$$w_{cr} = 12C_1 Dt/a^4, \tag{10.45}$$

whereas for a concentrated transverse force, F, acting at the center,

$$F_{cr} = 16C_1 Dt/a^2, \tag{10.46}$$

and so on.

Antisymmetric Mode. An analysis of an antisymmetric external action (load or temperature) simply mirrors the foregoing analysis, with equations (31) and (32) trading their roles. Upon setting to zero the work Q_2 in the right-hand side of equation (32), the latter is satisfied by either $\delta_2 = 0$ or $C_2^2 = G(\delta_1, \delta_2)$. In the first case, equation (31) becomes a cubic equation in the unknown δ_1, and the solution is obtained from (36) by switching the subscripts 1 and 2. Subscript switching is also all that is required to obtain the rest of the solution: it reproduces the results (37)–(40) for the critical deformations, displacements, and loads like, say,

$$Q_1^{cr} = 16C_2. \tag{10.47}$$

The load-displacement diagram is still given by the graph in Fig. 4. The critical load parameter for a given load pattern is evaluated from

$$Q_1^0 = \iint_A [w(x,y) \; xy/a^2] \, dA \tag{10.48}$$

in conjunction with (34). For example, an antisymmetric set of four corner loads P will produce a saddle-shape deformation until the critical value

$$P_{cr} = 4C_2 Dt/a^2 \tag{10.49}$$

is reached. At this load level, bifurcation takes place, the symmetric deflection component sets in and increases rapidly, with the resulting deformed shape approaching cylindrical. For an isotropic plate, the result (49) agrees with the numerical solution of Lee and Hsu (1971) to within 5%. Note that in the conventional plate theory, this load is statically equivalent to a self-balanced system of uniform twisting moments acting along the edges, $M = P/2$.

As is seen from the obtained results, the critical deformations and the corresponding loads strongly depend on the relation between the bending and extensional stiffnesses and approach zero for a truly inextensible plate. Such a plate can be visualized as made of a relatively flexible material but incorporating a very thin and very stiff midplane layer. For the limiting case of inextensible midplane, the effective thickness t approaches zero and, as a result,

$$\sigma_{cr} \to 0, \quad Q_{1(2)}^0 \to 0, \quad \delta_1 \to \pm \, \delta_2. \tag{10.50}$$

With the critical deformation and load very small, the deflected shape is cylindrical (isometric bending) practically from the onset of loading.

An interesting feature of the present analysis is that, because of a happy choice of variables, the normalized critical deflection and loads of a plate are determined only by the parameter C_1 (or C_2, as the case may be) defined in (33). The corresponding actual critical displacements and loads are then easily found with the aid of (29) and (34), depending on the plate composition and material properties. In view of this, the combination of simplicity and accuracy in the result (38) and, even more so, in the generic solution (39) for the critical curvature, is surprising. One can hardly expect more from a simple analytical model with only two degrees of freedom.

Imperfect Plates and Shallow Shells

Symmetric Loading. Now consider a geometrically imperfect, say thermally distorted, plate which is, in essence, a very shallow shell. The imperfection is in the form of a smooth camber and is characterized by a mean curvature H_0. This is the only parameter representing the surface geometry in the governing equations; it appears in a normalized form as δ_{20} defined in (29). At $H_0 = 0$ the surface is saddle-shaped; it is a minimal surface. Somewhat unexpectedly, its bending bifurcation behavior under both symmetric and antisymmetric loads is identical to the plate behavior. Therefore, in what follows, the mean curvature is assumed different from zero. In fact, only the case $H_0 > 0$ needs to be investigated, since the sign reversal, if necessary, can be achieved by a simultaneous reversal of the signs of the principal curvatures, and is a matter of sign convention. Note that a surface with nonzero mean curvature can be either convex or saddle-shaped (Fig. 3b and c).

The mean curvature appears only in the second of the simultaneous equations (31) and (32) for a shallow shell. This makes the response of a shallow shell to a symmetric load similar to the response of a plate. Initially, the shell deformation is symmetric:

$$\delta_1 = 0, \quad \delta_2 = [(Q_2 + C_2^2 \delta_{20})/2 + R]^{1/3} + [(Q_2 + C_2^2 \delta_{20})/2 - R]^{1/3}, \quad (10.51)$$

$$R = [(Q_2 + C_2^2 \delta_{20})^2/4 + (C_2^2 - \delta_{20}^2)^3/27]^{1/2},$$

but at

$$\delta_2 = \sqrt{C_1^2 + \delta_{20}^2} = \delta_2^{cr}, \quad Q_2 = 16\sqrt{C_1^2 + \delta_{20}^2} - C_2^2 \delta_{20} = Q_2^{cr}, \quad (10.52)$$

bifurcation occurs and antisymmetric deformation sets in. The subsequent displacements are

$$\delta_1 = \pm\sqrt{\delta_2^2 - C_1^2 - \delta_{20}^2}, \quad \delta_2 = (Q_2 + C_2^2 \delta_{20})/16. \quad (10.53)$$

The shell deformation corresponding to the critical displacement (52) is best quantified in terms of the resulting change in the total curvature of the surface. Since total curvature preserves in isometric bending, its incremental change characterizes extensional deformations of the shell. Under the simplifying assumption on the constancy of the principal curvatures for a shallow shell, the equation of the deformed surface can be written as

$$z(x, y) = \sigma_1 x^2/2 + \sigma_2 y^2/2 + d_1 xy/a^2 + d_2(x^2 + y^2)/2a^2. \quad (10.54)$$

The total curvature, in this approximation being also a constant, is evaluated with the aid of (29) and (33) at the point $x = y = 0$:

$$K = z_{xx}z_{yy} - z_{xy}^2 = K_0 - (t^2/a^4)G, \quad (10.55)$$

where $K_0 = \sigma_1 \sigma_2$ is the total curvature of the undeformed surface. Thus, the parameter G introduced in (18) and (33) is a measure of the change in the total curvature caused by the shell deformation. According to (33) and (51)–(52), the critical value of this parameter is

$$G_{cr} = -C_1^2. \quad (10.56)$$

From here and (55), the critical magnitude of an incremental change in the total curvature of the shell is found

$$(\Delta K)_{cr} = C_1^2 t^2 / a^4, \tag{10.57}$$

which coincides with the value (39) found earlier for a plate. This comes as no surprise since the total curvature of a surface is determined by its intrinsic geometry and the assumption on the shell shallowness amounts to identifying the intrinsic geometry of the middle surface with the plane geometry.

Antisymmetric Loading. The mentioned disparity in the composition of equations (31) and (32) makes the shell behavior under an antisymmetric load quite different from the plate behavior. The possibility of multiple deformation paths for a given load still exists, but the paths are isolated. In the absence of a bifurcation point where the paths cross, a switch between different deformation modes occurs as a snap through, i.e., as a dynamic transition with no interim equilibrium configurations. In such a case, the critical load for a shell is defined as one at which the solution to the governing equations becomes nonunique. This load, called the lower critical load, can be sought as an analytical minimum of Q_1 with respect to δ_2 from the equation

$$\frac{dQ_1}{d\delta_2} = \frac{\partial Q_1}{\partial \delta_2} + \frac{\partial Q_1}{\partial \delta_1}\frac{d\delta_1}{d\delta_2} = 0. \tag{10.58}$$

With all the necessary quantities obtained by differentiation from equations (31) and (32), and after some rearrangements, the last equation takes the form

$$16\delta_2^4 + C_2^2\delta_{20}(8 + C_2^2 - \delta_{20}^2)\delta_2 - 3C_2^4\delta_{20}^2/2 = 0. \tag{10.59}$$

It is expedient to take δ_{20} as a parameter describing the shallow shell geometry in the form $\delta_{20} = \varepsilon^3 C_2$, where ε is a small quantity such that ε^3 (but not ε^2) is negligible compared to 1. Of the four roots of equation (59) only two are real—one positive and one negative. The positive root is physically meaningless as it would lead to $\delta_1^2 < 0$. Taking into account (33), the approximate value of the negative root is found:

$$\delta_2^{cr} \approx -\varepsilon C_2(1 + \mu/2)^{1/3} \approx -\varepsilon C_2(1 + \mu/6) \approx -\varepsilon C_2. \tag{10.60}$$

The corresponding values of δ_1^{cr} and Q_1^{cr} evaluated to the same accuracy are

$$\delta_1^{cr} = C_2(1 + \varepsilon^2), \quad Q_1^{cr} = C_2[16 + 8(3 + \mu)\varepsilon^2] \approx 8C_2(2 + 3\varepsilon^2). \tag{10.61}$$

As it should be, at $\varepsilon \to 0$, the above approximate values tend to their respective counterparts for a perfect plate.

An accurate numerical solution of the governing equations with $\varepsilon^3 = 0.02$ is plotted in Fig. 5. It reflects the relative complexity of shallow shell behavior, although the graph still has enough semblance to that of Fig. 4 for a plate to appreciate their relation (recall the switched subscripts). The two additional, isolated branches for δ_1 and δ_2 pertain to a partially inverted (snapped through) shell, with the descending (inner) portion of the curves corresponding to unstable configurations. The critical load and displacement magnitudes in the graph match very closely those of the approximate solutions in (60) and (61).

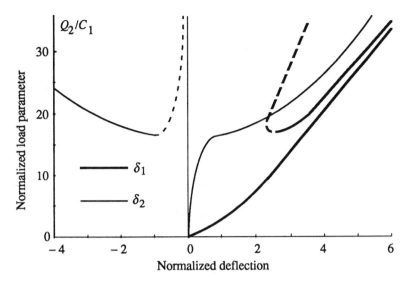

Figure 10.5. Bifurcation of a cambered plate (very shallow shell, $\delta_{20} = 0.02C_2$).

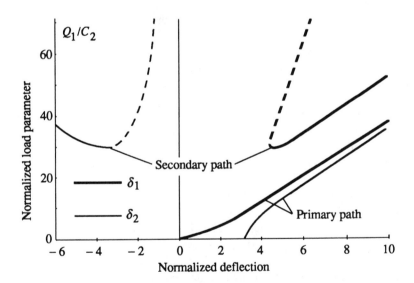

Figure 10.6. Bifurcation of a shallow shell with $\delta_{20} = 1.0\,C_2$.

Effect of Mean Curvature on Critical Load. Interestingly, the critical load for an imperfect plate is higher than the load (38) for an ideal plate and increases with the increasing mean curvature (Fig. 6). However, a numerical study shows that after attaining a maximum near $\delta_{20} = 1.2C_2$, this trend reverses and the bifurcation load gradually reduces all the way to zero. Figure 7 shows the relationship between the normalized critical load parameter in

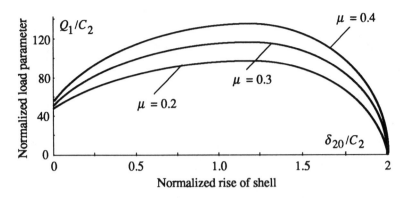

Figure 10.7. Normalized antisymmetric critical load versus mean curvature.

antisymmetric loading, Q_1^{cr}, and the normalized mean curvature of the shell (or cambered plate) δ_{20} for three values of Poisson's ratio μ.

The zero value for Q_1^{cr} means that the solution of the governing equations can be nonunique in the absence of loading. It is easily verifiable by substitution that for a shell with $\delta_{20} = 2C_2$ there are two possible states at $Q_1 = 0$: the original, stress-free state with $\delta_1 = 0$, $\delta_2 = \delta_{20}$; and a semi-inverted, self-stressed state with $\delta_1 = 0$, $\delta_2 = -C_2$. The load-displacement diagram for a shell with $\delta_{20} = 2C_2$ is presented in Fig. 8. At $\delta_{20} > 2C_2$ there is not one but two self-stressed states, but only one of them is stable and realizable. This latter state can be the starting point of a secondary load path, separate and independent of the primary path (the one originating from the natural state of the shell). This is illustrated by the load-displacement diagram (Fig. 9) resulting from an analysis for $\delta_{20} = 2.2C_2$.

For the same external load Q_1^0 the secondary path is always characterized by a higher total potential energy of the system than the primary path; hence the corresponding states are only locally stable. A perturbation of a sufficient magnitude in the presence of $Q_1 > Q_1^{cr}$ can transfer the shell from one path to the other, with a somewhat higher probability of settling on the primary path.

As follows from the symmetry properties of the obtained solution, a reversal of the external load direction is equivalent to the exchange of roles of the two diagonals of the shell. This leads to sign reversal of the first deflection mode δ_1, but is inconsequential for the second, symmetric mode δ_2. Accordingly, all of the related graphs can be extended to the half-plane $Q_1 < 0$ by point-symmetric extensions of paths δ_1 and reflection-symmetric images of paths δ_2.

The accuracy of a linear analysis of plates and shells is known to deteriorate with an increasing displacement magnitude. The above results indicate that for a plate or shell whose support conditions do not preclude isometric bending, the linear analysis not only suffers a quantitative error but, being oblivious to bending instability, may be even qualitatively inadequate.

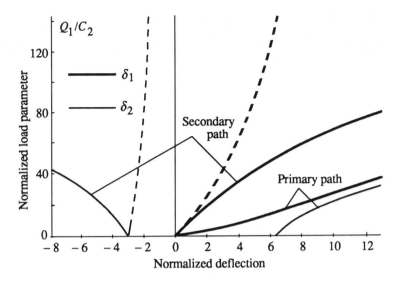

Figure 10.8. Bifurcation of a shallow shell with $\delta_{20} = 2.0\,C_2$.

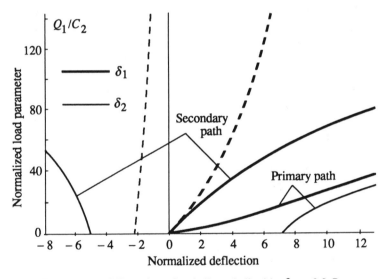

Figure 10.9. Bifurcation of a shallow shell with $\delta_{20} = 2.2\,C_2$.

The foregoing equations have been extended in Kuznetsov and Hall (1986) to describe the nonlinear vibrations of improperly supported plates and shallow shells. This problem is especially relevant for space structures, where bending stiffness is usually low in comparison with extensional stiffness. An amplitude-frequency relation has been obtained, and the role of extensional stiffness evaluated, for free vibrations of an unsupported square plate. This role is significant for plates with a high extensional stiffness but only moderate for isotropic plates. For a square isotropic plate, the nonlinear frequency of the saddle-shaped mode becomes equal to the frequency of the isometric cylindrical mode at the deflection amplitude of about seven thicknesses.

10.3 Shape Optimization
of a Membrane Shell of Revolution

Problem Formulation. The problem calls for the determination of the optimal shape of the meridian of a membrane shell of revolution under one or several load combinations. The optimality criteria can be rather diverse, e.g., a minimum weight or a uniform stress; minimum combined cost of the shell and the support ring, made of different materials; minimum internal pressure for an inflatable membrane enclosing a manned object of a given shape; and so on.

The optimization is subject to statical, geometric, and strength constraints and the optimized geometry is identified with the final, deformed shape of the shell. The shell is assumed of constant thickness and momentless but capable of supporting compression, although buckling is not considered. The strength condition determining the shell thickness is based on some local stress criterion, based on the maximum principal stress, maximum shear stress, or von Mises stress.

Two types of load are allowed: axisymmetric and antisymmetric (so-called "wind-type" load). As functions of the polar angle, these loads are described, respectively, by the first two terms in cosine series, with the resulting known closed-form solutions for the membrane forces. The problem statement, solution algorithm, and the obtained results will be illustrated in more detail as applied to the shape optimization of a dome (Kuznetsov and Ostrovski, 1971). The objective is to minimize the combined cost, C, of the materials of the shell and the supporting contour ring:

$$C = C_S S\, N_S/F_S + C_R\, 2\pi R T_R /F_R. \tag{10.62}$$

Here C_S and C_R are the unit costs of the shell and ring materials; F_S and F_R are their respective allowable stresses; N_S is the maximum force in the shell:

$$N_S = \max_r \max\, (|N_1(r)|,\ |N_2(r)|); \tag{10.63}$$

T_R is the tension force in the support ring which for a ring with a radius R is

$$T_R = N_1(R)\, R \cos \theta_R; \tag{10.64}$$

θ_R is the meridian slope at the edge; and, finally, S is the surface area of the dome; both θ and S are evaluated numerically in the process of optimization.

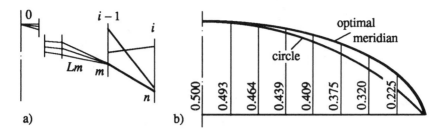

Figure 10.10. Shape optimization of a dome: a) analytical construction of optimal meridian; b) optimal and spherical dome profiles.

The meridional and hoop forces in an axisymmetric membrane shell are given in terms of the vertical force resultant, V, and the normal surface load, P:

$$N_1 = V/2\pi r \sin \theta, \qquad N_2 = (P - N_1 \sigma_1)/\sigma_2. \tag{10.65}$$

Additivity Condition and Recursive Procedure.   The presented statement of the problem renders it uniquely suitable for a solution by dynamic programming. Relative simplicity of numerical realization and a guaranteed global minimum are the decisive advantages of the method over the potential alternatives.   The necessary condition for the applicability of dynamic programming is the additivity (or separability) criterion for the objective function. In this case the criterion is satisfied by devising a multistep procedure where a contribution at each step to the final value of the objective function cannot affect the cumulative effect of the preceding steps.

The feasibility of such a procedure is due to the statical determinacy of a membrane shell. Its meridian can be constructed as a fine polygon starting at the apex and proceeding in small incremental steps towards the support ring (Fig. 10a). Because of the implicit presence, via (65), of the meridian curvature in the objective function, a modification of the conventional dynamic programming procedure must be used. In it, instead of examining designated points on the i-th applicate, the i-th step involves examining the set of admissible meridian segments in the interval $[i, i-1]$. Although this increases the problem dimension, several criteria for segment admissibility, reflecting geometric conditions such as meridian convexity and slope, help keep the volume of calculation in check.   Two inadmissible segments are shown in Fig. 10a; each has a slope incompatible with the overall profile of the shell.

The following relation, based on the additivity condition, governs the selection in the interval $[0, i-1]$, of the best meridional profile to append an admissible segment, mn, in the interval $[i-1, i]$:

$$t_i^{mn} = \min_{Lm} t_i^{Lmn} = \min_{Lm} \max \left(|N_i^{Lmn}|/F_s, \; t_{i-1}^{Lm} \right). \tag{10.66}$$

Here t_i^{mn} is the required thickness of the shell whose meridian includes segment mn and its optimal continuation in the interval $[0, i-1]$. The

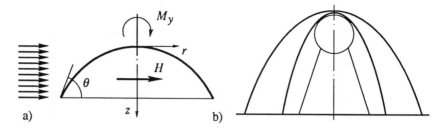

Figure 10.11. Optimization of an air-supported membrane: a) moment and force resultants of wind load; b) a membrane enveloping an object of a given shape.

continuation is chosen among the currently maintained optimized profiles, Lm, terminating at point m. As is seen from (66), the addition of segment mn may or may not require an increase in the shell thickness. Thus, at any step of the analysis, the number of maintained, tentatively optimized, profiles equals the number of admissible segments at the current interval.

The procedure easily accommodates additional geometric constraints, and may even benefit from them due to a reduction in the number of admissible segments (and the respective profiles) at each step. Such was the case when optimization was confined to domes with a fixed rise-to-span ratio (Fig. 10b). For illustration purposes, a spherical dome, corresponding to a very low unit price of the ring material, is compared to an optimal profile for the opposite extreme where the unit materials cost ratio is 50. Naturally, the second design has a steep meridian slope at the edge. The cost savings due to optimization were appreciable in spite of the seemingly insignificant difference between the two profiles.

Air-Supported Membrane under Wind Load. The only type of load other than axisymmetric, for which a closed-form solution is available, is an antisymmetric, "wind-type" load. Under such a load, the membrane forces are

$$N_1 = N_{1a} \cos \varphi, \quad N_2 = N_{2a} \cos \varphi, \quad S = S_a \sin \varphi. \qquad (10.67)$$

Their amplitude values (subscript a) are determined by the moment, M_y, and shearing force, H, resultants of the external load (Fig. 11a):

$$N_{1a} = M_y / \pi r^2 \sin \theta, \quad N_{2a} = (P - N_{1a}\sigma_1)/\sigma_2, \quad S_a = H/4r. \qquad (10.68)$$

Using these formulas, an optimization problem involving antisymmetric loads is solved in the same way as the foregoing axisymmetric problem.

One of the typical problems of this kind is the shape optimization of an air-supported membrane of revolution housing an object of a given shape (Fig. 11b). Here the objective has been to identify the membrane shape requiring a minimum air pressure necessary to withstand a given wind load. Two failure modes have been considered as constraints, both of them involving the principal stresses: the first is material failure in tension, the second the membrane disengagement resulting in wrinkling and flutter.

References

Calladine, C. R., and Pellegrino, S. (1991). First-order infinitesimal mechanisms. *Int. J. Solids Structures* **27**, 505–515.

Gerard, G. (1962). *Introduction to Structural Stability.* University Press, New York.

Gol'denveizer, A. L. (1953). *Theory of Elastic Thin Shells.* (In Russian) Gostechizdat, Moscow. English translation by Pergamon Press, New York, 1961.

Hilbert, D., and Cohn-Vossen, S. (1952). *Geometry and the Imagination.* Chelsea Publishing Company, New York.

Janiszewski, A. M., and Kuznetsov, E. N. (1988). Optimal reconfiguration of a thermally distorted wire mesh reflector. *Proc. 29th SDM Conf.*, Williamsburg, VA, pp. 1041–1047.

Kelvin, W., and Tait, P. G. (1890). *A treatise on Natural Philosophy,* 2nd ed. Cambridge University Press, Cambridge, U.K.

Kerr, A. D., and Coffin, D. W. (1990). On membrane and plate problems for which linear theories are not admissible. *ASME Journal of Applied Mechanics* **57**, 128–133.

Kötter, E. (1912). Uber die Moglichkeit, n Punkte in der Ebene oder im Raume... . *Festchrift Heinrich Muller-Breslau*, Kroner, Leipzig, 61.

Kuznetsov, E. N. (1969). *Introduction to the Theory of Cable Systems.* (In Russian) Stroiizdat, Moscow.

Kuznetsov, E. N. (1971). On the concept of statical determinacy and indeterminacy in structural mechanics. (In Russian) Stroiizdat, Moscow. *Trudy TsNIISK* **21**, 146–149.

Kuznetsov, E. N., and Ostrovski, A. Yu. (1971). Shape optimization of a membrane shell of revolution. (In Russian) Stroiizdat, Moscow. *Trudy TsNIISK* **21**, 149–156.

Kuznetsov, E. N. (1972a). Statical-kinematic analysis of systems involving unilateral constraints. (In Russian) *Studies in the Theory of Structures* **19**, 148–155.

Kuznetsov, E. N. (1972b). Problems in the statics of variant systems. *Soviet Mechanics of Solids* **8**, 108–112. (Available in Allerton Press translation.)

Kuznetsov, E. N. (1979). Statical-kinematic analysis and limit equilibrium of systems with unilateral constraints. *Int. J. Solids Structures* **15**, 761–767.

Kuznetsov, E. N. (1982). Edge effect in the bending of inextensible plates. *ASME Journal of Applied Mechanics* **49**, 649–651.

Kuznetsov, E. N. (1984a). Prestressed Cooling Tower. *United States Patent* 4,473,976, issued October 2, 1984.

Kuznetsov, E. N. (1984b). Statics and geometry of underconstrained axisymmetric 3-nets. *ASME Journal of Applied Mechanics* **51**, 827–830.

Kuznetsov, E. N., and Hall, W. B. (1986). Bending instability and vibration of plates and shallow shells. *Dynamics and Stability of Systems* **1**, 235–248.

Kuznetsov, E. N. (1988). Underconstrained structural systems. *Int. J. Solids Structures* **24**, 153–163.

Kuznetsov, E. N. (1989). The membrane shell as an underconstrained structural system. *ASME Journal of Applied Mechanics* **56**, 387–390.

Kuznetsov, E. N. (1991). Systems with infinitesimal mobility: Part I—Matrix analysis and first-order infinitesimal mobility; Part II—Compound and higher-order infinitesimal mechanisms. *ASME Journal of Applied Mechanics* **56**, 513–526.

Kydoniefs, A. D. (1969). Finite axisymmetric deformations of an initially cylindrical elastic membrane enclosing a rigid body. *Quart. J. Mech. Appl. Math.* **XXII**, Pt. 3, 319–331.

Landmann, A. E. (1985). Analysis of axisymmetric geodesic nets. Thesis, Dept. of General Engineering, University of Illinois/Urbana.

Lang, J. H., and Staelin, D. H. (1982). Electrostatically figured reflecting membrane antennas for satellites. *IEEE Transactions on Automatic Control* **AC-27**, 666–670.

Lee, S. S., and Hsu, C. S. (1971). Stability of saddle-like deformed configurations of plates and shallow shells. *Int. J. Nonlinear Mechanics* **6**, 221–236.

Levi-Civita, T., and Amaldi, U. (1930). *Lezioni di Meccanica Razionale*, 2nd ed. (In Italian) Vol. 1, Part 2. Zanichelli, Bologna.

Libai, A., and Simmonds, J. G. (1988). *The Nonlinear Theory of Elastic Shells*. Academic Press, New York.

Mann, H. B. (1943). Quadratic forms with linear constraints. *Amer. Math. Monthly* **50**, 430–433.

Marsden, J. E., and Hughes, T. J. R. (1983). *Mathematical Foundations of Nonlinear Elasticity*. Prentice-Hall, New York.

Maxwell, J. C. (1890). On the calculation of the equilibrium and stiffness of frames. *Scientific Papers of J. C. Maxwell*. Cambridge University Press, Cambridge, U.K.

Möbius, A. F., (1837). *Lehrbuch der Static*, Vol. 2. Leipzig.

Mohr, O. (1885). Beitrag zur Theorie des Fackwerkes. *Der Civilingenieur* **31**, 289–310.

Pipkin, A. C., and Rivlin, R. S. (1963). Minimum-weight design for pressure vessels reinforced with inextensible fibers. *ASME Journal of Applied Mechanics* **30**, 103–108.

Pipkin, A. C. (1968). Integration of an equation in membrane theory. *Z. angew. Math. Phys.* **19**, 818–819.

Pipkin, A. C. (1984). Equilibrium of Tchebychev nets. *Arch. Rat. Mech. Anal.* **85**, 81–97.

Pipkin, A. C. (1986). Continuously distributed wrinkles in fabrics. *Arch. Rat. Mech. Anal.* **95**, 93–115.

Read, W. S. (1963). Equilibrium shapes for pressurized fiberglass domes. *Journal of Engineering for Industry* **85**, 115–118.

Rivlin, R. S. (1955). Plane strain of a net formed by inextensible cords. *Journ. Rat. Mech. Anal.* **4**, 951–976.

Rivlin, R. S (1959). The deformation of a membrane formed by inextensible cords. *Arch. Rat. Mech. Anal.* **2**, 447–476.

Shulikovski, V. I. (1963). *Classical Differential Geometry*. (In Russian) Nauka, Moscow.

Shul'kin, Yu. B. (1984). *Theory of Elastic Bar Structures*. (In Russian) Nauka, Moscow.

Steele, C. R. (1964). Orthotropic pressure vessels with axial constraint. *AIAA Journal* **2**, 703–709.

Steigmann, D. J., and Pipkin, A. C. (1989a). Axisymmetric tension fields. *Z. Angew. Math. Phys.* **40**, 526–542.

Steigmann, D. J., and Pipkin, A. C. (1989b). Finite deformations of wrinkled membranes. *Quart. J. Mech. Appl. Math.* **42**, Pt. 3, 427–440.

Steigmann, D. J. (1990). Tension-field theory. *Proc. Roy. Soc. London* A **429**, 141–173.

Steigmann, D. J., and Pipkin, A. C. (1991). Equilibrium of elastic nets. *Proc. Roy. Soc. London* A (in press).

Strang, G. (1988). *Linear Algebra and its Applications*, 3rd ed. Harcourt Brace Jovanovich, New York.

Struik, D. G. (1961). *Lectures on Classical Differential Geometry*, 2nd ed. Addison-Wesley Reading, MA.

Tarnai, T. (1980). Simultaneous static and kinematic indeterminacy of space trusses with cyclic symmetry. *Int. J. Solids Structures* **16**, 347–359.

Tarnai, T. (1988). Bifurcations and singularities of compatibility (Abstract). 17th IUTAM Congress, Grenoble, France.

Vekua, I. N. (1959). *Generalized Analytic Functions.* (In Russian) Gostechizdat, Moscow. English translation by Pergamon Press, New York, 1962.

Wu, C. H. (1974). The wrinkled axisymmetric air bags made of inextensible membrane. *ASME Journal of Applied Mechanics* **41**, 963–968.

Zak, M. (1982). Statics of wrinkling films. *Journal of Elasticity* **12**, 51–63.

Index